T0418832

APPLICATIONS OF NUCLEAR AND RADIOISOTOPE TECHNOLOGY

APPLICATIONS OF NUCLEAR AND RADIOISOTOPE TECHNOLOGY
The Atom for Peace and Sustainable Development

Khalid Alnabhani
Centre for Risk, Integrity and Safety Engineering, Faculty of Engineering and Applied Science, Memorial University of Newfoundland, St. John's, Canada

Academic Press is an imprint of Elsevier
125 London Wall, London EC2Y 5AS, United Kingdom
525 B Street, Suite 1650, San Diego, CA 92101, United States
50 Hampshire Street, 5th Floor, Cambridge, MA 02139, United States
The Boulevard, Langford Lane, Kidlington, Oxford OX5 1GB, United Kingdom

British Library Cataloguing-in-Publication Data
A catalogue record for this book is available from the British Library

Library of Congress Cataloging-in-Publication Data
A catalog record for this book is available from the Library of Congress

ISBN: 978-0-12-821319-3

For Information on all Academic Press publications visit our website at https://www.elsevier.com/books-and-journals

Publisher: Joe Hayton
Acquisitions Editor: Maria Convey
Editorial Project Manager: Alice Grant
Production Project Manager: Prasanna Kalyanaraman
Cover Designer: Greg Harris

Typeset by Aptara, New Delhi, India

Contents

Preface

In spite of more than a quarter of a century since the tragic accident of "Chernobyl" that took place in 1986, the tragedy still evokes painful memories with any mention of the word "nuclear." In this sense, this has negatively affected plans to promote nuclear awareness about the peaceful application of nuclear energy. Arguably, nuclear technology plays a fundamental role in pushing the wheel of sustainable development in various sectors such as the economy, health, industry, and the environment. It is beyond the mere thought of an 'atomic bomb," and thus it is essential to dispute the idea of seeing nuclear technology as being as the "empty half cup" and instead we must look at it from the perspective of the "half-filled." Besides, it is essential for all stakeholders, politicians, and decision-makers to increase awareness of nuclear technology to ensure that governments, citizens, and nations can begin to benefit from the plethora of peaceful applications of nuclear and isotopic technology. Accordingly, the International Atomic Energy Agency under the nonproliferation of nuclear weapons treaty signed in 1970 among 191-member states has a unique role as the multilateral channel and the international safeguard inspectorate for the peaceful transfer of nuclear technology in relation to articles III and IV of the nonproliferation of nuclear weapons treaty. Accordingly, a lot of studies, efforts, and experiments have been made to harness the atom for peace and development. Since then, the world has witnessed a shift in peaceful nuclear energy applications in various sectors of sustainable development, which has grown exponentially. It includes but is not limited to clean nuclear power, health, efficient plant, and animal production, pest control, climate change reduction, improved water and soil quality, desertification control, wildlife improvement, sustainable environmental protection, nuclear forensic science in criminal investigations. In addition to many other industrial applications such as food irradiation to keep it fresh, oil spill tracking, characterization of the oil reservoir in the geological formation, and increasing the efficiency of oil production, sterilization of medical instruments, food imitation control, water desalination, and many other applications. This book discusses all those nuclear technologies and radioisotopes and their peaceful applications in advancing sustainable development and the science behind them in seven chapters.

This unique book will be a useful and important academic resource for scholars, researchers, graduate, and undergraduate students and policy-makers because this book will solve three major problems that have not yet been fully addressed:

- Enhancing nuclear awareness about nuclear peaceful applications.
- Explaining the theoretical and scientific facts behind nuclear peaceful applications in an easy and attractive manner supported by easy-to-understand examples and pictures.
- Correct prevailing perceptions about concerns related to nuclear safety and security that hinder nuclear development.

In the end, successfully completing many aspects of this work would not have been possible without the generous scientific and intellectual contributions by my fellow scientists from around the world. I would also like to take this opportunity to thank Memorial University for granting me the time and facilities to work on this book. Gratitude and appreciation are extended to Professor Faisal Khan who is one of the world's influential and leading scientists known for his landmark contributions to the sciences for his great effort and continued support. Tremendous gratitude and thanks are due to all MUN and C-RISE academics and professionals who offered me continued support that allowed me to overcome challenges and obstacles with determination and enthusiasm. I am also truly indebted to my beloved wife Aisha Al Salmi and my lovely three sons (Saif, Omar, and Sam) for their tireless support, encouragement, and sacrifice during this very hectic and challenging period.

Khalid Alnabhani

Dedication

In the name of Allah, the Most Gracious, the Most Merciful. I dedicate this book to all fellow atomic energy scientists from around the world for their unlimited support, generous advice, and exchanging of their information and expertise. This book presents cutting-edge research in the sciences of nuclear and has valuable information, theories, and modern scientific approaches which form a roadmap for more sophisticated scientific discoveries and applications in the future.

I also dedicate this book to professionals and students and those who seek to develop science and harness it for the service of mankind.

It is also a pleasure and an honor to dedicate this book to my country Oman and its faithful people, who have always been proud of me as the first Omani scientist among the world's influential and leading scientists known for their outstanding contributions to nuclear science.

Finally, I would like to dedicate this book as an expression of my sincere appreciation, thanks, and gratitude to my father "God bless his soul", my beloved wife "Aisha Al Salmi" and my lovely three sons "Saif, Omar, and Sam" for their tireless support, encouragement, and sacrifice during this very hectic and challenging period of our lives. I sincerely love you all more than you will ever know.

Khalid Alnabhani

CHAPTER

1

History of the atom and the emergence of nuclear energy

OUTLINE

Applications of Nuclear and Radioisotope Technology: The Atom for Peace and Sustainable Development
DOI: https://doi.org/10.1016/B978-0-12-821319-3.00008-7

1.1 The history of the atom and the beginning of the peaceful nuclear age

Atoms were first revealed in a very precise scientific way more than 1400 years ago, as it has been mentioned in several verses in the Holy Quran and the Hadith. For instance, Allah "May He be glorified and exalted" revealed in the Nobel Quran in Surah Younus, verse 61: "And not absent from your Lord is any [part] of an atom's weight within the earth or within the heaven or [anything] smaller than that or greater but that it is in a clear register."

Moreover, Al Nabhani and Khan (2019) argue that Allah "May He be glorified and exalted" has revealed prior 1400 years ago in the several verses in the Holy Quran and the Hadith that the atom has a weight, which has been confirmed recently by the modern science and said in the Nobel Quran in Surah Az-Zalzalah, verse 7:

"So, whoever does an atom's weight of good will see it."

The atom is one of the miracles of Allah mentioned in his holy book for more than 1400 years ago. It exists in both the earth and other planets. The Quran not only stresses the importance of the atom but also revealed the fact of the presence of the subatomic particles where many nuclear elements have not yet been discovered by a human. Furthermore, the Quran draws attention to the existence of the atom, the subatomic particles, and other science-related facts that modern physics has recently discovered. Bearing in mind, that during the period when Prophet Muhammad (PBUH) received the revelation, there was nobody has any idea of the atom or its subatomic particles and their compounds. This confirms the scientific miracle of the Qur'an.

Since then, the first theory of the atom was dated back to the fifth century BCE when Greek scientists and philosophers built their theory of atoms upon the work of ancient philosophers. They proposed the principle of identity that states that "matter was composed of atoms" (Ray and Hiebert, 1970). They have reached this conclusion primarily through deductive reasoning, logic, and mathematics but without conducting any experiments or providing concrete scientific proof. Accordingly, the term "atom" comes from the Greek word for indivisible. These theories were disregarded until the 16th and 17th centuries because religious intellectuals considered the theory to be a materialistic view of the world that denied the existence of spiritual forces (Ray and Hiebert, 1970).

At the beginning of the 19th century, scientists, such as John Dalton and Jöns Jakob Berzelius, revived the atomic theory by using quantitative and experimental data (Ray and Hiebert, 1970). For instance, John Dalton is a British chemist and considered as a pioneer in establishing the science of modern and quantitative chemistry, who has based his atomic theory on Democritus' ideas and believed that atoms are indivisible and indestructible. Additionally, he held a belief that different atoms form together to create all matter. In the new atomic theory, Dalton added his ideas that all atoms of a certain element are identical and combining in simple whole numbers. However, the atoms of one element will have different weights and properties than atoms of another element. Moreover, atoms cannot be created or destroyed.

Modern science came and proved the fact mentioned by Allah 1400 years ago in the holy book "The Quran," that atoms are not the smallest particles of matter and revealed that atoms are made from smaller subatomic particles such as the quark that has been scientifically discovered in 1939 by the German scientists Hahn and Strassmann (Al-Sheha, 2011). Modern science revealed that the center of an atom is the nucleus and contains protons and neutrons. Electrons are arranged around the nucleus in energy levels or orbits. Both protons and electrons have an electrical charge. The protons are positively charged while the electrons are negatively charged. The neutron is neutral, and the total number of electrons orbiting around an atom is always the same as the number of protons in that nucleus. The number of protons in an atom is used to refer to the atomic number. Atoms are arranged in the periodic table according to the increase in their atomic number.

Al Nabhani and Khan (2019) reveal that philosophers and scientists since then continued their studies to explore more about atoms until uranium was first discovered in 1789 by Martin Klaproth, a German chemist. It was not until a century later when Wilhelm Rontgen discovered ionizing radiation during his experimental research of passing electric currents through a vacuumed glass tube, thus generating continuous X-rays. A year later after this discovery, Henri Becquerel discovered that uraninite ore contained radium and uranium, consequently darkening the photographic plates due to the emission of beta radiation and alpha particles. Gamma rays were later found from the uraninite ore by another scientist known as Villard. Then later in 1896, Pierre and Marie Curie (Fig. 1.1) named the phenomena of radiation emission "radioactivity." In 1898, they became pioneers in isolating polonium and radium from the uraninite.

In the same year, 1898, Samuel Prescott discovered that radiation could destroy bacteria contained in food and he was credited for the evolvement of food radiation technology. Almost a century later, in 1902, Ernest Rutherford (Fig. 1.2) discovered that the radioactivity process occurs spontaneously by emitting either gamma rays or alpha or beta particles from the unstable nucleus, which ultimately transforms into a different element. Rutherford, through his research in 1919, was able to progress a broader consideration of the atoms when he fired alpha elements from a radium source into nitrogen, thus realizing the occurrence of new nuclear rearrangement with oxygen and proton formation.

FIGURE 1.1 Professor Marie Curie with her husband Pierre Curie in 1903 (From https://en.wikipedia.org/wiki/Marie_Curie#/media/File:Pierre_Curie_(1859-1906)_and_Marie_Sklodowska_Curie_(1867-1934),_c._1903_(4405627519).jpg).

FIGURE 1.2 Ernest Rutherford at McGill University in 1905 while conducting his research and experiments to identify the thorium emanations. (From: https://en.wikipedia.org/wiki/Ernest_Rutherford#/media/File:Ernest_Rutherford_1905.jpg).

FIGURE 1.3 The Danish physicist Niels Bohr who reveal the atomic structure and quantum theory (From: https://en.wikipedia.org/wiki/Niels_Bohr).

In the 1940s, another scientist called Niels Bohr (Fig. 1.3) contributed to advancing our comprehension of atoms, including the organization of electrons around the nucleus.

However, earlier in 1911, Frederick Soddy had already learned that naturally radioactive elements had numerous various isotopes with the same chemical properties. During the same year, George de Hevesy indicated that radionuclides were instrumental as tracers, as minute amounts were readily detectable with simple instruments. These findings are the secret behind the emergence of contemporary radiotracer technologies.

Ernest Rutherford continued his nuclear discoveries and in 1911 discovered the atomic nucleus. Between the years 1911 and 1920, he concluded that the protons and neutrons had almost similar mass while conducting his famous experiment with cathode-ray tubes (Charlie Ma and Lomax, 2012). Rutherford further theorized that there was a neutral particle in the nucleus. The nucleus is joined together by a strong force that aims to overcome the repulsing electrical energies between protons. Based on the mass of the nucleus, some atomic nuclei are unstable because the binding force contrasts in different atoms. These unstable atoms then decay into other elements to get rid of the excess energy in the form of particle emission or energy emission, thus becoming more stable.

A sizeable amount of energy is contained in an atom, which is scientifically considered the smallest particle of an element. Most of this energy is found inside different isotopes of individual elements. Variants of the same element existing in the same physical state but exhibiting different chemical properties and with the equivalent number of protons and a diverse number of neutrons are regarded as "isotopes." Emission of energy occurs during nuclear fission, which takes place when neutrons split or during nuclear decay. In this context, and specifically in 1932, the neutron was discovered by James Chadwick. In the same year, Cockcroft and Walton produced nuclear transformations through the bombardment of atoms with accelerated protons. In 1934, Irene Curie and Frederic Joliot

FIGURE 1.4 Enrico Fermi, the first founder of the process of nuclear fission reaction and won Nobel Prize in 1938 for his scientific contributions. (From: https://en.wikipedia.org/wiki/Enrico_Fermi#/media/File:Enrico_Fermi_1943-49.jpg).

concluded that some bombarded reactions may result in the forming of new, artificial radionuclides.

Sime (2014) argues that according to the historical context, the process of nuclear fission was first discovered and launched by Enrico Fermi in 1934 (Fig. 1.4), a physicist from the University of Chicago, who found that neutrons could split an atom into numerous other small atoms. After a year, Fermi further found that a more considerable variety of new artificial radionuclides could be created when neutrons were used in bombardment as an alternative for protons. Fermi is therefore considered as the first scientist to succeed in creating a fissionable nuclear reactor and developing the first self-sustaining nuclear chain reaction.

Based on the results obtained from Fermi's experiments, Otto Hana and Fritz Strassman who were both German scientists concluded from their experiment in which they fired at a neutron with uranium from beryllium and radium source, that the resulting materials were half lighter in atomic mass compared to reactant uranium (Kirklan, 2010). The difference in the atomic mass between the product materials and reactant material was in the form of heat that resulted from the vigorous splitting of the atom, which is called fission (Kirklan, 2010). The resulting heat could be massive and therefore be used in the generation of nuclear power.

Consequently, Lanouette (1992) argues that when Bohr arrived in New York, he announced the news of his success in performing a self-sustaining chain fission reaction at the Kiser Wilhelm Institute. Two years later, Leo Szilard and Fermi made a breakthrough through the development of real-time self-sustaining self-reaction through a uranium chain reactor (Lanouette, 1992). Bohr went on to share these findings with Einstein, which led to an important discovery of critical mass that brought a noteworthy change in the scientific world. A team of scientists that included Fermi, Bohr, and Szilard wrote a letter signed by Einstein to President Roosevelt (Fig. 1.5) illustrating the dangers of nuclear weaponry (Alan, 2015). The aim of this letter was only to intimidate and prevent the Germans from constructing the nuclear bomb.

However, this never stopped the great Manhattan project from commencing at a full pace, which resulted in the construction of two atomic bombs. According to Hassan and Chaplin (2010), unfortunately, the first bombs were made based on the principle discovered by Bohr, Leo Szilard, and Fermi of a self-sustaining fission chain reaction with enriched Uranium-235 through neutron bombardment as per Eq. 1.1:

$$^{235}_{92}U + ^{1}_{0}n \rightarrow ^{141}_{56}Ba + ^{92}_{36}Kr + 3^{1}_{0}n \tag{1.1}$$

The continued studies and research on nuclear reactions resulted in the production of numerous forms of energies- approximately 8.3×10^7 KJ, which became the working principle of the atomic bomb that was later dropped in Hiroshima in Japan on August 6, 1945, at 8:16 local time. This bomb was referred to as the "Little Boy" (Hassan and Chaplin, 2010). While the second atomic bomb made from the fusion of plutonium was known as the "Fat Man" and was dropped in Nagasaki on August 9, 1945, at 11:02 local time (Hassan and Chaplin, 2010). These two attacks caused the deaths of more than 250,000 people.

After these tragedies, nuclear scientists devoted their efforts, scientific studies, and research to harness atomic and nuclear technology for peaceful applications. Since then, the first application that has been focused on was the generation of electricity through the exploitation of the huge amounts of heat produced from the fission reaction.

1.2 The birth of peaceful nuclear applications

The birth of peaceful nuclear applications was announced by President Dwight D. Eisenhower on December 8, 1953, through his speech in front of the United Nations council (Fig. 1.6) under the title of "Atoms for Peace" (Kupp, 2005). He stated "in fact, we did no more than crystallizing a hope that was emerging in many minds in many places… the split of the atom may lead to the unifying of the entire divided world" (IAEA, 2020a). Through this speech, the President was attempting to diminish the nuclear arms race between the United States and the Soviet Union. Eisenhower

Albert Einstein
Old Grove Rd.
Nassau Point
Peconic, Long Island

August 2nd, 1939

F.D. Roosevelt,
President of the United States,
White House
Washington, D.C.

Sir:

Some recent work by E.Fermi and L. Szilard, which has been communicated to me in manuscript, leads me to expect that the element uranium may be turned into a new and important source of energy in the immediate future. Certain aspects of the situation which has arisen seem to call for watchfulness and, if necessary, quick action on the part of the Administration. I believe therefore that it is my duty to bring to your attention the following facts and recommendations:

In the course of the last four months it has been made probable - through the work of Joliot in France as well as Fermi and Szilard in America - that it may become possible to set up a nuclear chain reaction in a large mass of uranium,by which vast amounts of power and large quantities of new radium-like elements would be generated. Now it appears almost certain that this could be achieved in the immediate future.

This new phenomenon would also lead to the construction of bombs, and it is conceivable - though much less certain - that extremely powerful bombs of a new type may thus be constructed. A single bomb of this type, carried by boat and exploded in a port, might very well destroy the whole port together with some of the surrounding territory. However, such bombs might very well prove to be too heavy for transportation by air.

-2-

The United States has only very poor ores of uranium in moderate quantities. There is some good ore in Canada and the former Czechoslovakia, while the most important source of uranium is Belgian Congo.

In view of this situation you may think it desirable to have some permanent contact maintained between the Administration and the group of physicists working on chain reactions in America. One possible way of achieving this might be for you to entrust with this task a person who has your confidence and who could perhaps serve in an inofficial capacity. His task might comprise the following:

a) to approach Government Departments, keep them informed of the further development, and put forward recommendations for Government action, giving particular attention to the problem of securing a supply of uranium ore for the United States;

b) to speed up the experimental work,which is at present being carried on within the limits of the budgets of University laboratories, by providing funds, if such funds be required, through his contacts with private persons who are willing to make contributions for this cause, and perhaps also by obtaining the co-operation of industrial laboratories which have the necessary equipment.

I understand that Germany has actually stopped the sale of uranium from the Czechoslovakian mines which she has taken over. That she should have taken such early action might perhaps be understood on the ground that the son of the German Under-Secretary of State, von Weizsäcker, is attached to the Kaiser-Wilhelm-Institut in Berlin where some of the American work on uranium is now being repeated.

Yours very truly,
A. Einstein
(Albert Einstein)

FIGURE 1.5 A copy of the letter signed by Einstein and sent to President Roosevelt to the United States President Franklin D. Roosevelt on August 2, 1939. (From https://en.wikipedia.org/wiki/Einstein%E2%80%93Szil%C3%A1rd_letter).

FIGURE 1.6 President Eisenhower announcing in front of the United Nations council on December 8, 1953 "Atoms for Peace" (From: https://www.flickr.com/photos/nrcgov/15856679667/in/photostream/).

believed that all nations in the world should have access to nuclear technology for peaceful usage through the right to obtain information in relation to advances in nuclear medicine, nuclear reactors to produce energy, and many other peaceful applications.

President Eisenhower aimed to create a central agency called the "uranium bank" that would collect, store, and distribute radioactive materials to all nations, so the United States and Soviet Union would no longer have a monopoly on nuclear applications (Lavoy, 2003). This "uranium bank" would allow nonnuclear countries to obtain fissionable content for medicine, energy, and other peaceful applications. According to this vision, in 1957, the United Nations established the International Atomic Energy Agency (IAEA), which functions as the world's nuclear regulator (Fischer, 1997). In this regard, IAEA Director-General Dr. Mohamed ElBaradei stated on March 5, 1999, that IAEA has three main objectives, which are: "to assist Member States, particularly developing countries, in the use of nuclear technology; to promote radiation and nuclear safety; and to ensure to the extent possible that pledges related to the exclusively peaceful use of nuclear energy are kept "(IAEA, 2020b).

In March 1970, the IAEA was put in charge of enforcing and monitoring the adherence to the nuclear nonproliferation treaty (NPT) (Sharp, 1996). NPT includes both nuclear and nonnuclear states. Nuclear states or nuclear club states or P-5 comprise of the five main states, which are the United States, Russia, United Kingdom, France, and China (Rajagopalan and Mishra, 2015). All other states besides these are non-nuclear states. The nuclear club states that are permanent members of the Security Council and the IAEA are mainly responsible for enforcing the NPT. Scholars such as Müller (2016) agree that this treaty ensures three main areas which are: (1) all non-nuclear states shall not build, develop or buy nuclear weapons; (2) all nuclear states that own nuclear weapons shall work toward disarmament; (3) all countries will be supported with the information required for peaceful applications of nuclear technology. Almost half a century after the NPT entered into force, about 190 countries from all over the world had signed the NPT (Müller, 2016). Chapter 2 will shed light in more detail about the NPT and will discuss systematically how effective it is in promoting nuclear peaceful applications and minimizing nuclear proliferation.

About more than half a century ago, the scientific research of isotopic and nuclear science technology has continued to contribute to the advancement of nuclear technology through the revolution in the ionizing capabilities of isotopic radiations and their interactions with materials. This advancement has contributed to the emergence of a wider range of nuclear peaceful applications.

1.3 Basics in the sciences of nuclear radioactive materials

1.3.1 Radiation

Radiation simply refers to the emitted energy that comes in different forms as a result of the kinetic energy of mass in motion. Radiation may comprise atomic or subatomic particles (electrons, protons, and neutrons). The other type of radiation may comprise electromagnetic radiation including radio waves, microwaves, infrared, ultraviolet, visible light, X- and gamma-rays, where the energy is transported through space by oscillating electrical and magnetic fields at different wavelengths and frequency (Cherry et al., 2012). When electromagnetic radiation, such as visible light and radio waves, interact with matter, they behave in the form of wave-like phenomena. However, when electromagnetic radiation interacts with individual atoms, it can also take the form of discrete packets that are scientifically called photons (quanta). Photons have no mass or electrical charge but move at the speed of light. Such physiognomies differentiate them from other forms of particulate radiation. Fig. 1.7 illustrates diverse photon energies of the electromagnetic spectrum.

It's worth mentioning that X-rays and γ-rays tend to have the highest energy but with the shortest wavelength and the highest frequency at the end of the spectrum. Thus the energy of X-ray and γ-ray photons may range from few Kiloelectronvolts (KeV) up to Megaelectronvolts (MeV),

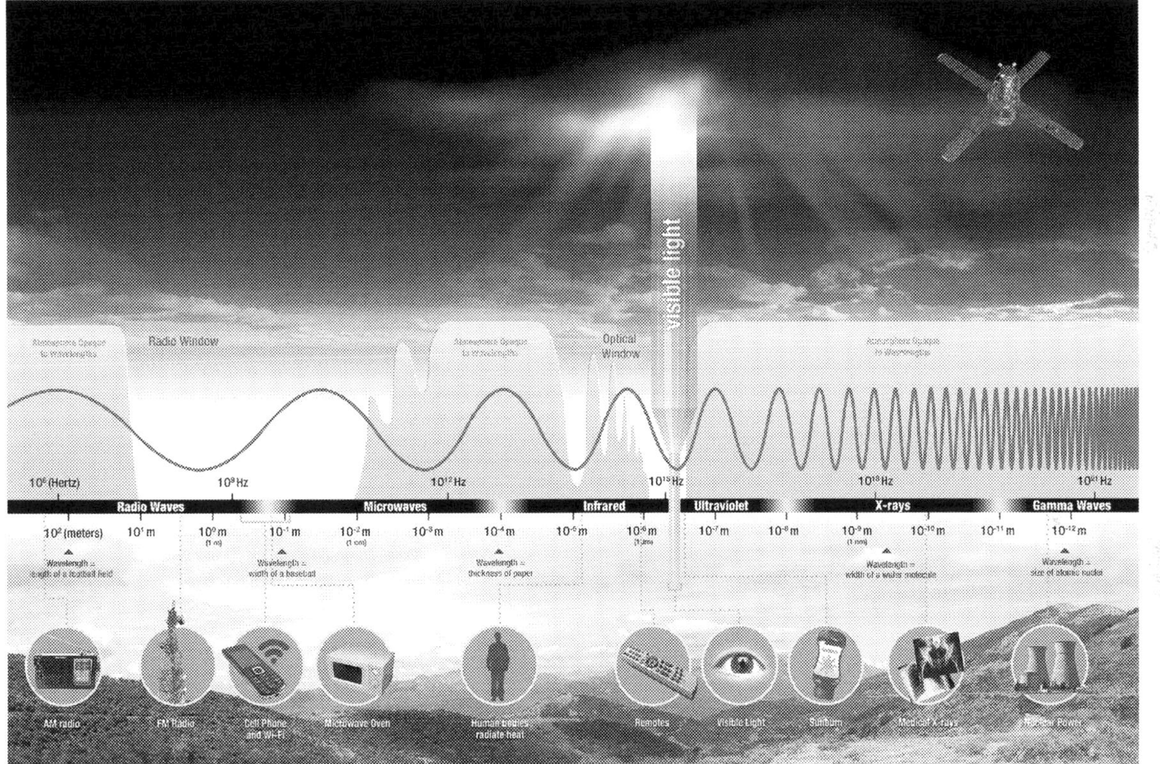

FIGURE 1.7 Different types of electromagnetic spectrum. (From: https://www.jpl.nasa.gov/edu/news/2020/1/23/nasa-says-goodbye-to-space-telescope-mission-that-revealed-a-hidden-universe/ - Image policy for public https://www.jpl.nasa.gov/imagepolicy/).

or Teraelectronvolts (TeV), respectively. According, to their high energies and short wavelengths, X-rays and γ-rays often interact with matter quite inversely from other, more acquainted types of electromagnetic radiation (Cherry et al., 2012). On the other hand, visible light photons have low energy that ranges to only a few electron volts.

1.3.2 Radioactivity

Radioactivity was discovered in 1896 by Henri Becquerel while studying whether natural phosphorescent materials emit similar rays or not. He found that the uranium salts emitted rays that could penetrate through a metal sheet or thin glass (L'Annunziata, 2016). Further, Becquerel proved scientifically that uranium salts gave off less intense radiation than uranium metals. This radiation causes ionization, which is used to measure the intensity of radioactivity. Accordingly, Becquerel is considered to be the first who provided evidence that some of the radiation emitted by uranium and its salts were similar in properties to electrons (L'Annunziata, 2012).

In 1898, Marie Curie discovered that it was not only uranium that gave off the mysterious rays discovered by Becquerel, but thorium did as well. Pierre and Marie Curie observed that the intensity of the spontaneous rays emitted by uranium or thorium increased as the amount of uranium or thorium increased. They concluded that these rays were a property of the atoms of uranium and thorium; they decided to consider these substances as radioactive materials. The emanation of the spontaneous rays from atoms would now be referred to as "radioactivity" (Al Nabhani and Khan, 2019).

Additionally, they found that another radioactive element with chemical properties like bismuth was present in pitchblende. She named this new element, polonium. They found a second new radioactive element in the pitchblende ore with chemical properties close to that of barium, and they named that new element "radium," from the Latin word radius meaning "ray." It is worth mentioning that Underhill (1996) stated, "Radium is of primary concern not only because it is radioactive, but also because it is chemically toxic" (Al Nabhani and Khan, 2019).

Radium may be almost as toxic as polonium and plutonium, the most toxic elements known to man. (It is estimated that one teaspoon of plutonium could kill 100,000 people through its chemical toxicity alone.) Due to its chemical properties, radium is termed a "bone seeker." Kumar and Dangi (2016) stated that in 1900 the French chemist and physicist Paul Villard discovered while studying the radiation emitted by radium, highly penetrating radiation in the form of electromagnetic waves that is consisting of photons.

In 1903 Ernest Rutherford was the first to name that highly penetrating radiation discovered by Villard as gamma rays (Al Nabhani and Khan, 2019). A few years prior to Villard's discovery, Rutherford in 1899 had already named two types of nuclear radiation as "alpha" and "beta," which he characterized based on their relative penetrative power in that the alpha radiation would be more easily absorbed by the matter than beta radiation. In harmony with this nomenclature, Rutherford assigned the term gamma rays to the more penetrating radiation (L'Annunziata, 2012). Therefore, radioactivity is the emission of radiation originating from a nuclear reaction or because of the spontaneous decay of unstable atomic nuclei.

Al Nabhani and Khan (2019) defined radioactive decay as the process that occurs when an unstable atomic nucleus has excess energy and therefore decays to lose the excess energy by the emission of elementary particles (such as alpha particles, beta particles, neutrons) or as gamma-rays in the form of photons or electromagnetic waves to reach stability. These types of radioactive decays are categorized under the ionized radiation category. The energy emitted by these radiations is often enough to ionize biological cells and cause damage to it, and therefore it is a serious health risk. Accordingly, radiation is a serious safety issue (Kumar and Dangi, 2016).

The law of radioactivity is based on the fundamental of the probability of decay per nucleus per unit of time in which a parent nucleus decays at a certain time into a daughter nucleus. This process called disintegration (decay) and mathematically it is expressed as per Eq. 1.2:

$$\mathrm{dN} = \lambda \mathrm{N.dt} \tag{1.2}$$

when

$$\lambda = (-\mathrm{dN}/\mathrm{dt})/N$$

where N is the number of radioactive nuclei, $-\mathrm{dN}/\mathrm{dt}$ is the decrease (negative) of this number per unit of time, and λ is the probability of decay per nucleus per unit of time. The decay constant λ is different for each decay mode and for each nuclide.

However, it is a common practice to use the half-life ($T_{1/2}$) instead of the decay constant (λ) for indicating the degree of the decay rate of a particular radioactive nuclide. Therefore, the half-life is defined as the period of time in which half of the radioactivity has disintegrated, and mathematically it is expressed s per Eq. 1.3:

$$T_{1/2} = (1/\lambda)/\ln(1/2) \tag{1.3}$$

Thus, λ can be expressed as:

$$\lambda = \ln 2/T_{1/2}$$

$$\lambda = 0.693/T_{1/2}$$

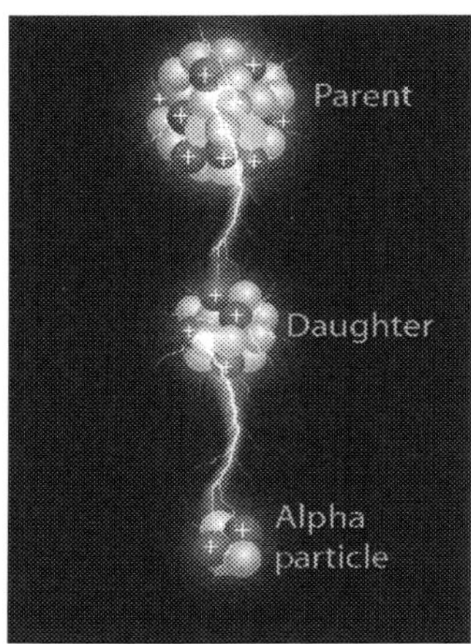

FIGURE 1.8 Alpha decay.

1.3.3 Modes of radioactive decay

There are several types of particles or electromagnetic waves that may result from the decay of an unstable atom to reach stability by losing its excess energy in form of particles or energy emission that is called ionizing radiation (Al Nabhani and Khan, 2019). There are many modes of radioactive decays but the most common ones are (1) alpha radioactivity; (2) beta radioactivity; and (3) gamma radioactivity.

The first and most common form of ionizing radiation is alpha radiations (Fig. 1.8) that comprise a helium nucleus, which are made up of two protons and two neutrons. Alpha radiation is always associated with the release of an energy (Q) and most alpha particles have energies between 4 and 6 MeV (IAEA, 2008). This energy is equivalent to kinetic energy of alpha particles, the energy of recoil daughter nucleus, and the energy lost as gamma radiation from the daughter nucleus when it is at excited status as a result that helium nucleus is tightly bound. Alpha decay chemically represented by Eq. 1.4:

$$ {}_{Z}P^{A} \rightarrow {}_{Z-2}D^{A-4} + {}_{2}He^{4} + Q_{\alpha}\left(Q_{\alpha} = K_{T} = E_{\alpha} + E_{recoil} + E_{\gamma}\right) \tag{1.4} $$

where P is the parent nuclide, D is daughter nuclide, Q is the equivalent to the total kinetic energy during the alpha particles decay process.

For instance, the decay of alpha particles from Th^{228} will release energy equivalent to 5.5 MeV as shown in Eq. 1.5:

$$ {}_{88}Th^{228} \rightarrow {}_{88}Ra^{224} + {}_{2}He^{4} + Q_{\alpha}\left(Q_{\alpha} = K_{T} = E_{\alpha} + E_{recoil} + E_{\gamma} \cong 5.5\,MeV\right) \tag{1.5} $$

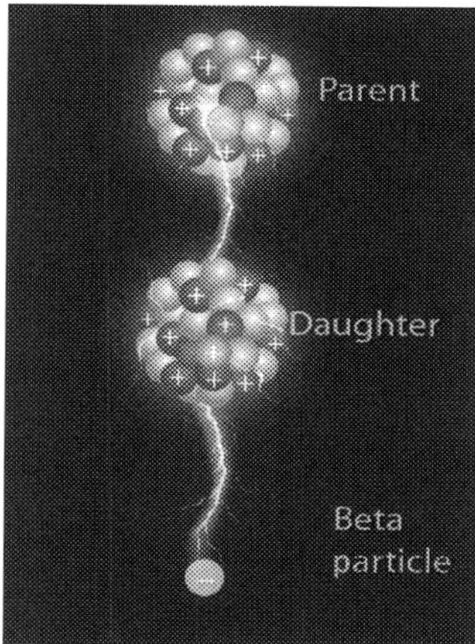

FIGURE 1.9 Beta decay ($\beta-$ decay).

Using Einstein's Eq. 1.6 of mass and energy we can calculate the energy equivalence to mass loss as the following:

$$E = (\Delta m) c^2 \tag{1.6}$$

Therefore, large atoms often decay by discharging an energetic alpha particle. Alpha particles are relatively large and positively charged; and for that reason, do not penetrate through matter easily. Nevertheless, these particles can cause extreme damage to materials by dislocating atoms as they slow.

The second form of radioactive decay is the beta particle radiations that are high-energy and high-speed electrons. They are emitted from the nucleus-decayed neutron when the neutron to proton ratio is too high. Conservation of charge requires a negatively charged electron to be emitted because neutrons are neutral particles and protons are positive, and this is known as negative beta decay ($\beta-$ decay) (Fig. 1.9) and chemically represented by Eq. 1.7. Some isotopes decay by means of converting a proton to a neutron, in consequence emitting a positron ($\beta+$ decay) as shown in Eq. 1.8:

$$^{A}P_{Z} \rightarrow {}_{Z+1}D^{A} + {}_{-1}\beta^{0} + v + Q_{\beta-}(\text{Negatron}) \tag{1.7}$$

$$\left(Q_{\beta} = K_{T} = E_{\beta} + E_{recoil} + E_{v} + E_{\gamma}\right)$$

where P is the parent nuclide, D is daughter nuclide, ${}_{1}\beta^{0}$ is negative beta particle, v is antineutrino, and K_T is the total kinetic energy released during

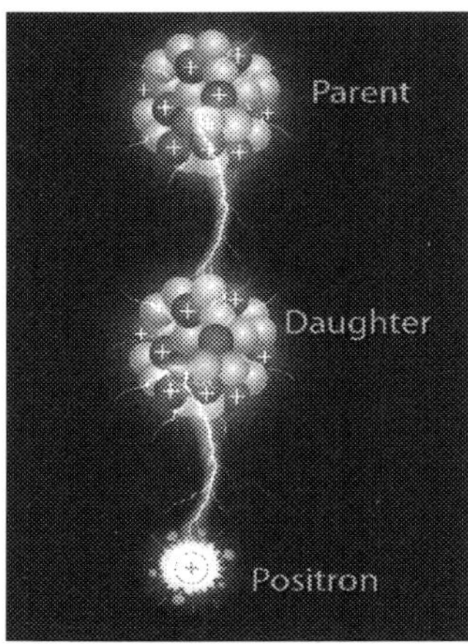

FIGURE 1.10 Beta decay (β+ decay).

beta decay:

Example: ${}_{92}U^{235} \rightarrow {}_{93}Np^{235} + {}_{-1}e^0 + v + Q_{\beta-}$

$$^{A}P_{Z} \rightarrow _{Z-1} D^{A} + _{+1}\beta^{0} + v + Q_{\beta+} \text{ (Positron)} \tag{1.8}$$

$$\left(Q_{\beta} = K_{T} = E_{\beta} + E_{\text{recoil}} + E_{v} + E_{\gamma}\right)$$

where P is the parent nuclide, D is daughter nuclide, ${}_{1}\beta^{0}$ is a positive beta particle (Fig. 1.10), v is the neutrino, and K_T is the total kinetic energy released during beta decay.

$$\text{Example: } {}_{19}K^{38} \rightarrow {}_{18}Ar^{38} + {}_{+1}\beta^{0} + v + Q_{\beta+}$$

The third type of radioactive decay is gamma rays (Fig. 1.11), which are photons that are emitted from the nucleus when it is in an excited state, and often an atom will de-excite by emitting an electromagnetic wave that is chemically represented by Eq. 1.9. IAEA (2008) stated that "gamma rays are more energetic, and their energies range from ten thousand to ten million electron volts and can easily penetrate through most of the materials." This radiation requires lead shielding because they have no charge and can penetrate most matter easily. Therefore, people who are exposed directly to either alpha, beta, and gamma radiations through direct skin contact or internal contamination through ingestion or inhalation, are at an extremely serious health risk that may develop into cancerous diseases:

$$^{A}P{*}_{Z} \rightarrow _{Z} D^{A} + {}^{0}\gamma_{0} \text{ (Gamma emission)} \tag{1.9}$$

where P* is an excited parent nuclide, D is daughter nuclide, and γ is gamma emission.

$$\text{Example: } {}_{92}U^{235} \rightarrow {}^{0}\gamma_{0} + {}_{92}U^{235}$$

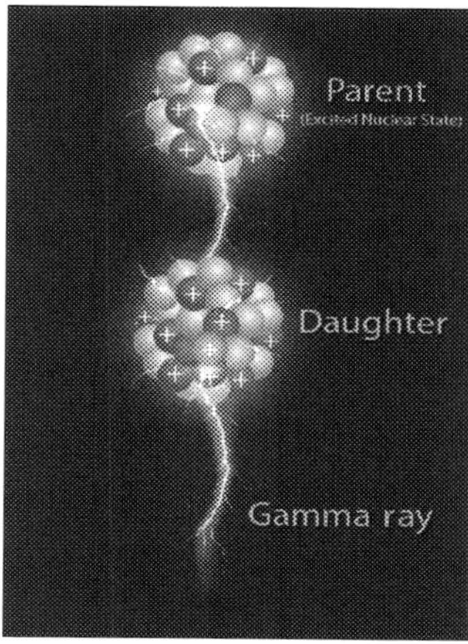

FIGURE 1.11 Gamma decay.

1.3.4 Isotopes

The term isotope was first introduced by the British chemist Frederick Soddy from the Greek words iso (4σος "equal") and topos (τόπος "place"). This means that different isotopes of a single element take up the same place on the periodic table (Nagy, 2009; Davisson, 2002). In the late 19th century, chemists discovered new elements with specific atomic numbers, which were categorized and arranged in the periodic table designed by Mendeleev by their atomic weight at the beginning (Eric, 2009).

In clarifying mass differences, Frederick Soddy of 1914 proposed that elements encompassed "isotopes" of the same atoms. Evidence for the actuality of isotopes was first observed by Sir Joseph John Thomson who, through the use of essentially an adapted cathode ray tube and concluded that "neon is not a simple gas but a mixture of two gases, one of which has an atomic weight about 20 and the other about 22" (Maher et al., 2015). Accordingly, Mr. Joseph John Thompson is considered one of the first to discover isotopes and is credited with the inventing of the mass spectrometer. Moreover, Davisson (2002) argues that Ernest Rutherford is recognized for determining the size and electronic state of atoms. However, it was not until 1932 that another researcher, James Chadwick, while working in Rutherford's lab, discovered that neutrons are neutral particles, and this explains the inconsistencies between atomic mass (neutrons + protons) and atomic number (protons). This also explains why atoms with higher mass are unstable and exhibit spontaneous radioactive decay. Therefore, isotopes can be defined as atoms that have the same number of protons and electrons, but different amounts of neutrons and therefore having different physical properties.

Isotopes can either naturally occur or be artificially produced. Most elements have naturally arising isotopes with a total of 100% percent natural abundances (Enghag, 2008). Hydrogen, for instance, has three naturally occurring isotopes, which are ^{1}H (protium), ^{2}H (deuterium or D), and ^{3}H (tritium or T) (Yan and Hino, 2016). Their natural abundances percent are 99.9885%, 0.0115%, and 10^{-18} %, respectively, thus all together should form 100% percent natural abundances. The sum of the percentage of natural abundances of all the isotopes of any given element should always add up to 100%. On the other hand, other isotopes such as ^{4}H to ^{7}H are not naturally available but can be artificially produced in the laboratory.

In the case of hydrogen isotopes, tritium is a radioactive isotope decaying into Helium 3 and releasing beta particles with a half-life of 12.32 years (Luo and Hong, 2017). While all heavier isotopes are artificially made and ^{5}H is the most stable and the least stable isotope is ^{7}H (Gurov; et al. 2004). Usually, the unstable isotopes have excess energy and therefore become more radioactive and spontaneously decay to get rid of that excess energy by emitting it in form of alpha, beta, or gamma until it reaches stability and eventually forms a new, different element. Bismuth ($^{209}{}_{83}Bi$) has the highest atomic and mass number of all the stable nuclides. But most of the nuclides with atomic number and a mass number higher than 83 and 209, correspondingly, are radioactive, as shown in Fig. 1.12.

The discovery of isotopes played a significant role in the emergence of contemporary theories related to fission, and fusion, which contributed to the harnessing of the atom in a wide range of peaceful applications. Many significant uses of isotopes today include power generation using ^{235}U fissile isotope; nuclear medicine and radiation therapy; industrial applications; food and agriculture applications; archaeology applications; forensic evidence; aerospace applications, and many other applications.

1.4 Isotopes separation methodologies

Power generation from nuclear power plants is among the most important and contemporary peaceful nuclear applications. Nuclear power plants often make use of ^{235}U fissile isotopes to generate the required energy. Unfortunately, the abundance of ^{235}U isotopes in nature is only 0.7% of natural uranium, while ^{238}U abundance in nature is about 99.3%. (Sivasankar, 2008). Commonly, nuclear power reactors such as light water reactors require ^{235}U isotope enrichment between 3% and 5% (Tsoulfanidis, 2012). The processes used in raising the ^{235}U content to 3%–5% is known as isotope separation or uranium enrichment (Tsoulfanidis, 2012). Laughter (2009) reported that in the early 1980s, Russia, France, the United States, Germany, the Netherlands, and the United Kingdom are known countries

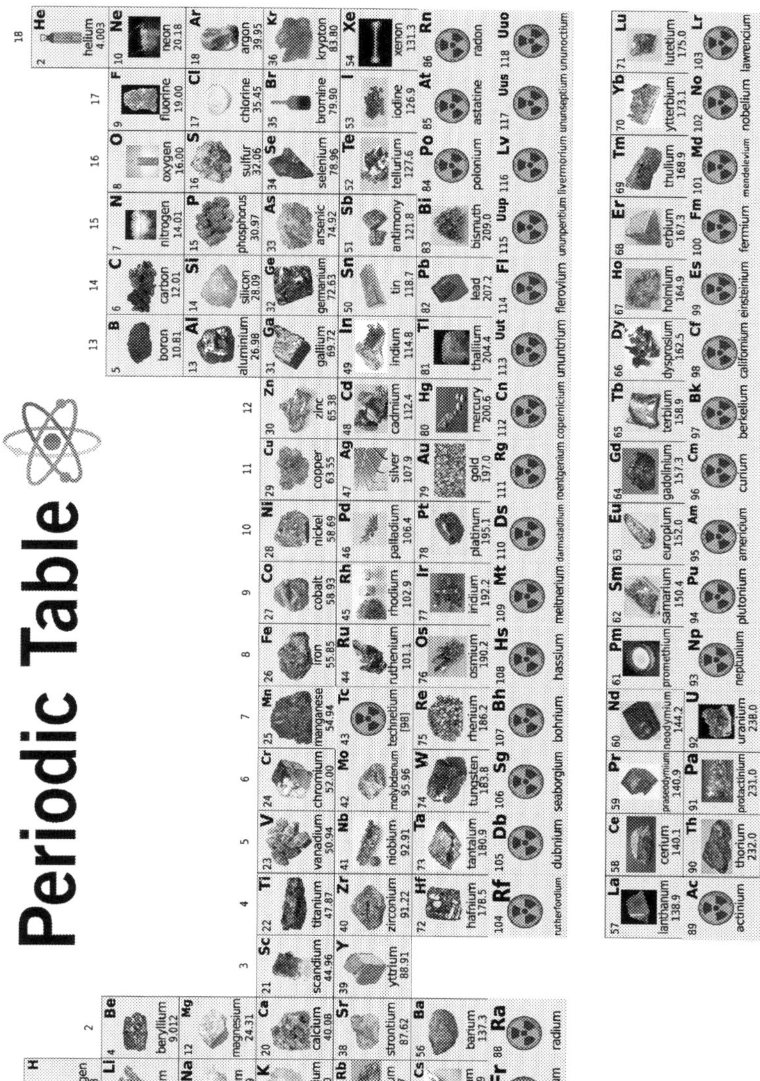

FIGURE 1.12 The periodic table.

for their usage of gas separation methods for the enrichment of Uranium-235 fissile isotope.

1.4.1 Gaseous diffusion approach

According to El-Sharkawi (2015), gaseous diffusion was first developed by Jesse Wakefield in 1934 during the development of the Manhattan Project. In 1940, the gaseous diffusion technology was further developed by German scientist Francis Simon and Hungarian scientist Nicholas Kurti at the Clarendon Laboratory of the United Kingdom. Later, researchers working on the Manhattan Project in Oak Ridge developed several different methodologies to separate Uranium-235 from Uranium-238. Gaseous diffusion is one of these methodologies and was historically used during the Manhattan project for building the nuclear bomb Second World War. The "Little Boy" bomb was an outcome of this methodology that was dropped in Japan in the city of Hiroshima on August 6, 1945, during the Second World War. The gaseous diffusion methodology is a process for separating and enriching isotopes based on the molecular diffusion of a gaseous isotopic mixture through porous membranes (Philippe and Glaser, 2014).

Working principle: Uranium-235 is the only naturally occurring fissile nucleus and its natural abundance is only 0.72%. ^{235}U is the only fissile isotope that supports a continuous nuclear fission chain reaction for power generation in nuclear power reactors if its enrichment reached about 3%–5% (Brown et al., 2013). In order to reach the required enrichment percentage, a gaseous diffusion method is used. According to many scholars such as Zumdahl and DeCoste (2012), it is based on Graham's law of diffusion or effusion that was formulated in 1848 by Thomas Graham, which states that the rate of diffusion of a gas is inversely proportional to the square root of its molecular mass and mathematically expressed by Eq. 1.10:

$$\text{Rate of diffusion for gas}_1/\text{Rate of diffusion for gas}_2 = \sqrt{}(M_2/M_1) \quad (1.10)$$

where:

$Rate_1$ is the rate of diffusion for the first gas.
$Rate_2$ is the rate of diffusion for the second gas.
M_1 is the molar mass of gas 1
M_2 is the molar mass of gas 2

During this process, Uranium-238 must be converted into a gaseous state so uranium oxide must be mixed with fluorine, and the solid mixture is then heated to a temperature above 56°C so that it can be processed as gas and forming uranium hexafluoride (Oxtoby et al., 2015). Thus, uranium hexafluoride is the only compound of uranium sufficiently volatile and can be used in the gaseous diffusion process. Fluorine in uranium hexafluoride consists of a single isotope, ^{19}F. The 1% variance in molecular weight between $^{235}UF_6$ and $^{238}UF_6$ is often due to the difference in masses of the uranium isotopes, using Graham's Law in uranium hexafluoride equation, as shown in Eq. 1.11. During this process, $^{235}UF_6$ will move faster than the heavier molecules of $^{238}UF_6$ and therefore pass easily the membranes compared to the $^{238}UF_6$ (Oxtoby et al., 2015):

$$\text{Rate 1/Rate 2} = \sqrt{}(M2/M1) = \sqrt{}(352.041/349.034) = 1.00 \quad (1.11)$$

where:

Rate of diffusion for gas_1, is the rate of effusion of $^{235}UF_6$

Rate of diffusion for gas_2, is the rate of effusion of $^{238}UF_6$

M_1 is the molar mass of $^{235}UF_6 = 235.043930 + 6 \times 18.998403 = 349.034348$ g/mol

M_2 is the molar mass of $^{238}UF_6 = 238.050788 + 6 \times 18.998403 = 352.041206$ g/mol

According to Cotton (2013), uranium hexafluoride gases (UF_6) during the gaseous diffusion process are forced through porous membranes usually of nickel or aluminum with a pore size of 10–25 nm, which is less than one-tenth the mean of the UF_6 molecule. These membranes are constructed of sintered nickel or aluminum because UF_6 is a highly corrosive substance. Since the molecular weight of $^{235}UF_6$ and $^{238}UF_6$ is virtually equal, minimal separation of the ^{235}U and ^{238}U is affected by a single pass through the barrier. According to Oxtoby et al. (2015), the lighter molecules of ^{235}U pass out of the porous membranes more rapidly than the denser particles of ^{238}U (Fig. 1.13). Moreover, to generate the desired level of enrichment, it is necessary to connect a significant number of diffusers in a sequence of stages called a cascade. During this process, all components of a diffusion plant need to be maintained at an optimal temperature and pressure to ensure that the UF_6 remains in the gaseous phase. Finally, the gas exiting the diffusion container is relatively enriched in the lighter molecules, while the heavier molecules are residual gas that gets depleted. The process is continuing, where the gas is required to be compressed at each stage to make up for the loss in pressure through the diffuser. This leads to the compression heating of the gas, which is then cooled before inflowing into the diffuser. The process of pumping and cooling makes diffusion plants huge consumers of electric power. Due to this, Gibson (2011) argues that gaseous diffusion is viewed as the most expensive to operate and required massive amounts of infrastructure compared to the second-generation gas centrifuge plant that will be discussed in the coming section.

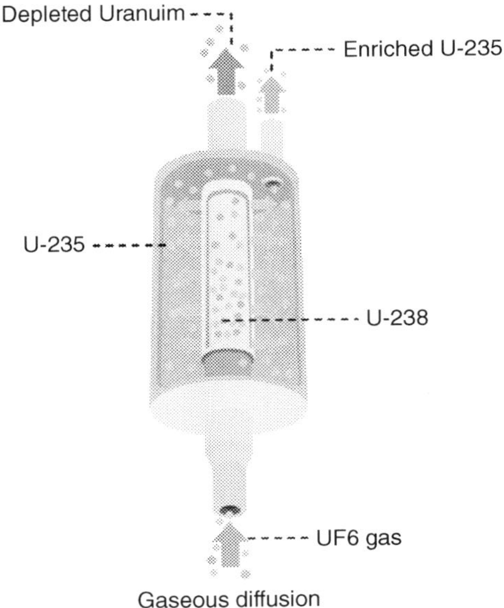

FIGURE 1.13 Gaseous diffusion device.

1.4.2 Gas centrifugation

Many scholars such as Gibson (2011) and Kemp (2012) argue that the gaseous diffusion method is viewed as the most expensive method used for producing enriched uranium that became obsolete and was replaced by the second-generation gas centrifuge technology. The second-generation gas centrifuge technology is more economical because it requires far less electric power to produce the same amounts of the separated Uranium-235 (Kemp, 2012). Historically, centrifuge technology was the first device used to separate chemical isotopes used by scholar Jesse Beams from the University of Virginia, to separate chlorine-35 from chlorine-37 in 1934 (Kemp, 2009). Since then, the centrifuge technology was considered by American physicists to be the best solution for large-scale enrichment.

The main working principle of this technique is based on the mass differential between the two isotopes that are required to be separated from each other (Fig. 1.14). Since Uranium ^{238}U possesses three more neutrons in its nucleus, it means that it has a higher mass compared to ^{235}U. Accordingly, Gibson (2011) argues that all uranium separation methods depend on the fact that ^{235}U is about 1.26% lighter than ^{238}U. To separate Uranium-235 isotope from Uranium-238 isotope using the gas centrifugation method, it is required to first be converted into gaseous form; it is therefore converted into uranium hexafluoride (UF_6), which is used to enrich uranium

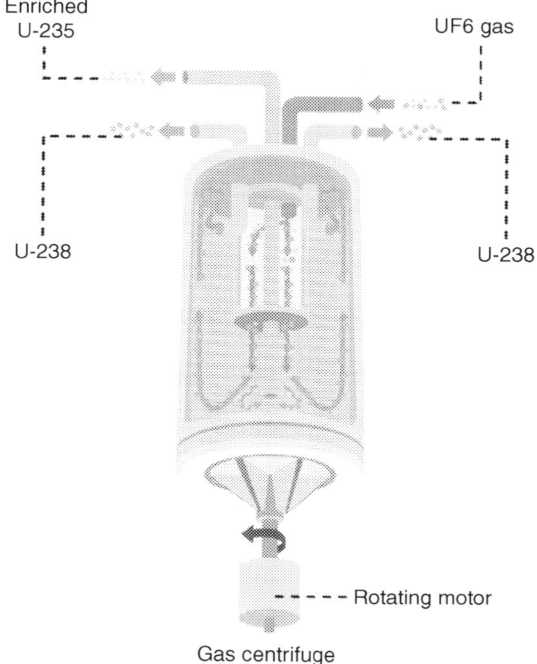

FIGURE 1.14 Gas centrifugation device.

(Gibson, 2011). UF_6 gas is entered into a centrifuge cylinder and rotated at high speed. Accordingly, a strong centrifugal force is created that forces more of the denser gas molecules comprising the ^{238}U toward the wall of the cylinder, while the lighter gas molecules of ^{235}U gathered near the center. The slightly enriched stream in ^{235}U is removed and fed into the subsequent higher stage, while the slightly depleted stream gets recycled back into a lower stage.

From the economic point of view, the gas centrifuge method is known to be more economical compared to the gaseous diffusion method because it needs 96% less electric power compared to a gaseous diffusion plant of the same separate work capacity. In this context, Roberts (1974) reported that "A gaseous diffusion plant of 8.75 million SWU capacity requires about 2400 megawatts of electric power. A gas centrifuge plant of that same capacity requires substantially less power, within about 10% of that power."

1.4.3 Laser isotope separation

All isotopes of the same element are different in the atomic mass number due to the variance in the number of neutrons in the nucleus, despite having had almost identical chemical properties but different physical properties (Quagliano, 1969; Duke and Williams, 2007). Accordingly, isotopes can be separated based on the differences in the mass of the atoms. According to USNRC (2008), each isotope has a unique signature of spectroscopic that their atoms can absorb certain electromagnetic radiation (light) at an individual and well-defined wavelength specific to each atomic or molecular species. This causes atoms to become excited to higher vibrational, rotational, or electronic energy levels if they are exposed to a particular wavelength.

Since each isotope has a unique signature of spectroscopic, then each isotope will absorb light slightly differently from other isotopes and when the targeted atoms are being ionized, they become excited, their state will change and may enter preferentially into chemical reactions. The excited positively charged atoms are ultimately collected electrostatically on a cathode and this method is called "photochemical isotope separation" (Ghiassee, 1979). The first photochemical isotope separation method first theoretically emerged in 1920. The first successful experiment involving isotope separation was conducted in 1932, where the regular light source was used as a source of excitation. According to USNRC (2008), during the early 1970s, scholars from Lawrence Livermore National Laboratory were the pioneers that investigated the use of lasers for isotope separation instead of the normal light. In 1974, after a lot of research and experiments, Lawrence Livermore National Laboratory's scholars managed to introduce a new concept for isotopes separation based on laser technology that is called "atomic vapor laser isotope separation (AVLIS)" (USNRC, 2008). This approach has been used for Uranium-235 separation from Uranium-238 (uranium enrichment) and became one of the standard enrichment approaches in the United States. Scholars such as Fuss (2015) and Greenland (1990) argue that this technology drastically reduces the energy, time, and physical footprint of an enrichment facility. Not only that the LIS offers the possibility of a higher degree of enrichment in a single step, while in old-style centrifugal or diffusion isotope separation processes, up to a thousand stages of centrifugation was required to attain a weapons-grade advancement. Accordingly, scientists from the former Soviet Union, the United States, and China have further developed this technology and become predominant in the field of uranium enrichment to reach the required level of enhancement. On the other hand, the cheaper and more effective LIS technology can cause a serious proliferation threat, because it can allow nations to hide their clandestine enrichment operations from the global community more efficiently (Fuss, 2015).

There are currently two standard LIS processes used in the industry. These include AVLIS and molecular laser isotope separation (MLIS).

1.4.3.1 Atomic vapor laser isotope separation

Isotopes are differentiated by the difference in the number of neutrons and therefore each isotope has a certain hyperfine structure different from others, which influences the nuclear magnetic dipole (Greenland, 1990). Furthermore, Zare (1977) argues that these differences make each isotope only absorb a certain spectrum, and therefore a precisely tuned laser can be used to excite one specific isotope and not the other isotope. Paisner (1988) describes that in the AVLIS process where the material is converted first into an atomic vapor stream using multisteps photoionization. Then, a precisely tuned laser is used to separate isotopes of those materials. A laser is a device that can produce large numbers of photons compose of electromagnetic radiation and all having almost the same rate. USNRC (2008) outlines that this technique depends on the quantum mechanical connection between energy and frequency in an atom's electrons. AVLIS achieved separation of Uranium-235 from Uranium-238 based on the energy of transition difference between the two isotopes. For instance, the energy of transition difference between the Uranium-235 and Uranium-238 is about $4.2 \times 10^{\times 5}$ eV (USNRC, 2008). Moreover, the USNRC recommended that in order to excite and ionize Uranium-235 isotopes without affecting the other Uranium-238 isotopes, the laser beam must be tuneable to an accuracy of 1 part of 10^5. Once the Uranium-235 isotopes' atoms are excited, they become positively charged and are then attracted to a negatively charged plate called the cathode through an electromagnetic field.

1.4.3.2 Molecular laser isotope separation

MLIS is the second type of isotope separation method using laser technology. According to Kok (2009), this method was first developed at the Los Alamos National Laboratory in the early 1960s and shaped out late 1970s. MLIS is based on a specially tuned laser to separate the targeted isotope using selective ionization of hyperfine transitions of that particular isotope. It is almost similar to AVLIS, but it operates in a cascade setup like the gaseous diffusion process and requires low energy consumption. In the case of Uranium isotopes separation, the MILS working medium is uranium hexafluoride stream instead of vaporized uranium to separate Uranium-235 isotopes from Uranium-238 isotopes (Kok, 2009). The expanded and cooled stream of UF_6 is irradiated with an infrared laser and commonly the carbon dioxide laser with transitions of around 16 μm (Jensen et al.,1982). The photons are absorbed by the excited $^{235}UF_6$ and cause its photolysis to $^{235}UF_5$. The precipitated $^{235}UF_5$ is enriched with ^{235}U, which will then be fed into the next stage of the cascade for further enrichment, which is eventually collected through cathodes as shown in Fig. 1.15.

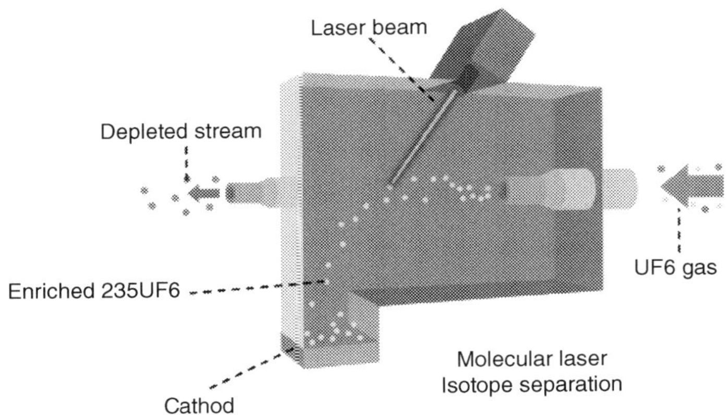

FIGURE 1.15 Molecular laser isotope separation (*MLIS*).

LIS technology might increase the difficulties of policing nuclear non-proliferation (Fuss, 2015).

1.4.4 In-situ Uranium-235 recovery based on prompt fission neutrons technology

Al Nabhani and Khan (2019) argues that the available geological and geochemical studies, collected seismic data, oilfield well-logging correlation data, and collected coring and cutting samples from the different fields including oilfields, provides an excellent indication about the possible quantities of uranium and thorium in the reservoir that can be extracted. These data help to identify exactly at what depths these uranium formations are located and what are their intervals lengths. Not only that they found that with the support of "prompt fission neutrons (PFNs) technology" it became easier to determine the quantity and the abundance of ^{235}U before the extraction process takes place, which saves a lot of time, effort, and cost in subsequent enrichment operations.

During the in-situ Uranium-235 recovery drilling activity, the drilling cuttings are collected at the surface by the geologist for further geological analysis. The results obtained from the geological analysis help to further identify with some certainty the geological formation where the uranium, thorium, and any other nuclear fission materials are deposited or located. However, these results cannot tell us the exact depth, the quantities, exact isotopes, and other important geological parameters related to targeted uranium, thorium, and any other nuclear fission isotopes. Thus, to confirm the depth and to quantify the quantity of Uranium-235, isotopes, for instance, a measurement while drilling techniques can be used as well as a direct uranium logging technique called PFNs (Al Nabhani and Khan, 2019). The PFN logging technique comprises of a pulsed source of neutrons flux. In the pulsed source of neutrons flux, a high voltage of 14 MeV pulses at 1000 cycles per second, and emits about 108 neutrons per second that accelerate and excite deuterium ions into tritium as expressed in Eq. 1.12:

$$H_2 + H_3 \rightarrow n + 4He \tag{1.12}$$

Al Nabhani and Khan (2019) explained that the PFNs logging tool emits neutrons flux to the targeted geological formations, which collide with only ^{235}U isotopes and lead to slow-neutron-induced fission of ^{235}U in the formation. Epithermal neutrons and thermal neutrons returning from the formation after the collision following fission of natural ^{235}U isotopes in the rock formations are counted separately in detectors in the logging tool called thermal/epithermal neutron detector. The thermal/epithermal neutron detector gives the percentage ratio of ^{235}U where the ratio of epithermal to thermal neutrons is directly proportional to ^{235}U isotopes.

According to Givens and Stromswold (1989), the time-gated ratio of epithermal to thermal neutron counts provides a measure of uranium content. Uranium content measurements obtained from PFNs logs have shown good agreement with core measurements, so this technique provides a major data source for delineation and exploitation of uranium isotopes mineralization. The PFN technique provides a precise direct measurement of in-situ uranium recovery, in particular the ^{235}U isotopes over even very narrow intervals. Not only that, the PFN technique is also able to identify other fissionable material such as plutonium isotopes by detecting their thermal neutrons.

Al Nabhani and Khan (2019) explained that once uranium, thorium, and other fissionable material zones are identified, the cased hole section must be perforated, and subsequently, multiple zone completion equipment are installed in order to extract nuclear materials from more than one geological pay zone.

Once the drilled well is classified as economically feasible and ready for production, a recovery project starts. This includes drilling several injection wells that pump a lixiviant solution, which is exactly like the solution used for the enhanced oil recovery process. The lixiviant solution is a groundwater solution mixed with oxygen or sodium bicarbonate (leachate) or CO_2 or Sulfuric acid leach, and possibly other chemical additives if needed, that is pumped into sandstone or shale containing uranium, thorium, and other nuclear radioactive isotopes. The dissolved oxygen in the lixiviant solution oxidizes and dissolves uranium, which is then lifted to the surface through the recovery drilled well. Since thorium is less soluble, its particles are lifted to the surface in the form of suspended particles in the produced formation water (lixiviant). The solution that contains uranium, thorium, and other nuclear radioactive isotopes is collected via a special multiple zone completion equipment. These materials are then pumped from the wellhead directly to the Uranium-235 in-situ recovery processing plant. Fig. 1.16 depicts a diagram of the Uranium-235 recovery process flow.

Al Nabhani and Khan (2019) highlight that the in-situ Uranium-235 Recovery process comprises of four stages, which are the ion exchange unit, elution unit, precipitation unit, and filtering and drying unit. The ionic exchange unit contains millions of positive ion resins, which attract and binds the uranium negatively charged particles. The resins will then be transferred to the washing unit (Elution unit) and be washed by the brine solution that creates a reduced environment. This, therefore, increases the concentration of the uranium solution, which based on PFN measurements contains more Uranium-235 isotopes. A uranium solution will then enter the precipitation unit where an acid, such as hydrochloric acid, is injected into the solution to modify the PH. Hydrogen peroxide will be added at this stage to stimulate uranium to precipitate more and fall. The uranium solution is then dewatered and dried. Finally, the uranium is collected in the form of powder. The collected powder is then packed in drums and transported to the nuclear facilities where it will be subjected to further refining and enrichment processes to the required levels using any of the previously mentioned techniques. Finally, once uranium, thorium, and other nuclear materials are separated, the remaining leachate from the dewatering process is then passed into a chemically designed circuit to release the solid particles, chemically reconditioned, and subsequently prepared to be reinjected back into the in-situ Uranium-235 recovery field. Fig. 1.17 depicts a diagram of the proposed Uranium-235 recovery process flow.

1.5 Production of radionuclides

It is very imperative to produce artificially made radionuclides that are considered important for diagnostic and therapeutic nuclear medicine treatments as well as in many industrial applications. This is due to the naturally occurring radionuclides often being unsuitable for such applications due to their characteristically long half-lives and unideal physical or chemical characteristics. Artificial radionuclide production involves the nuclear reaction of changing the composition of protons and neutrons within the nucleus. Many scholars such as Saha (1984) reported that the most common methods of radionuclide production include nuclear reactions, such as fission, fusion, neutron activation using cyclotron, reactors, and radionuclides generators. In 1934, Irene Curie and her husband, Frederic Joliot, discovered the artificial production of radioisotopes, and they were the first ones who managed to produce artificial unstable radionuclide that decayed through positron emission (Leroy, 2003; Hurst, 1997).

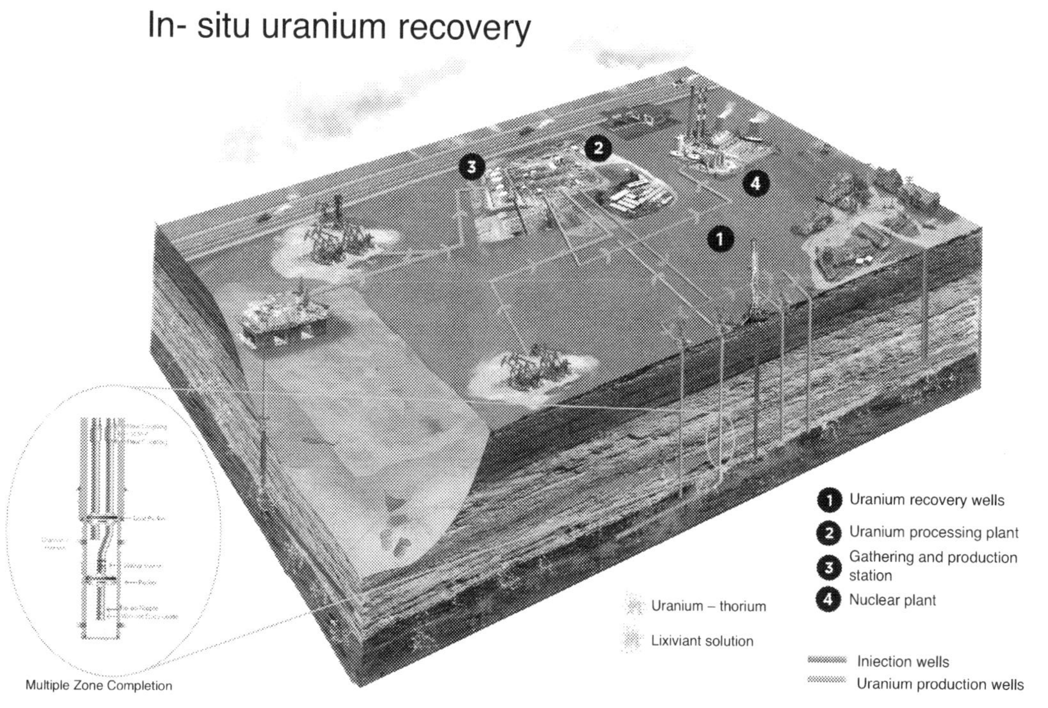

FIGURE 1.16 An overview of the integrated in-situ Uranium-235 recovery process.

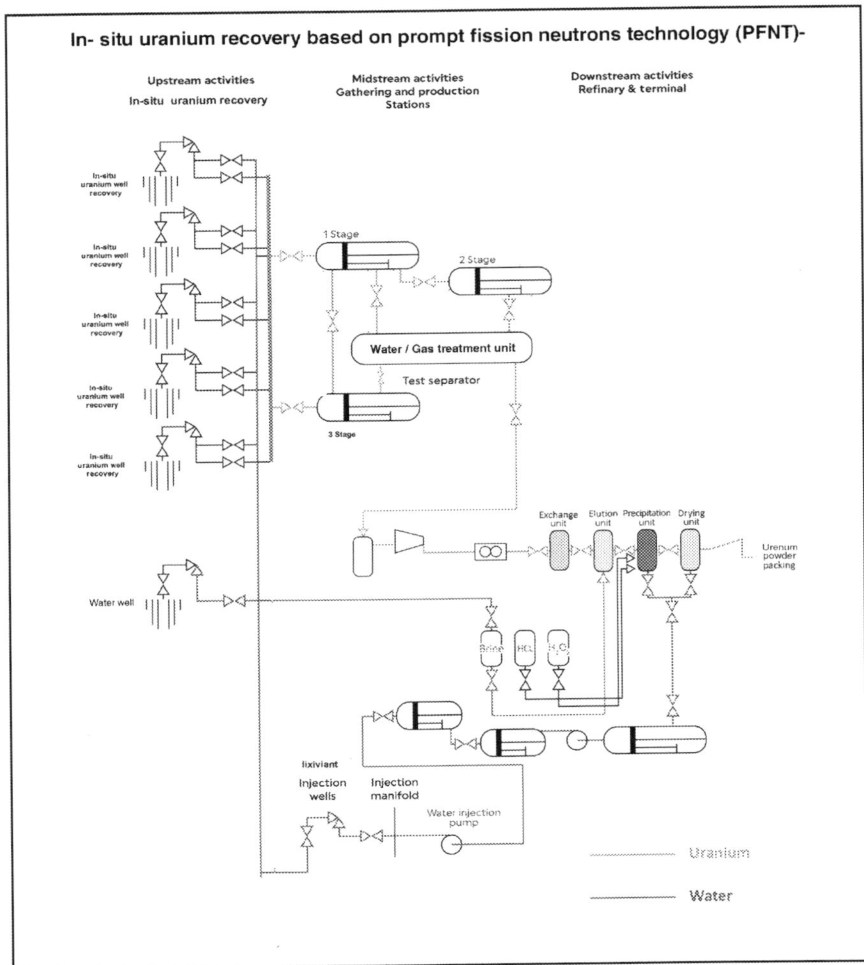

FIGURE 1.17 In-situ Uranium-235 recovery process flow chart.

The next section will shed light on the most common methodologies used for artificial radionuclides production.

1.5.1 Induced fission

A nuclear fission reaction is splitting off a heavy nucleus by bombarding it by a projectile into two lighter atoms and consequently producing new lighter nuclei called binary fission, accompanied by the release of huge thermal energy, kinetic energy, gamma radiation, and free neutrons typically two or three neutrons (Sanctis et al., 2016). Fission of low energy neutrons commonly takes place in those isotopes whose nuclei contain odd numbers of neutrons such as Uranium-233, Uranium-235, Plutonium-239, Plutonium-241 because the nuclear binding forces are relatively weak in an odd number of neutrons in a nucleus compared to the even number of neutrons in a nucleus (Hore-Lacy, 2010). These isotopes are not naturally occurring but are artificially produced from the fertile nuclei of ^{232}Th, ^{238}U, and ^{240}Pu in a nuclear reactor. On the other hand, the world nuclear association (2020) reported that scientists managed to introduce fission in nuclei containing an even number of neutrons by bombarding them with neutrons that have energy over one million electron volts (MeV) and move at the speed of light.

When neutrons are used to bombard fission nuclides such as Uranium-235 or plutonium-239 in a nuclear reactor, then the fission process starts and often it results in the splitting of the heavy nucleus into two lighter weight of the fission fragments called fission product along with the release of two or three neutrons, and huge energy accompanied with highly radioactive materials such as Cesium-137, Iodine-131, and Strontium-90 (Medalia, 2011). The fission chain reaction is commenced when some beryllium is mixed with polonium, radium, or any other alpha emitter in the reactor that will be used to bombard the Uranium-235 nucleus (Hore-Lacy, 2010). The Uranium-235 nucleus absorbs a neutron and converts into Uranium-236, which becomes comparatively unstable, and it will likely attempt to become stable by splitting into two fragments of around half the mass followed instantaneously by the emission of a number of neutrons. It is worth mentioning that the U.S. Congress Joint Committee on Atomic Energy (1967) reported that, during Uranium-235 fission reactions in a thermal reactor, typically 2 or 3 neutrons are produced (or on average of 2.45 neutrons) while the probability of fission from Plutonium-239 produces an average of 2.9 to 3 neutrons per fission reaction (Cochran, 1998). Conversely, this is with a smaller fission cross-section value compared to Uranium-235 and accordingly has a lower critical mass than ^{235}U. The chain reaction can only be self-sustaining if a sufficient amount of fissionable material is used (Shipman et al., 2012). Scholars such as Brown et al. (2013) argue that scientifically, the fission process depends on the level of the "critical mass," as shown in Fig. 1.18. Therefore, if there is a less critical mass, then it is called subcritical mass, and cannot sustain a chain reaction because neutrons end up flying out into space instead of crashing with nuclei and the rate of neutron loss is greater than the rate of neutrons created through fission (Brown et al., 2013). However, if the rate of neutron loss is equal to the rate of neutrons created through fission, then it is called "critical mass," while if it is more than the critical mass it is called "supercritical mass" (Brown et al., 2013). This can generate enough energy for a massive explosion where the rate of neutrons' loss is less than

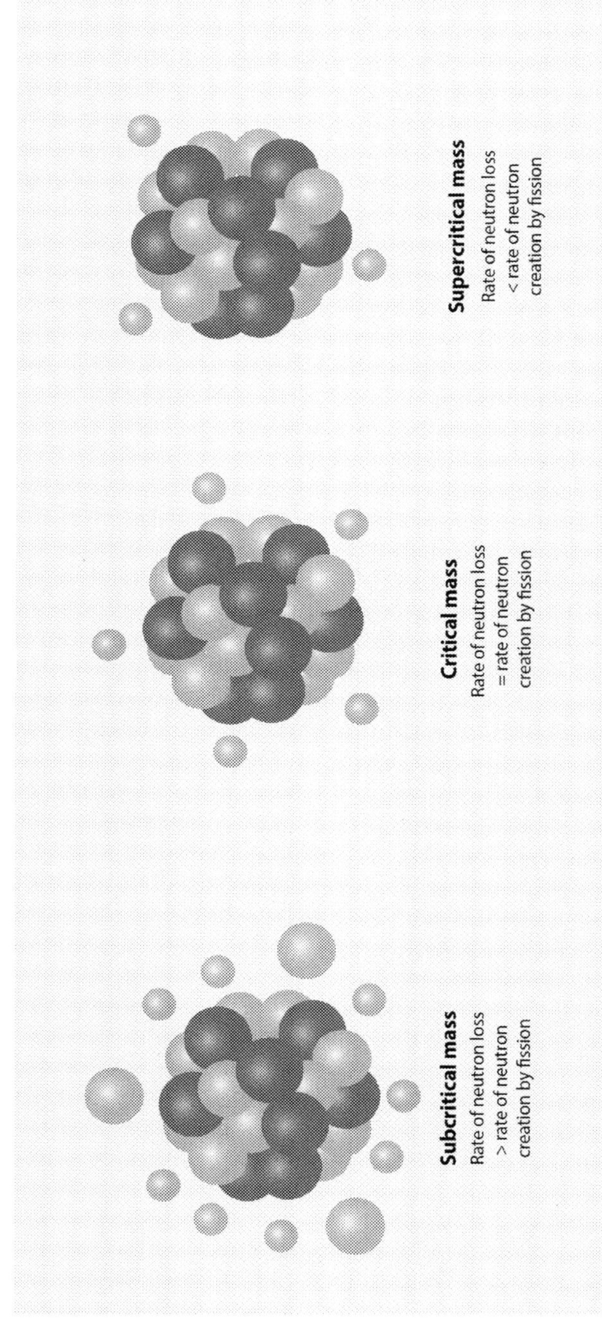

FIGURE 1.18 The role of critical mass in a chain fission reaction.

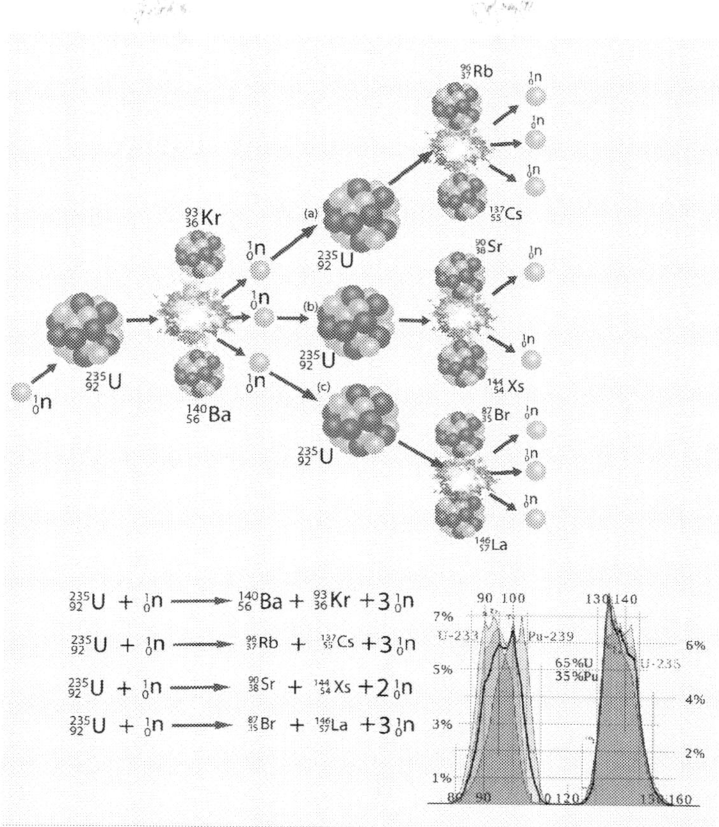

FIGURE 1.19 Fission reaction of Uranium-235 and the produced fission fragments.

the rate of neutrons creation by fission. The critical mass often depends on the type of material: its purity, the temperature, the shape of the model, including how the neutron reactions are controlled.

According to Joyce (2017), approximately 85% of the energy (~200 MeV) released from a single nuclear fission reaction is kinetic energy from the fission fragments. The rest of the energy comes from gamma radiations emitted as well as the kinetic energy of the neutrons. According to the conservation law, the total energy is conserved in any reaction. Therefore, in the fission reaction of Uranium-235, it will produce new fission fragments such as Ba, Kr, Sr, Cs, I, Te, Rb, La, Zr, and Xe with atomic masses distributed around 95 and 135 (Fig. 1.19) (Hore-Lacy, 2010).

In this context, Breeze (2014) argues that Barium and krypton are the most frequent fission reaction's products from the fission reaction of ^{235}U,

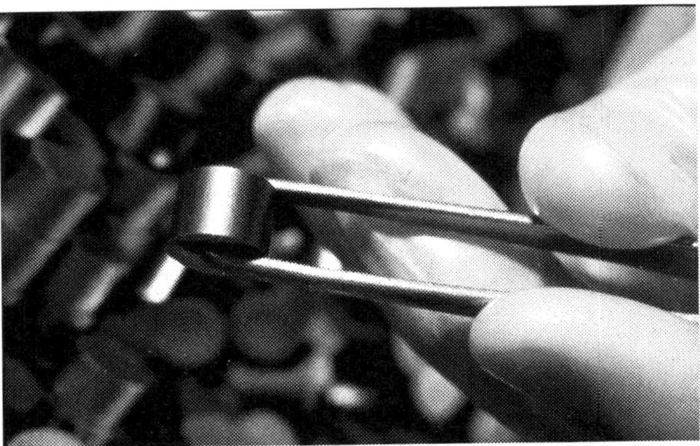

FIGURE 1.20 Uranium fuel pellet size (From: https://en.wikipedia.org/wiki/Nuclear_fuel#/media/File:Fuel_Pellet.jpg).

as illustrated in Eq. 1.13:

$$^{235}_{92}\text{U} + {}^{1}{}_{0}\text{n} \rightarrow {}^{144}_{56}\text{Ba} + {}^{89}_{36}\text{Kr} + 3^{1}_{0}\,\text{n} + 210\,\text{MeV} \tag{1.13}$$

Both the barium and krypton isotopes decay to form more stable isotopes such as neodymium and yttrium and emit many electrons from the nucleus in the form of beta decay accompanied by gamma rays, which make the fission products extremely radioactive. However, the radioactivity intensity diminishes with time due to their half-lives but is still considered very harmful. Fission reaction is not only related to the production of new products, but it is also associated with the production of colossal energy.

According to conservation laws, nucleons (protons + neutrons) are conserved during any nuclear reaction. Referring to Uranium-235 fission nuclear reactions, shown in Eq. 1.13, we can notice that the number of neutrons in the reactants side (235 +1) equal to the number of neutrons in the products side (141 + 92 + 3), which both sides equal to 236 neutrons. Not only that but it has been also noticed that there is a small loss in atomic mass that may be revealed to be comparable to the energy released. This energy is a result of the difference between total binding energy released in the fission reaction of an atomic nucleus and from the specific break-up energy, which is estimated to be on averages of 200 MeV for a single ^{235}U nuclear fission reaction (Joyce, 2017). Therefore, the energy released from the fission of 1 kg of Uranium-235 can produce about 2.5 million times as much energy as is created by burning 1 kg of coal. While one uranium fuel pellet that has a size of 0.3-inch diameter by 0.5-inch long (size of a fingertip Fig. 1.20) can generate the equivalent energy of 17,000 cubic feet of natural gas, 1780 pounds of coal, or 149 gallons of oil (Hordeski, 2009).

Accordingly, nuclear fission is an overwhelming scientific revolution in the field of energy that not only produces new products of nuclides but also enormous energy from the fission process. This energy can be used in many peaceful applications such as electricity generation and also unfortunately in nonpeaceful applications such as military applications including the fission nuclear bombs and warheads.

1.5.2 Fusion

Scholars such as Serway and Jewett (2010) defined the fusion as the process of converting lighter nuclei into heavier nuclei accompanied by the conversion of the mass into massive amounts of energy. Scientifically, atoms in the nuclei are connected firmly with a nuclear bond force. To produce a new nucleus, individual atoms need to undergo a fusion process, and this occurs only when certain atoms are given enough energy to reach another atom overcoming the electric repulsive force and breaking the bonding force between the atoms to fuse. This energy can be achieved in two ways: the first way comprises speeding up atoms in a particle accelerator; the second way comprises heating them at high temperatures, which is also called the thermonuclear reaction (Haider, 2019). This type of reaction gives the nuclei enough kinetic energy to overcome the strong repulsive forces ensuing from the positive charges in their nuclei. This is called the energy barrier or "Coulomb barrier" given in Eq. 1.14 so that they can get close enough to collide and undergo nuclear fusion reaction. The Coulomb barrier proportionally increases as the atomic numbers of the colliding nuclei increase (Cerny, 2012):

$$U_{Coul} = k(q_1 q_2 / r) \tag{1.14}$$

where

k is the Coulomb's constant, which is 8.9876×10^9 N m^2/C^2.
q_1, q_2 are the charges of the interacting particles.
r is the interaction radius.

When an atom is heated beyond its ionization energy, electrons of an atom are excited and become ionized, forming a cloud of ions. This cloud is known as plasma. It is electrically neutral but conductive medium and magnetically manageable (Dendy, 1990). Several fusion reactors take advantage of this to regulate the particles as they are heated. The sun, for instance, is a real example of a net fusion reaction of hydrogen to form helium that takes place at a solar-core temperature of 14 million kelvin (Stokes, 2019). The net outcome of such fusion reaction is the fusion of four protons into helium (Fig. 1.21), with the discharge of two positrons, two neutrinos, and convert two of the protons into neutrons accompanied by the release of massive energy.

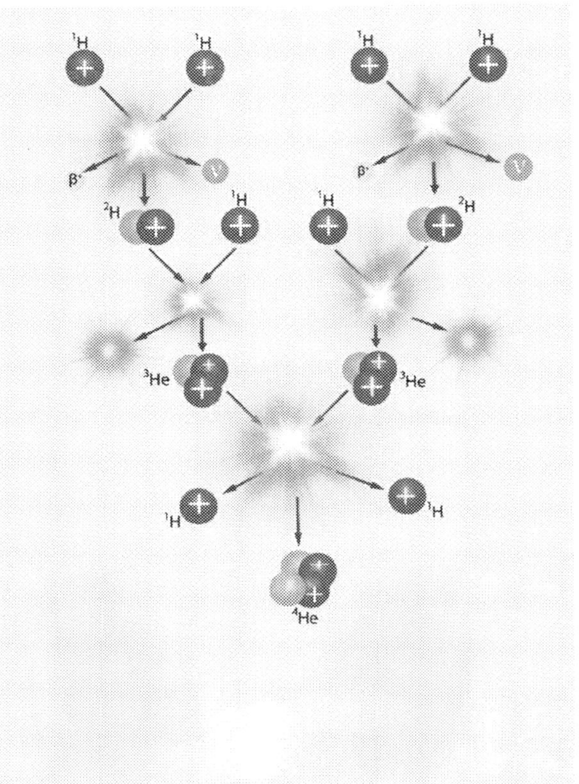

FIGURE 1.21 Fusion reaction of four hydrogen atoms into a helium.

Helium is produced when four hydrogen atoms (the heavy isotopes of hydrogen, which are a deuteron $^{2}{}_{1}H$, and a triton $^{3}{}_{1}H$) undergo thermonuclear fusion reaction at extremely high temperatures, as shown in Eq. 1.15 and Fig. 1.22. Karttunen et al. (2013) argue that such fusion reaction causes the new nucleus to have about 0.7% mass less than that of four hydrogen nuclei. This indicates that the fusion reaction does not only produce new nuclei but also produces huge energy. Therefore, according to conservation laws, the lost mass from helium is turned into energy during the fusion cross ponding to an energy release of 6.4×10^{14}J per kilogram of hydrogen (Karttunen et al., 2013):

$$ {}_{1}^{2}H + {}_{1}^{3}H \rightarrow {}_{2}^{4}He + 2{}_{1}^{0}n \quad (1.15) $$

1.5.3 Neutron activation

Neutron activation is another approach for radionuclides production. Not only that, neutron activation has a wide application in peaceful

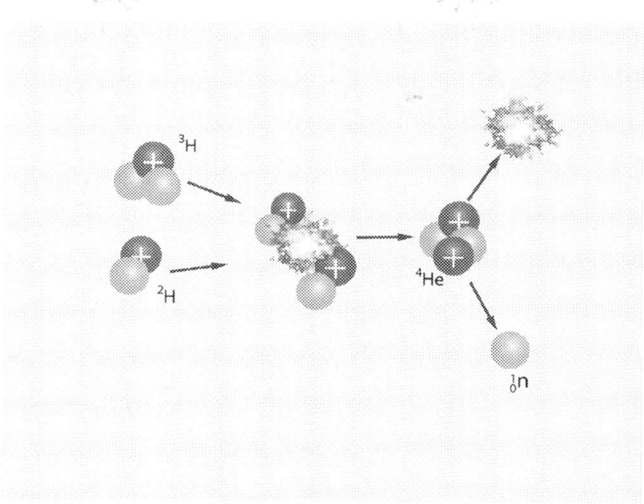

FIGURE 1.22 D–T fusion reaction.

nuclear applications such as elements analysis in any environment under investigation such as monitoring the environment for metallic pollutants and many other useful applications (Robertson and Carpenter, 1974). The basic premise of neutron activation is based on the principle that neutrons have no net electrical charge. Hence, they are neither attracted nor repelled by the atomic nuclei (Cherry et al.,2012). This process takes place when neutrons hit the targeted atomic nuclei, then some of the neutrons are captured by the targeted nuclei of the target atoms. After capturing neutrons, the targeted nucleus becomes heavier than before, unstable and entering excitation states. Eventually, the unstable radioactive nucleus will try to become more stable by decaying its excess energy in the form of emitting gamma rays, or particles such as alpha or beta. The unstable radioactive nuclei may have half-lives ranging from seconds to many years. It is worth mentioning that the neutron activation reaction is a very precise process that must be carefully handled otherwise it may end up in a fission reaction. All this depends on the kinetic energy of the neutron.

A number of scientists such as Podgorsak (2016) identified that there are two common types of neutron activation reactions. According to Podgorsak (2016), the first one is known by (n, γ) reaction. In this reaction, the targeted nucleus (^A_ZX) captures the activated neutron and become neutron activated nucleus ($^{A+1}_Z$X) in the excited state, which undergoes de-excitation to its ground state by emitting a prompt γ rays [^A_ZX (n, γ) $\longrightarrow$ $^{A+1}_Z$X]. Such as ^{41}K(n, γ) $\longrightarrow$ ^{42}K; ^{50}Cr(n, γ) $\longrightarrow$ ^{51}Cr; ^{124}Xe (n, γ) $\longrightarrow$ ^{125}Xe $\longrightarrow$ ^{125}I. These new nuclei represent the isotopes of the same chemical element. While the second type of neutron activation reaction is the (n, p) reaction. In this case, the target nucleus (^A_ZX) is activated by the neutron and promptly ejects a proton [$^{A}_{Z}X$ (n, P) $\longrightarrow$ $_{Z-1}^{A}X$] such as the case with ^{14}N (n, p) $\longrightarrow ^{14}C$; ^{32}S(n, p) $\longrightarrow ^{32}P$; ^{35}Cl(n, p) $\longrightarrow ^{35}S$. Note that, the new nuclei result from (n, p) neutron activation reaction does not represent the same chemical element.

Cobalt-60 is a good example of neutron activated isotope that is considered as a valued source of gamma radiation for radiotherapy generated through the Neutron activation process by bombarding Cobalt-59 with neutrons (Baba et al., 2013). Accordingly, Cobalt-60 is widely produced using the neutron activation reaction process in a nuclear reactor, which eventually decays by the emission of a beta particle and gamma rays and form Nickel-60 (6^0Ni_{28}). This reaction is known to have a half-life of about 5.27 years with a decay rate of 1.089% per month (Baba et al., 2013).

Moreover, the neutron activation process is a very useful method for tracing element analysis, nondestructive analysis, water and minerals abundances, and many other applications.

1.5.4 Cyclotrons and synchrotron

1.5.4.1 Cyclotrons

The production of radionuclides involves the conversion of the original atoms into another atom that may have similar chemical properties or different. According to IAEA (2008), the exact type of reaction required for a certain radionuclide production depends on a number of parameters, and the most important of them is the intensity and the type of energy of the bombarding particle that may result in the emission of particles or not. For instance, if the targeted nucleus is bombarded by a neutron that has no charge and has a certain type of energy, this refers to the free neutron's kinetic energy and given in electron volts that range from cold neutrons to thermal neutrons, epithermal neutrons, slow neutrons, fast neutrons, and ultrafast neutrons. Then the reaction will continue without the emission of any particles and eventually the newly induced nuclide will carry the same chemical properties as those of the original nuclide (IAEA, 2008). If, however, the target nucleus is bombarded by a charged particle that has more energy than the binding energy of nucleons in the nucleus, which is of the order of 8 MeV on average, then the reaction will emit particles because the incoming projectile will force other particles in the target nucleus to be ejected (IAEA, 2008). Eventually, the newly induced nuclide will not have the same chemical properties as those of the original nuclide. Accordingly, such a conversion process is usually accompanied by a change in the number of protons and/or neutrons in the targeted nucleus. Therefore, by cautiously selecting the target nucleus, the bombarding particle, and its energy, it is conceivable to produce the required radionuclide (IAEA, 2008).

Ernest Lawrence is considered as the pioneering American nuclear scientist who first to invent the first cyclotron in 1932 at the University of California and in 1939 he has been rewarded the Nobel Prize in Physics for his invention of the Cyclotron (Brown et al., 1995). Since then and until the emergence of synchrotron during the 1950s, cyclotrons were the dominant technology for particle acceleration. Despite the emergence of the synchrotron, cyclotrons are however still used in physics, nuclear medicine, and nuclear research labs for the production of particle beams. They are also widely used as the first stage of the multistage particle accelerators.

Cyclotrons work on the principle of the use of magnetic force to bend a moving charged particle into a semicircular path due to the effect of the Lorentz force that is perpendicular to the horizontal electric field (Peach et al., 2011). When the moving charged particle is at the center gap between the two dees, it is accelerated by the electric field created between the two dees that are placed in the magnetic field region in the vacuum chamber, as depicted in Fig. 1.23. The charged particle then moves in a spiral path outward from the center gap between the dees that mainly accelerate the moving charged particle toward the dees (Peach et al., 2011). During this process, the perpendicular magnetic field that is created by the two magnets located at the top and at the bottom of the two dees continues bending the movement of the moving particle. This process continues until the moving charged particle reaches the final loop in the dees and finally leaves the cyclotron at a very high-speed via an evacuated beam tube to hit the targeted nucleus. At this point, a nuclear reaction occurs due to the collisions that will create the new desired particle.

Cyclotrons are widely used for the production of different types of radionuclides that are used for different applications. They are widely used for the production of radionuclides that are used in nuclear medicine and radiopharmaceuticals applications, including positron emission tomography (PET) and single-photon emission computed tomography (SPECT) (IAEA, 2008). Nearly most of the PET's and SPECT's radionuclides are produced via using negative ion cyclotrons that are able to accelerate charged particles to higher energies. There are different types of cyclotrons of different capacities. For instance, according to IAEA (2008), level I cyclotron of 10 MeV proton beam delivery can be used to produce the conventional PET's isotopes, such as ^{11}C, ^{11}N, ^{15}O, and ^{18}F. While a higher energy particle beam cyclotron Levels II or III is suitable for the production of SPECT's isotopes such as ^{201}Tl, ^{67}Ga, ^{123}I, and ^{111}In. Therefore, it is worth mentioning to select the correct accelerator to deliver required particles such as protons, deuterons, and others. The energy required for a certain nuclear reaction is the most important criterion of Cyclotron's selection.

Currie et al. (2011) outlined that, in the cyclotrons, there are commonly four nuclear reactions, which are protons (p), deuterons (d), triton (t), and

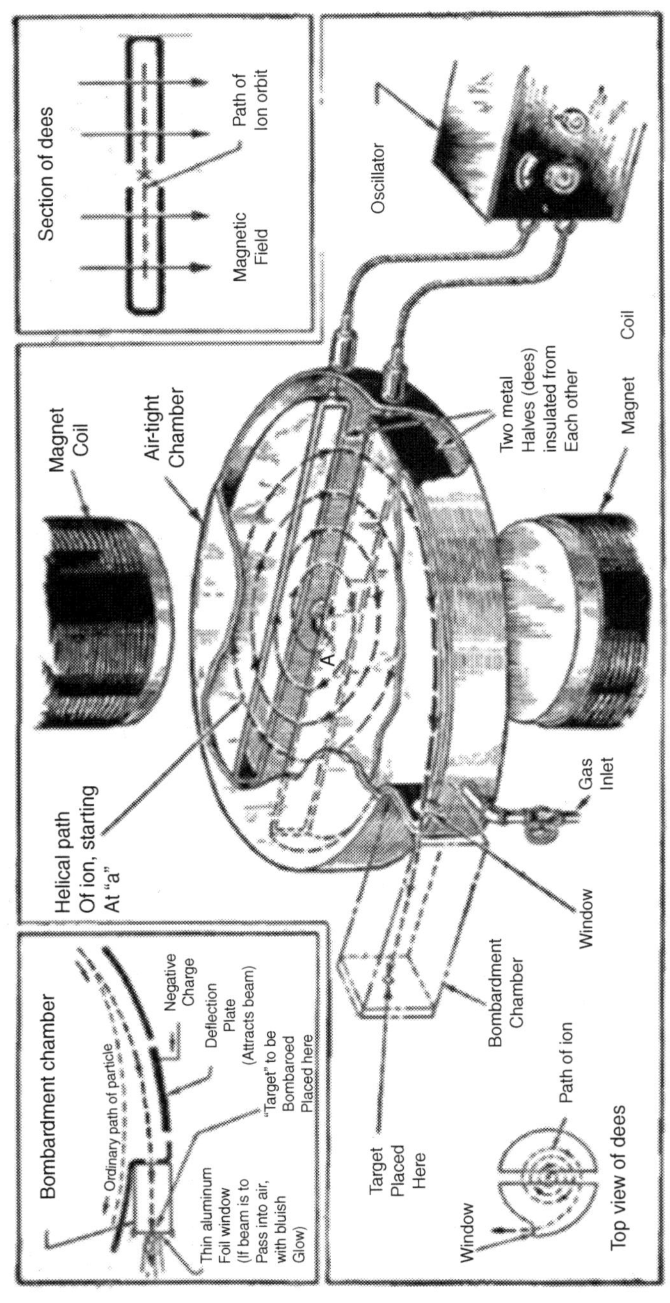

FIGURE 1.23 Cyclotron (From: https://en.wikipedia.org/wiki/Cyclotron).

alpha (α) particles to accelerate a projectile particle to very high velocities so that it can penetrate the orbital electrons of the target nucleus.

1.5.4.2 Synchrotron

According to Fernandez and Ripka (2013), the synchrotron's concept was first introduced by the Australian physicist Oliphant who was a professor at the University of Birmingham in 1943. This concept then was further developed by the Russian Physicist Vladimir Veksler in 1944. In 1945, the American physicist Edwin McMillan built the first new accelerator and he called it "Syncro-cyclotron" which later changed into a synchrotron. Since then and until recent years, scientists have been working in developing the capabilities of the synchrotron further, whose acceleration has been dramatically increased from 8 MeV to about 6.5 TeV. The Large Hadron Collider, which is shown in Fig. 1.24 (CERN Accelerator Complex) is considered as the world's largest and fastest-particle accelerator (Armitage, 2013). The Large Hadron Collider was built in 2008 by the European Organization for Nuclear Research (CERN). It is located near Geneva, Switzerland, with a circumference of about 27 km. It can hasten beams of protons, electrons, positrons, antiprotons, alpha particles, oxygen, and sulfur nuclei to an energy of 6.5 TeV (CERN 2020).

The basic working principle of the synchrotron is almost similar to the principle of work the cyclotrons in terms of using the magnetic field. Accordingly, scientists argue that the synchrotrons are a result of the extensive development of the classical cyclotrons and their successor isochronous cyclotron (Livingston and Howard, 1958). However, synchrotron science differs from the science of cyclotrons only in one physical phenomenon, which is that electromagnetic poles are controllable to change the direction of moving electron and therefore their wavelengths controllable, whereas in the cyclotron the magnetic field is static and subsequently the RF frequency remains constant (Joho, 1984). The sudden changes in the direction of the moving electrons by controlling the magnetic poles and their wavelengths make the electrons to emit the desired energy.

The synchrotron works by injecting an electron beam from the electron gun with an initial energy of approximately 20 MeV (or any other charged particles) into an ultra-high vacuum stainless steel tube. Then it is further accelerated into a linear accelerator called LINAC, which will increase their energy from 20 to 250MeV using a power amplifier. At this stage, the electrons are traveling at 99.9998% of the speed of the light. Then the accelerated electron beam is transferred from the linear accelerator into the poster ring that also contains a power amplifier, which accelerates them further from 250 MeV to about 2900 MeV until they reach almost the speed of light (Canadian Light Source Outreach, 2012). The poster ring also contains a series of magnets that maintain the movement of the accelerated electron beam in a circular motion. Once the electron beam

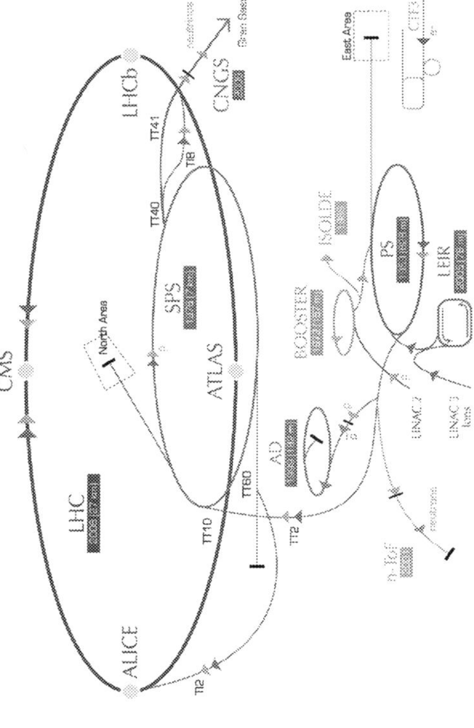

FIGURE 1.24 CERN accelerator complex (From: http://cds.cern.ch/record/1260465).

reaches almost the speed of the light, the electron beam exits the poster ring to the storage ring. This ring is about 171 m and comprises of a series of 12 straight sections to make the beam moves in a straight line but by the end of each section, there are two dipole magnets at each section and a series of four to six-pole magnets. Their main function is to bend the beam again into a circular narrow motion as well as play a significant role in controlling the wavelengths via the monochromator (Canadian Light Source Outreach, 2012). As mentioned earlier, the sudden changes in the direction of the moving electrons and their wavelengths in the storage ring, make the electron beam to emit the desired energy that is eventually collected at different end stations located at the edges of the storage ring. Furthermore, this highly accelerated particle can also be used to bombard a targeted atom and convert it into the required isotope. The emitted energy is at the range of the infrared wavelength and the X-ray wavelength. Fig. 1.25 depicts the main components of a Synchrotron, which comprises of the electron gun, linear accelerator (LINAC), booster ring, storage ring (or rings), bending magnets, beamlines, and the end stations.

1.5.5 The radionuclide generator

A radionuclide can also be produced using the radionuclide generator, which is a device that easily separates and extracts the daughter nuclide from its parent nuclide that is contained in this device and widely used in the nuclear medicine (Shultis and Faw, 2016). The basic working principle of the radionuclide generator is that a required parent nuclide is placed in the generator. The parent nuclide will continuously decay and will produce a new daughter that will be extracted continuously and collected, repetitively (Shultis and Faw, 2016). This method is widely used to generate radionuclides that are used in medical diagnoses and imaging applications, such as ^{99}Mo -^{99m}Tc that is used widely as an imaging label agent in nuclear medicine because it emits γ rays of a relatively low radiation dose and has a half-life of 6 hours, which can be easily identified with a gamma camera because of its single peak at 141 KeV (Blowera, 2015).

According to many scholars including Zolle (2007), all agreed that technetium-99Mo generator is commonly used to extract the metastable isotope of technetium—^{99m}Tc from the decay of its parent nuclide ^{99}Mo. This process depends on the chromatography column, where the parent nuclide (^{99}Mo) in the form of molybdate ion, MoO_4^{2-} is adsorbed onto acid alumina (Al_2O_3). The parent nuclide ^{99}Mo will decay and produce its daughter nuclide in the form of Sodium pertechnetate (TcO_4^-) that has a single charge and therefore bounded loosely to the alumina. Finally, the saline solution will elute the soluble ^{99m}Tc from the column and form a saline solution containing the ^{99m}Tc.

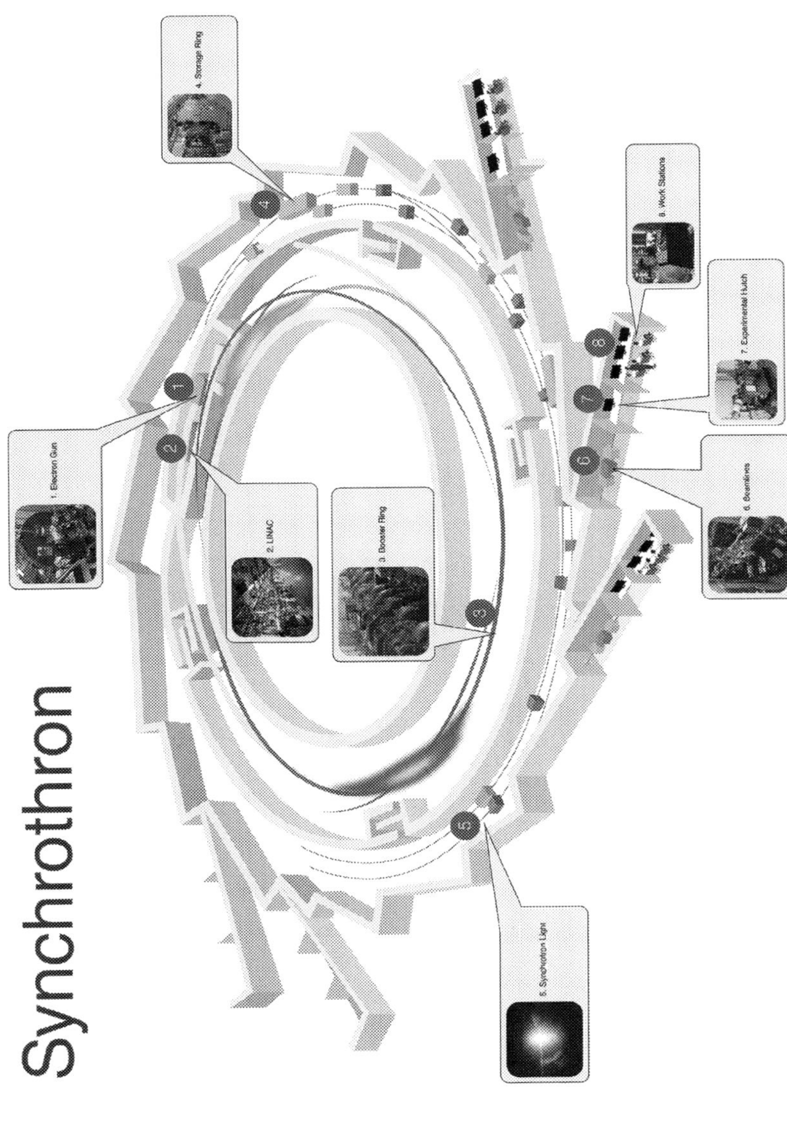

FIGURE 1.25 Synchrotron.

FIGURE 1.26 The calutron developed by Ernest for the Manhattan Project during the Second World War for separating and enrichment isotopes of uranium. (From: https://en.wikipedia.org/wiki/Calutron#/media/File:Alpha_calutron_tank.jpg).

1.5.6 The mass spectrometer

The mass spectrometer is another methodology used in the separation of the isotopes according to their mass-to-charge ratio, which was first invented in 1886 by the German physicist Eugen Goldstein (Cheng, 2017). A calutron shown in Fig. 1.26 is a type of sector mass spectrometer that was developed by Ernest O. Lawrence for the Manhattan Project during the Second World War for separating and enrichment isotopes of uranium (William, 2005). Since then, the mass spectrometers have been used widely in many industrial, health, science, space exploration including but are not limited to isotopic composition ratio determination, uranium enrichment, isotope dating and tracing, trace gas analysis, atom probe, pharmacokinetics, protein characterization, and many other applications.

A number of scholars such as Hoffmann and Stroobant, (2013) and many others have described the basic theory and principles of mass spectroscopy (Fig. 1.27). The key working principle of the mass spectrometer is that the required sample to be analyzed is placed in the mass spectrometry in the vacuum tube. The sample is first converted into vapor (gas phase ions) by heating. Then, the sample atoms are ionized using the electron

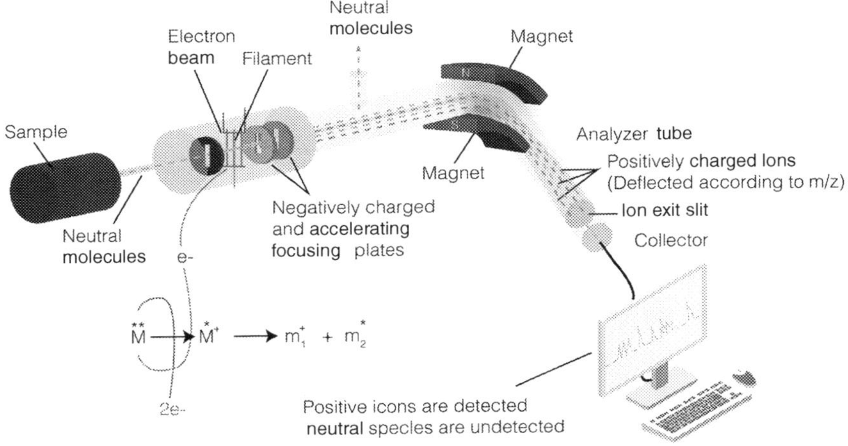

FIGURE 1.27 The mass spectroscopy.

ionization technique via the electron gun that generates a high energy beam of electrons, which will displace the electron from the organic molecule and form a radical cation known as the molecular ion. As the electron beam moves toward the molecule ions, it causes molecules' ions to become unstable and therefore will collide with other ions and become small fragments. This fragmentation may result in either radical ions with an even number of electrons or a molecule with new a radical cation as shown in Eqs. 1.16–1.18:

$$M + e^{-}M^{\cdot +} + 2e^{-} \tag{1.16}$$

$$M^{+}EE^{+}\,(\text{EvenIon}) + R \tag{1.17}$$

$$M^{+}OE^{+}(\text{Odd Ion}) + M\,(\text{Molecule}) \tag{1.18}$$

These ions are then accelerated by an electric field, fragmented into smaller ions while moving toward the detector. The detector then separates these fragments based on mass to charge ratio (m/z), and neutral molecules are not detected. The mass spectrometer has wide industrial and scientific applications, such as substances chemical analysis, food flavor analysis, and many other applications.

1.6 Conclusion

The first theory of the atom was dated back to the fifth century BCE when Greek scientists, and philosophers built their theory of atoms upon the work of ancient philosophers. These theories were disregarded until the 16th and 17th centuries because religious intellectuals considered the theory to be a materialistic view of the world that denied the existence of spiritual forces. At the beginning of the 19th century, scientists such as John Dalton and Jöns Jakob Berzelius revived the atomic theory by using quantitative and experimental data. Since then, philosophers and scientists continued their studies to explore more about atoms until the mystery of fission and fusion reactions were solved and showed their destructive power during the Second World War. Since then, nuclear scientists have devoted their efforts and scientific studies and research to harness atomic and nuclear technology for peaceful applications. Accordingly, the applications have been focused on the generation of electricity through the exploitation of the huge amounts of heat produced from the fission reaction.

The birth of peaceful nuclear applications was announced by President Dwight D. Eisenhower on December 8, 1953, through his speech in front of the United Nations council entitled "Atoms for Peace." Through this speech, the president was attempting to diminish the nuclear arms race. According to Eisenhower 's vision, in 1957, the United Nations established the IAEA. IAEA functions as the world's nuclear regulator assist all associated countries in developing nuclear technology for peaceful usage. On July 1, 1968, NPT was declared one of the important international conventions that and entered into force on March 5, 1970, and has three main objectives, which are: (1) nonproliferation of nuclear weapons; (2) the peaceful use of nuclear energy; (3) and disarmament. Since then and for more than five decades, the scientific research of isotopic and nuclear science technology has continued to contribute to the advancement of nuclear technology through the revolution of isotopes separation and radionuclide production based on the ionizing capabilities of isotopic radiations and their interactions with materials. Nuclear and isotopic technologies have facilitated a plethora of peaceful applications in the energy, medical, scientific, industrial, and agricultural sectors, which will be discussed in more detail in the coming chapters.

References

Al-Sheha, A., 2011. The key to understanding Islam. Osoul Center, Riyadh, Saudi Arabia, pp. 1–251.

Al Nabhani, K., Khan, F., 2019. Nuclear radioactive materials (TENORM) in the oil and gas industry. Safety, Risk Assessment Manag. Elsevier Inc, Amsterdam, The Netherlands, pp. 1–328.

Alan, H., 2015. Never Lose Your Nerve!. World Scientific Publishing Co, Singapore, pp. 1–288.

Armitage, J., 2013. The Virilio Dictionary—Great Accelerator. Edinburgh University Press, UK, pp. 1–264.

Baba, M., Mohib-ul-Haq, M., Khan, A., 2013. Dosimetric consistency of Co-60 teletherapy unit—a ten years study. Int J Health Sci 7 (1), 15–21.

Blowera, P., 2015. A nuclear chocolate box: the periodic table of nuclear medicine. Dalton Trans 44 (11), 4819–4844.

Breeze, P., 2014. Power generation technologies, 2nd Edition. Newnes and Elsevier, UK, pp. 1–379.

Brown, T., et al., 2013. Chemistry: the Central Science. Pearson. Canada, pp. 1–1323.

Brown, L., Pippard, B., A, P., 1995. Twentieth-Century Physics. IOP Publishing Ltd, CRC Press, UK, pp. 1–644.

Canadian Light Source Outreach, 2012. Synchrotron science. Classroom resources. What Is a Synchrotron? 1–52. https://www.lightsource.ca/ckfinder/userfiles/files/SSCR2013Jul23-Formatted-for-Web%20(1).pdf. Accessed 6 June, 2020.

CERN, 2020. Official website. https://home.cern/science/acceleratorss. Accessed 6 June, 2020.

Cerny, J., 2012. Nuclear Spectroscopy and Reactions 40-A, Part 1. Academic Press, New York and London, pp. 1–538.

Charlie Ma, C.-M., Lomax, T., 2012. Proton and Carbon Ion. Therapy Imaging in Medical Diagnosis and Therapy. CRC Press, Boka Raton, USA, pp. 1–256.

Cheng, G., 2017. Mechanistic Studies on Transition Metal-Catalyzed C–H Activation Reactions Using Combined Mass Spectrometry and Theoretical Methods. Springer, Singapore, pp. 1–126.

Cherry, S., Sorenson, J., Phelps, M., 2012. Physics in Nuclear Medicine. Elsevier, Saunders, Philadelphia, pp. 1–544.

Cochran, T., 1998. Technological Issues Related to the Proliferation of Nuclear Weapons. In: Strategic Weapons Proliferation Teaching Seminar, pp. 1–43.

Cotton, S., 2013. Lanthanide and Actinide Chemistry. John Wiley & Son, England, UK, pp. 1–272.

Currie, G., Wheat, J., Davidson, R., Kiat, H., 2011. Radionuclide production. J. Radiographer. 58 (3), 46–52.

Davisson, L., 2002. Isotopic Tracers in Surface Water. Awwa Research Foundation and American Water Works Association, USA, pp. 1–115.

Dendy, R., 1990. Plasma Dynamics. Clarendon Press, Oxford University Press, New York, USA, pp. 1–161.

Duke, C., Williams, C., 2007. Chemistry for Environmental and Earth Sciences. CRC Press, Boca Raton, USA, pp. 1–248.

ElSharkawi, M., 2015. Electric Energy: An Introduction, 3rd Edition. CRC Press, Boca Raton, USA, pp. 1–569.

Enghag, P., 2008. Encyclopedia of the Elements: Technical Data - History - Processing - Applications. John Wiley & Sons, Weinheim, Germany, pp. 1–1309.

Eric, S., 2009. Selected Papers on the Periodic Table. Imperial College Press, London, UK, pp. 1–156.

Fernandez, B., Ripka, G., 2013. Unravelling the mystery of the atomic nucleus: A sixty year journey 1896–1956. Unravelling the Mystery of the Atomic Nucleus: A Sixty Year Journey 1896-1956. Springer, New York, NY (pp. 1–530) https://doi.org/10.1007/978-1-4614-4181-6 .

Fischer, 1997. History of the International Atomic Energy Agency: the first forty years. https://wwwpub.iaea.org/MTCD/publications/PDF/Pub1032_web.pdf. Accessed 20 May, 2020.

Fuss, W., 2015. Laser isotope separation and proliferation risks. Max-Planck-Institut fuer Quantenoptik. MPQ- 346, Germany, pp. 1–28.

Ghiassee, N., 1979. The Spectroscopy and Photolytic Decomposition of Some Volatile Uranium Compounds. PhD Thesis, Imperial College of Science & Technology. London. https://spiral.imperial.ac.uk/bitstream/10044/1/35085/2/Ghiassee-N-1979-PhD-Thesis.pdf.

Gibson, G., 2011. Energy and Us: Sources, Uses, Technologies, Economics, Policies and the Environment. Xlibris Corporation, USA, pp. 1–363.

Givens, W.W., Stromswold, D.C., 1989. Prompt fission neutron logging for uranium. Nuclear Geophys. (International Journal of Radiation Applications and Instrumentation, Part E), UK 3 (4), 299–307.

Greenland, P., 1990. Laser isotope separation. Contemp. Phys. (6) 405–423.

Gurov, Y., 2004. Spectroscopy of superheavy hydrogen isotopes in stopped-pion absorption by nuclei. Physics of Atomic Nuclei 68 (3), 491–497.

Haider, Q., 2019. Nuclear fusion: holy grail of energy. In: Igor, G. (Ed.), Nuclear Fusion—One Noble Goal and a Variety of Scientific and Technological Challenges. IntechOpen https://doi.org/10.5772/intechopen.82335. Accessed 3 June, 2020.

Hassan, Y., Chaplin, R., 2010. Nuclear Energy Materials and Reactors. EOLSS Publications Co. Ltd, Oxford, UK, pp. 1–428.

Hoffmann, E., Stroobant, V., 2013. Mass Spectrometry: Principles and Applications, 3rd Edition. John Wiley & Sons, UK, pp. 1–502.

Hordeski, M., 2009. Hydrogen and Fuel Cells-Advances in Transportation and Power. River Publisher, Denmark, pp. 1–287.

Hore-Lacy, I., 2010. Nuclear Energy in the 21st Century. World Nuclear University Press and Elsevier Inc., UK and USA, pp. 1–168.

Hurst, D., 1997. Canada Enters the Nuclear Age: A Technical History of Atomic Energy of Canada Limited as Seen from Its Research Laboratories. McGill-Queen's Press, Canada, pp. 1–434.

IAEA, 2008. Cyclotron Produced Radionuclides: Principles and Practice. Technical Reports Series No. 465. IAEA, Vienna, Austria, pp. 1–215.

IAEA, 2008. Radiotracer Residence Time Distribution Method for Industrial and Environmental Applications. IAEA, Vienna, Austria, pp. 1–153.

IAEA, 2020a. Official website. https://www.iaea.org/about/overview/history. Accessed 20 May, 2020.

IAEA, 2020b. Official website. News center/statements. https://www.iaea.org/newscenter/statements/peaceful-usesnuclear-energy. Accessed 20 May, 2020.

Jensen, R., Judd, O., Sullivan, A., 1982. Separating isotopes with lasers. Los Alamos Science, USA, 4, 1–33.

Joho, W., 1984. Interfacing the Sin Ring Cyclotron to a Rapid Cycling Synchrotron with an Acceleration and Storage Ring Astor. IEEE, USA, pp. 611–614.

Joyce, M., 2017. Nuclear Engineering: A Conceptual Introduction to Nuclear Power. Butterworth-Heinemann and Elsevier Ltd., UK, pp. 1–420.

Karttunen, H., Kr?ger, P., Oja, H., Poutanen, M., Donner, K., 2013. Fundamental Astronomy. Springer Science & Business Media, p. 1.

Kemp, R., 2012. Centrifuges: A New Era for Nuclear Proliferation. In: Sokolski, Henry (Ed.), Nuclear Nonproliferation: Moving Beyond Pretense. Nonproliferation Policy Education Center, Washington, D.C., USA Avilable from http://www.npolicy.org/userfiles/image/oving%20Beyond%20Pretense%20web%20version.pdf#page=58.

Kirklan, K., 2010. Physical Sciences: Notable Research and Discoveries. Frontiers of Science. Facts on File Inc., New York, USA, pp. 1–189.

Kok, K., 2009. Nuclear Engineering Handbook, 1st Edition. CRC Press, Boca Raton, USA, pp. 1–786.

Kumar, A., Dangi, V., 2016. Electromagnetic spectrum and its impact on human life. International Journal of All Research Education and Scientific Methods (IJARESM) 4 (8), 67–72.

Kupp, R., 2005. A Nuclear Engineer in the Twentieth Century. Trafford Publishing, Victoria, British Columbia, Canada, pp. 1–372.

L'Annunziata, M., 2012. Handbook of Radioactivity Analysis. Elsevier Inc., USA, pp. 1–1379.

LAnnunziata, M., 2016. Radioactivity: Introduction and History, from the Quantum to Quarks, 2nd Edition. Elsevier B.V. Amsterdam, The Netherland, pp. 1–932.

Lanouette, W., 1992. Ideas by Szilard. Physics by Fermi. Bulletin of the Atomic Scientists. 48 (10), 16–23.
Laughter, M., 2009. Profile of world uranium enrichment programs-ORNL/TM-2009/110. Oak Ridge National Laboratory for U.S. Department of Energy, USA, pp. 1–41.
Lavoy, P., 2003. The enduring effects of atoms for peace. Arms Control Association, Washington, D.C., USA, pp. 1–9. Available from https://core.ac.uk/download/pdf/36731195.pdf.
Leroy, F., 2003. A Century of Nobel Prize Recipients: Chemistry, Physics, and Medicine. Marcel Dekker - CRC Press, Boca Raton, USA, pp. 1–300.
Livingston, R., Howard, F., 1958. The Oak ridge relativistic isochronous cyclotron. U.S. Atomic Energy Commission, Washington, D.C., USA, pp. 1–129. Available from https://www.osti.gov/servlets/purl/4275955.
Luo, F., Hong, Y., 2017. Renewable Energy Systems: Advanced Conversion Technologies and Applications. CRC Press, Bock Raton, USA, pp. 1–880.
Maher, S., Jjunju, F.P.M., Taylor, S., 2015. Colloquium: 100 years of mass spectrometry: Perspectives and future trends. Rev. Modern Phys. 87 (1). doi:10.1103/RevModPhys.87.113.
Medalia, J., 2011. Japanese Nuclear Incident: Technical Aspects. Congressional Research Service, Washington, D.C., USA, pp. 1–18.
Müller, H., Müller, D., 2016. WMD Arms Control in the Middle East: Prospects, Obstacles and Options. Routledge, London and New York, pp. 1–358.
Nagy, S., 2009. Radiochemistry and Nuclear Chemistry, Vol. 1. EOLSS Publishers, Oxford, UK, pp. 1–430.
Oxtoby, D., Gillis, H., Butler, L., 2015. Principles of Modern Chemistry. 8th Edition, Cengage learning, Boston, USA, pp.1–1264.
Paisner, J.A., 1988. Atomic vapor laser isotope separation. Appl. Phys. B Photophys. Laser Chem. 46 (3), 253–260. doi:10.1007/BF00692883.
Peach, K., Wilson, P., Jones, B., 2011. Accelerator science in medical physics. Br. J. Radiol. 84 (1), S4–S10. doi:10.1259/bjr/16022594.
Philippe, S., Glaser, A., 2014. Nuclear archaeology for gaseous diffusion enrichment plants. Sci. Global Secur. 22 (1), 27–49. doi:10.1080/08929882.2014.871881.
Podgorsak, E., 2016. Radiation Physics for Medical Physicists. Springer, Switzerland, pp. 1–906.
Quagliano, J., 1969. Chemistry, 3rd Edition. Prentice-Hall Inc., Englewood Cliffs, New Jersey, USA, pp. 1–844.
Rajagopalan, R., Mishra, A., 2015. Nuclear South Asia: Keywords and Concepts. Routledge, India and The UK, pp. 1–326.
Ray, H., Roselyn, H., 1970. Atomic pioneers: from ancient Greece to the 19th Century. United States Atomic Energy Commission. Division of Technical Information, USA, pp. 1–65.
Roberts, J.T., 1974. Uranium enrichment: supply, demand and costs. IAEA Bull. 16 (1–2), 94–104.
Robertson, D., Carpenter, R., 1974. Neutron activation techniques for the measurement of trace metals in environmental samples. Technical Information Center, Office of Information Services, United States Atomic Energy Commission, USA, pp. 1–79.
Saha, G., 1984. Fundamentals of Nuclear Pharmacy. Production of radionuclide, 2nd Edition. Springer Science+Business Media, LLC, New York, USA, pp. 1–277.
Sanctis, E., Monti, S., Ripani, M., 2016. Energy from Nuclear Fission: An Introduction. Springer, Switzerland, pp. 1–278.
Scott Kemp, R., 2009. Gas centrifuge theory and development: a review of U.S. programs. Sci. Global Secur. 17 (1), 1–19. doi:10.1080/08929880802335816.
Serway, R., Jewett, J., 2010. Physics for scientists and engineers. Cengage Learn. (5) 1–368.
Sharp, J., 1996. About Turn, Forward March with Europe: New Directions for Defence and Security Policy. Institute for Public Policy Research and Rivers Oram Press, UK and USA, pp. 1–321.

Shipman, J., Wilson, J., Higgins, C., 2012. An introduction to physical science. Cengage Learn. 1–792.

Shultis, J., Faw, R., 2016. Fundamentals of Nuclear Science and Engineering. CRC Press, Boca Raton, USA, pp. 1–660.

Sime, R., 2014. Science and politics: the discovery of nuclear fission 75 years ago. Physics Forum, Ann. Phys. Berlin. 526 (3-4), A27–A31.

Sivasankar, B., 2008. Engineering Chemistry. Tata McGraw-Hill Education, New Delhi, India, pp. 1–557.

Stokes, N., 2019. Artifact Collective: An Attempt to Consciousness. Nick Stokes. Tacoma, USA, pp. 1–481.

The World Nuclear Association, 2020. Nuclear fission, 2020. https://www.world-nuclear.org/informationlibrary/nuclear-fuel-cycle/introduction/physics-of-nuclear-energy.aspx. Accessed 1 June, 2020.

Tsoulfanidis, N., 2012. Nuclear Energy: Selected Entries from the Encyclopedia of Sustainability Science and Technology. Springer Science & Business Media. New York, USA, pp. 1–530.

Underhill, 1996. Naturally Occurring Radioactive Materials: Principles and Practice. St. Lucie Press, CRC Press, USA, pp. 1–160.

U.S. Congress Joint Committee on Atomic Energy, 1967. Reactors systems. Committee prints. 90th Congress. US Government Priniting Office, Washington, USA, pp. 1–370.

USNRC, 2008. In: Uranium Enrichment Processes: Laser Enrichment Methods (AVLIS AND MLIS) Module 3.0. Rev 03. Directed Self-Study, pp. 3-1–3-62.

William, P., 2005. The uranium bomb, the calutron, and the space-charge problem. Phys. Today, USA 58 (5), 45–51.

Yan, X.L., Hino, R., 2016. Nuclear Hydrogen Production Handbook. Nuclear Hydrogen Production Handbook. CRC Press, Boca Raton, USA, pp. 1–878. https://www.crcpress.com/Nuclear-Hydrogen-Production-Handbook/Yan-Hino/p/book/9781439810842. Accessed 25 May, 2020.

Zare, R., 1977. Laser separation of isotopes. Scientific American Inc., USA, pp. 86–98. Available from https://web.stanford.edu/group/Zarelab/publinks/zarepub132.pdf.

Zolle, I., 2007. Performance and quality control of the 99Mo/99mTc generator. Technetium-99m Pharmaceuticals: Preparation and Quality Control in Nuclear Medicine. Springer, Berlin, pp. 77–93. doi:10.1007/978-3-540-33990-8_5.

Zumdahl, S., DeCoste, D., 2012. Chemical principles, 7th Edition. Cengage Learn, USA, pp. 1–1200.

CHAPTER 2

The role of international law via NPT in promoting nuclear peaceful applications

OUTLINE

2.1 Introduction

Nuclear technology is considered as one of the most sophisticated discoveries in the 20th century. This technology can be used in many different applications, either for the development and advancement of humanity

Applications of Nuclear and Radioisotope Technology: The Atom for Peace and Sustainable Development
DOI: https://doi.org/10.1016/B978-0-12-821319-3.00003-8

such as, power generation, diagnosis, and treatment of intractable diseases such as cancer and heart, monitoring environmental pollution, improving agricultural and livestock production (IAEA, 2012), or for the inevitable destruction of humanity through developing of the weapons of mass destruction such as the nuclear and radiological bombs that threaten the existence of humanity and civilization as experienced in Hiroshima and Nagasaki. The importance of investigating nonproliferation treaty (NPT) lies in the risk associated with the development and proliferation of nuclear weapons that may arise as a result of the weakness of some articles' formulations, which may grant the right to some countries to develop nuclear weapons. Therefore, it is the responsibility of the international community to bridge any gaps in the current NPT and to create global awareness about the urgent need for nuclear states to obey and cooperate with the restrictions placed by the international law on controlling the proliferation and development of nuclear weapons programs and eventually to abandon their ongoing nuclear programs for the sake of achieving the total world peace.

In this chapter, the controversial efforts made to the nonproliferation of nuclear weapons and the impact of the nuclear countries on the role of the international agencies involved in the nonproliferation of nuclear weapons will be examined from the liberalism and realism perspectives, which are one of the most famous political philosophies followed in developed countries advocate for human rights and the environment. Thus, in this chapter, an attempt will be made to shed the light on what has the international community done so far to hinder the development of nuclear weapons that threaten international peace and security from liberalism and realism perspectives? Have the mechanisms introduced by the international community such as the International Atomic Energy Agency (IAEA), Security Council, and the International Court of Justice to achieve these goals been successful? and what are the arguments of liberalists and realists in this regard?

A theoretical answer to the raised questions can be derived from the hypothesis, which proposes that lasting global peace is achievable if the NPT mandates all the countries permitted to develop nuclear weapons to stop and abandon their nuclear programs without any biases in view of achieving equality among all countries and reconsidering the privileges granted by the NPT to the nuclear countries. Rather a concerted effort should be directed at achieving and sustaining international peace and security as well as, the call for the nonproliferation of nuclear weapons and the adherence to the Nuclear-Test-Ban Treaty are the two sides of the same coin that will potentially enhance the authority and the effectiveness of the NPT and other related agencies such as the Security Council, the International Court of Justice, and IAEA in sustaining global peace.

In this investigation, the analytical methods will be used in line with the international relations theory; they include argumentative, comparative, and political analysis and will be validated by highlighting empirical evidence derived from international historical events, the role of the international community, their activities such as the proclamation of international treaties and the level of adherence, the role of other international safeguards. Furthermore, and in order to have a better understanding, this investigation will begin with an overview of nuclear weapons proliferation, the potential impacts of nuclear weapons, and the need to create policies aimed at nonproliferating nuclear weapons. This will help to have a better understanding of the progress of the investigation on how effective the role of NPT and the nuclear weapon states (nuclear club states), which are considered as the key driver in stopping the nuclear proliferation conflict and instead promoting nuclear peaceful applications by revealing the conflicting views and stances of realists and liberals countries regarding the nonproliferation of nuclear weapons given NPT and the role of other international safeguards such as Security Council and the international court of justice.

2.2 What are nuclear weapons?

The National Research Council, Policy and Global Affairs, Committee on International Security and Arms Control (2005) attempted to define nuclear weapons based on available treaties and other scientific recourses as explosive materials made of nuclear fuel or radioactive isotopes (plutonium or highly enriched uranium) capable of creating an uncontrollable nuclear exponential fission chain reaction subsequently causing the massive destruction of human lives, animals, the environment, and the properties. Furthermore, many scientific studies and international treaties also categorized nuclear and radiological weapons also under mass destruction weapons besides chemical, biological due to the level of destruction that these weapons can cause that can last for many years due to the half-life of the used radiological substances such as the case in Hiroshima and Nagasaki.

2.3 An overview of nuclear weapons proliferation

The history of nuclear technology development became apparent after the Cold War when powerful countries realized the need to have more powerful and destructive weapons to discourage and resist any foreign invasion. The studies presented by the Preparatory Commission for The Comprehensive Nuclear-Test-Ban Treaty Organization (2018) and

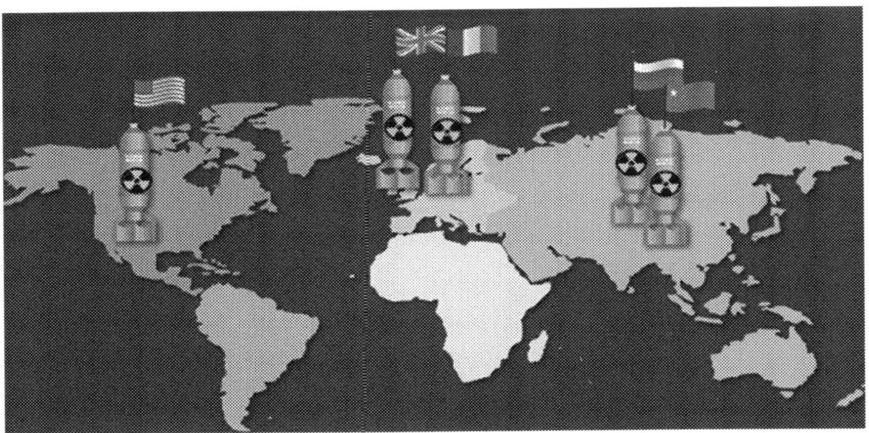

FIGURE 2.1 The five nuclear weapon states (nuclear club states).

Cavendish (2003) indicate that many countries have successfully developed nuclear weapons that can be categorized into two main categories: (1) fission atomic bombs (Enriched Uranium 235); and (2) fusion hydrogen bombs (Plutonium 239). Furthermore, nuclear countries have been categorized into three main categories, which are: (1) nuclear weapon states that have been given permission by the international law to develop nuclear weapons and they are the United States, Russia, United Kingdom, France, and China that collectively referred to as the nuclear weapon states or sometimes called nuclear club states, as shown in Fig. 2.1 (Griffiths and Ameri, 2013).

These states are permanent members of the Security Council (Cimbala, 2012); (2) nuclear states operating outside the nuclear club such as India, Pakistan, and North Korea that were against the exclusive selection of the nuclear club states, which subsequently have withdrawn from the NPT (Schaffer and Schaffer, 2016); (3) nuclear threshold states such as Iran (Rublee, 2010)

The five nuclear weapon states have however been granted the total proliferation of nuclear programs that has been defined by Sidel and Levy (2007) in two main categories: (1) horizontal proliferation of nuclear weapons (includes a total proliferation of nuclear weapons in non-nuclear weapons states); and (2) vertical proliferation (involves the mass production of nuclear weapons in quantity and quality in nuclear weapon states as a deterrent to compel other non-nuclear weapon states to abandon their nuclear weapon plans). The available studies reveal that the main reasons behind the proliferation of nuclear weapons could be attributed to the following: (1) the nuclear states can use it to intimidate other states during disputes; (2) science and technology advancement; (3) nuclear weapon states have the right to horizontally and vertically proliferate to protect their political interests in the region; (4) the nuclear black market that supports states operating outside the nuclear club.

2.4 An overview of the nonproliferation treaty

According to article 38 of the Statute of the International Court of Justice, international conventions remain the primary source of general international laws (Thompson, 2015, International Court of Justice, 2019a). Thus, NPT is one of the important international conventions that was declared on July 1, 1968, and became effective on March 5, 1970, after it has been signed by the three main countries that controlled advanced nuclear power (the United States, the Soviet Union, and the United Kingdom) NPT has three crucial objectives: (1) nonproliferation of nuclear weapons; (2) the peaceful use of nuclear energy; (3) and disarmament (United Nations Office for Disarmament Affairs, 2019).

In principle, the treaty aims at achieving global nonproliferation of nuclear weapons as well as encouraging all participating countries to embrace the safe use of nuclear technology given promoting global peace and protecting humanity (United Nations Office for Disarmament Affairs, 2019). But realists and liberals have a different view about the NPT in this regard, and this is what the following sections try to reveal after understanding the main NPT strengths and weakness:

2.5 Nonproliferation treaty's strengths and weakness

The primary objective of the nonproliferation of nuclear weapons treaty (NPT) is to discourage the proliferation of nuclear weapons that considered as one of its main strengths. However, the main gap of NPT is that this objective is only applicable to non-nuclear weapon states that signed this convention, and it excludes the nuclear weapon states that had the authority to manufacture and detonate nuclear weapons or nuclear explosive devices before 1967 despite articles 1 and 2 of the treaty contradict with its main purpose, which stipulated that the prohibition was intended to curb the use of nuclear weapons and nuclear explosive devices. This contradiction made three states, which are India, Israel, and Pakistan, refusing to adhere to the NPT and in 2003 North Korea, has announced its withdrawal from the NPT (Mishra, 2008).

While there have been positive achievements brought about by the treaty, there have been many criticisms in its application. For instance, the nuclear weapon states have in many ways contributed to the breach of the provisions specified by the treaty and provided support in term of weapons and technologies to states regarded as their allies in view of enhancing the military power in these strategic states such as the United States–India nuclear collaboration, Pakistan–China nuclear collaboration (Carranza, 2016). Another example illustrated by Sarram (2015) regarding the plans made by the American government to support Israel's nuclear program, which contradicts the provisions of the treaty, and this shows the double standards of the American government that also publicly stresses the need for all countries to abandon their nuclear programs, while continue breaching the treaty by overlooking proposals to create some nuclear weapon-free zones similar to the zones located in the Middle East, Northern Europe, and South-East Asia. Some nuclear states such as the United States and China remain unwilling to sign the treaties that will declare these areas as nuclear weapon-free zones due to their political interests in these regions (Sarram, 2015).

2.6 Nonproliferation treaty from realism theory

The leading proponents of the realism theory include Hans Morgenthau, John Herz, and Hennery Kissenge who regarded the international system as a collection of sovereign states that evolved by Thucydides after the treaty in Westphalia was signed. Realists believe that it is crucial that countries have the means to defend their borders, interests, and citizen. Many studies argue that realist scholars have not much of confidence in the effectiveness of international institutions. Consequently, these beliefs could lead to a preliminary conclusion that the world will turn into anarchy system. According to the study presented by Ahmed (2017), this could be obvious by tracing back the early beginnings of global nuclear war development to the nuclear programs that started by the Soviet Union in response to the achievements made by the United States in the development of nuclear technology. Subsequently, other states such as France and the United Kingdom launched similar nuclear projects as a deterrent to potential invasion from the Soviet Union. Other states such as China, Israel, India, and Pakistan have launched nuclear weapon development programs majorly citing the need to protect their international borders, and their interests in the region are the main reason for their involvement with the development of nuclear weapons (Rooth, 2015; Ahmed, 2017). Thus, it can be stated from the realism perspective that nuclear weapons are essential as a deterrent to foreign invasion, and possession of nuclear weapons gives countries a certain level of recognition and respect exactly like what North Korea is trying to do. This led other countries such as Iran and Pakistan have initiated nuclear weapon programs with the backing of some nuclear states in view of protecting their international borders (Ahmed, 2017).

Based on the realism theories, analysts propose that embracing these theories could lead to an ignorance of crucial issues regarding international security that may result in the global anarchy system, for instance, the rising tensions between the United States and North Korea or Iran that could escalate into a Third Nuclear World War.

2.7 Nonproliferation treaty from liberalism theory

The liberalism ideology advocates equality and the use of cooperation solutions through international institutions and states to achieve international peace (Rousseau and Walker, 2012). Peace, therefore, can be achieved through international treaties such as NPT, TBT, and international institutions such as the Security Council, the UN, and the International Court of Justice to resolve international disputes. Many states, which hold liberalist views, believe in the role of international institutions and aimed at achieving world peace through dialogue and cooperation rather than aggression. The liberalism ideology suggests that there can be sustainable peace and fewer conflicts between democratic states when there are cooperative solutions and all the participating states adhere to the regulations created by relevant international institutions such as the treaties and international safeguard institutions which can be called "democratic peace theory" (Rousseau and Walker, 2012); however, with regards to nuclear nonproliferation, there have been some allegations by liberalists of bias leveled at the international institutions that are seemingly under the influence of the five powerful nuclear states. These allegations have been cited as reasons why countries such as India, Pakistan, and Israel have refused to join the NPT and claim NPT is an injustice. The lack of cooperation and equality among participating states to a large extent will hinder the efforts of the international treaties (e.g., NPT, TBT, Convention for the Suppression of Acts of Nuclear Terrorism, etc.) and international institutions (e.g., Security Council and The International Court of Justice) to achieve the world peace.

The following section will reveal the effectiveness of the international treaties and institutions to achieve the nonproliferation of nuclear weapons from the liberalist and realist's perspective:

2.7.1 International treaties (nonproliferation of nuclear weapons treaty and banning nuclear weapon tests treaty)

The nonproliferation of nuclear weapons treaty (NPT) was established to propagate the need to stop the proliferation of nuclear weapons. Articles I and II of the treaty indicate that it is aimed at discouraging the development and use of nuclear weapons (United Nations Office for Disarmament Affairs, 2019). The liberalists have however criticized NPT's objectives and the decisions made to grant authority to the five nuclear states to develop nuclear weapons and for this reason, countries like India, Israel, and Pakistan started to develop their nuclear program outside the NPT because of the absence of equality. Lately, in 2003, North Korea announced its withdrawal from the NPT (Mishra, 2008) for the same criticism. The lack of cooperation by some states is due to the biased conditions included in the NPT which some states claimed was not in line with creating equality or moral cooperation among all participating states.

There have been allegations of connivance between some nuclear states and specific non-nuclear states to develop nuclear weapons (Carranza, 2016). Sarram (2015) pointed out that the American governments' decision to support Israel's nuclear program was in contradiction to the NPT. The alliance between the United States and Israel in the development of nuclear weapons is aimed at preventing other states in the Middle East from embarking on similar nuclear weapons projects. The political interest of the United States in supporting Israel's nuclear programs is evident in its refusal to sign a treaty that will declare the entire Middle East a nuclear weapon-free zone as this will require Israel to abandon its nuclear programs. This is also the case in Asia where China refused to declare South-East Asia a nuclear weapon-free zone. (Sarram, 2015).

The liberalists have also criticized the inability of the international institutions to enforce the NPT and TBT, which ban all states from carrying out nuclear tests. This treaty was created with the primary intention of stopping the proliferation of nuclear weapons, but the reality is different. Also, the liberalists have frowned at the obvious negligence and the breach of the statutes of these treaties by countries such as the United States and France.

2.7.2 International institutions (Security Council, The International Court of Justice, and the International Atomic Energy Agency)

Security Council is labeled crucial judicial authority that was formed in 1945 under UN charter, article 23 under Chapter V with a primary objective to maintain international peace and security (United Nations, 2019a,b). Security Council given the responsibility of stopping the banning the proliferation of nuclear weapons with binding resolutions and sanctions according to the UN Charter, article 39 under chapter VII (United Nations, 2019b; Vredenburgh, 1991). The liberalists recognize the functions of the Security Council, which include implementing the regulations of the UN charter chapter VII aimed at nonproliferation of nuclear weapons

(United Nations, 2019a). There have also been allegations of bias leveled against the Security Council. For instance, the Security Council failed to compel Israel to sign the treaty declaring the entire Middle East region a nuclear weapon-free zone; Israel supposedly has the backing of the United States in this decision. This event affirms the doubts of liberals regarding interference of powerful states in contradicting the decisions made by the Security Council thus conforming to a realism perspective while limiting the functions of the Security Council, the alliance between the United States and Israel is a good example in this regard.

The International Court of Justice has functions, which include providing support to international institutions to achieve the nonproliferation of nuclear weapons globally. The United Nations established the International Court of Justice immediately after the Second World War in 1945 (International Court of Justice, 2019b). The responsibilities of the International Court of Justice include passing resolutions on all issues presented to it regarding the imposing of sanctions on states that breach the treaties aimed at achieving the nonproliferation of nuclear weapons. The liberals have however identified two key cases of criticisms where the International Court of Justice has failed to prove its commitment to stop the proliferation of nuclear weapons. The first case is the inability of the International Court of Justice to unequivocally and explicitly rules that the use or proliferation of nuclear weapons in any state is illegal and forbidden (International Court of Justice, 2019 c). The second case is the inability of the International Court of Justice to enforce the Ban on Nuclear Tests. For instance, a case was raised in 1973 to the International Court of Justice by Australia and New Zealand requesting sanctions against France for illegally performing nuclear tests in the Pacific Ocean was struck out when it was clearly declared by the International Court and France that the court had no legal jurisdiction over such issue (International Court of Justice, 2019d). Accordingly, the liberals criticized in the International Court of Justice that give a special privilege and rights to the nuclear state while non-nuclear states are prohibited from having the same rights. These cases affirm the doubts of liberals regarding the interference of powerful states in contradicting the decisions made by the International Court of Justice thus conforming to a realism perspective.

Finally, the IAEA, which is an intergovernmental body that has been established in 1957 by the United Nations to evaluate and develop methods through which nuclear energy can be safely used and controlled to achieve peace and sustainability development in many areas, such as in health, agriculture, electricity (IAEA, 2019a). IAEA has contributed positively in making NPT as well as regulating the safe use of nuclear reactors used in many countries for power generation. IAEA under the nonproliferation of nuclear weapons (NPT) has a significant role as the multilateral channel and the international safeguard inspectorate for the peaceful transfer of nuclear technology in relation to articles III and IV of the NPT that state, respectively, "the IAEA administers international safeguards to verify that non-nuclear weapon states party to the NPT fulfill the nonproliferation commitment they have made, with a view to preventing diversion of nuclear energy from peaceful uses to nuclear weapons or other nuclear explosive devices. Article III" and "The Agency facilitates and provides a channel for endeavors aimed at "the further development of the applications of nuclear energy for peaceful purposes, especially in the territories of non-nuclear weapon States Party to the Treaty, with due consideration for the needs of the developing areas of the world. Article IV." (United Nations Office for Disarmament Affairs, 2019).

Moreover, at the global and regional level; there are many worldwide atomic energy agencies collaborating with the IAEA in the same scope of work such as the Arab Atomic Energy Agency, which was established in 1989 under the auspices of the League of Arab States whose primary objectives include the development of a strategic vision to make sure that nuclear technology is used for achieving peace and sustainability development in the Arabic region (AAEA, 2019).

2.8 History of the emergence of peaceful nuclear applications

More than 250,000 citizens were killed as a result of the atomic nuclear bombs that were dropped in Hiroshima, Japan, on August 6, 1945, and on Nagasaki on August 9, 1945. These two major nuclear disasters in the history of humanity have left widespread angry reactions at the international level until today. Since then, scientific studies and research have devoted their efforts to harnessing atom and nuclear technology for peaceful applications. Precisely, the first application focused on the generation of electricity through the exploitation of enormous amounts of heat emitted from the fission reaction.

As mentioned earlier in the first chapter, the speech made then by the president to the United States, Eisenhower played a crucial role in the commencement of peaceful nuclear applications and declared the birth of peaceful nuclear applications officially. The thematic concern in the speech was "Atoms for Peace" effectively delivered to the General Assembly on December 8. After that, this was accompanied by the creation of the (IAEA) in 1957 to respond to the critical issues raised from the application of nuclear technology that led to the Hiroshima and Nagasaki tragic endings and the subsequent destruction of life and property. Importantly, the ratification of the statute by President Eisenhower in 1957 played a significant role in the official rebirth of peaceful nuclear applications whereby his excellency claimed that "In fact, we did no more than

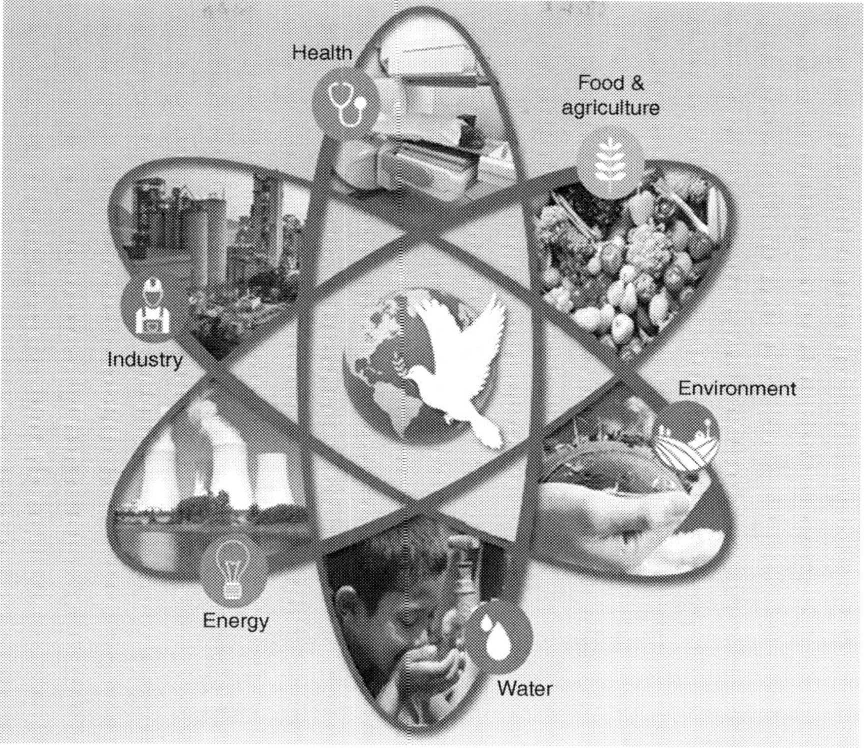

FIGURE 2.2 Atom for peace and sustainable development.

crystallizing a hope that was developing in many minds in many places ... the splitting of the atom may lead to the unifying of the entire divided world (IAEA, 2019b)."

Research in both isotopic and nuclear science technology has continued to contribute to the advancement of nuclear technology for more than three decades especially through the breakthrough in the ionizing capability of isotopic radiations through their interactions with materials. This technology facilitates the breakdown of molecules resulting in a plethora of effects in the materials thereby leading to their applications in the energy, medical, scientific, industrial, and agricultural sectors (Fig. 2.2).

For instance, nuclear and isotopic technology have contributed toward improving farming methods and increasing productivity within the food and agricultural industries. Notably, through nuclear mutation technology, they can naturally develop hybrid vigor crops with characteristics that make them fit to adapt to harsh environmental conditions such as desertification, drought, extreme temperatures, and increased productivity.

Similarly, the application of ionizing radiation technology has also eased the mitigation of human, animal, and plant pest causing diseases such as bird flu, malaria, parasites, worms, and palm.

Additionally, nuclear, and isotopic technologies, through the help of Caesium-137 and carbon-13, also assist farmers in the determination of the quality of soils suitable for cultivation. Nitrogen-15 isotope probe technology a modern technique used to facilitate "smart farming" which is largely applied in optimum to rationalize the consumption of water and fertilizers. Therefore, this method plays a key role in measuring the soil moisture rate and at the same time determine the proportion of water needed for the plants about their optimum absorption rates.

On the other hand, the application of nuclear and isotopic technology in the health sector has played a key role in the treatment and control of many intractable diseases such as heart, cancer, and blood vessel diseases. Significantly, isotopic radiotherapy and the three-dimensional diagnostic techniques immensely contributed to the early detection of diseases. Moreover, using irradiation technology, it is also recognized for its role in making highly effective immunosuppressants.

Notwithstanding, the use of nuclear and isotopic technology has been a key driver in the advancement of the industrial sector. For instance, through the use of radiotracers such as Xenon-133, Argon-41, and Borm-82 including other relevant radiotracers, they have been utilized widely in the oil and gas industry to facilitate the tracking of oil leakages and measuring the size and properties of the oil receiver during the geological formation to improve the efficiency of its production. Furthermore, they are used in food processing to kill pathogens and microorganisms, and the polymerizing effect of the radiation materials has primarily been applied in changing the chemical, physical, and biological properties of the elements to generate new materials with desirable qualities. Archaeologists are also relying on the radioactive dating of the isotopes to accurately dating prehistoric objects and to locate structural defects in buildings, antiques, and statues through radiation polymerization. In engineering, radioisotopes are playing a key role in facilitating the detection of irregularities in metal casings, measurement of microscopic thickness, and testing welds. Briefly, nuclear research is contributing considerable benefits to humankind in many plentiful ways. The coming chapters aim at presenting them in more detail and will shed light on how peaceful nuclear and radioisotopes technologies can benefit humanity and pushing the sustainable development wheel.

2.9 Conclusion

Nuclear weapons pose a serious threat to international security, stability, and global peace. The need to stop the proliferation of nuclear weapons has been the focus of international communities. The international community has taken measures such as the enactment of international law, organizing international conventions, and establishing institutions to create a safeguard and framework through which the proliferation of nuclear weapons can be controlled and eventually stopped. Despite this, the efforts of international law have been hindered by some nuclear weapon states, which have breached NPT agreement implicitly and, in some cases, aided other states to develop nuclear weapons. Subsequently, states such as Israel, India, and Pakistan that are not recognized as major nuclear weapon states have successfully initiated nuclear weapon programs aided by the main nuclear weapon states. Countries such as the United States and China have also refused to sign the treaties that aim to make regions such as the Middle East, Northern Europe, and South-East Asia as nuclear weapon-free zones and instead they have been reported to have aided the nuclear weapons programs in Israel, India, and Pakistan. In addition, there are criticisms that have been raised against the United Nations and its legal institutions in the treatment given to all the participating states that are regarded as biased, and this can hinder the international efforts aimed at stopping the proliferation of nuclear weapons globally. The five nuclear weapon states have been criticized for rejecting the signing of a Nuclear-Test-Ban Treaty that will increase the influence of the NPT under international laws. Thus, this raises suspicions regarding its original purposes, which many states fear is to protect the status of the nuclear states.

It is evident that adopting the realist views in the affairs of all the states involved will result in lasting tensions that the international institutions may not be able to handle and therefore lead to global anarchy and nuclear wars. Nuclear war should never be promoted, and all efforts should be made to prevent the occurrence of any situation that could potentially lead to hostilities between nuclear states or between a nuclear state and a non-nuclear state. It is, therefore, crucial to address first all criticisms that have been raised by non-nuclear states while ensuring that a genuine effort is made to eliminate the threats that make the other states compelled to initiate nuclear weapon programs. This level of international politics will encourage every state to have confidence in the affirmations given by their counterparts to obey the resolutions of the NPT, the Security Council, and the International Court of Justice that eventually lead to the world to avoid anarchy that may result from the realism ideology. Therefore, the goals of the international law can also be achieved when non-nuclear states such as Israel, Pakistan, and India feel confident enough to trust the decisions arrived at by the international institutions in view of achieving long-lasting world peace and also to stop their nuclear development programs.

Accordingly, the international institutions and treaties designed for the nonproliferation of nuclear weapons can only be successful with the cooperation of all the member states as emphasized by the liberalism ideology. In view of achieving global peace, then it is important that all forms of threats should be eliminated to make all existing states feel safe through trust-building and serious cooperation to get rid of all nuclear weapons. Therefore, the five main nuclear powers are primarily responsible for cooperation to ensure that nuclear energy is safely used for positive purposes. The cooperation of the five nuclear states will enhance the authority of the international treaties and institutions such as the NPT, the Security Council, and the International Court of Justice to make them more effective in their duties. Nuclear states should take the initiative to utilized enriched uranium in nuclear warheads into reactor fuel to generate electricity that benefits people rather than killing them. In 1994, the Megatons to Megawatts treaty with Russia to reuse nuclear warheads into reactor fuel is a good example of how to harness the atom for peace and sustainable development. Not only that, during the Obama administration progress was made in the United States nuclear plans, where 10% of US electricity comes from dismantled nuclear weapons.

Therefore, more efforts are needed first from the nuclear states to abandon their nuclear program to become a real model strive for international peace and security so that NPT and nuclear weapons tests treaty becomes effective to stop and control any nuclear proliferation. Subsequently, other non-nuclear states such as Iran, North Korea, Israel, India, Pakistan will become convinced to be part of NPT and will ultimately help in avoiding any possible nuclear war in the coming future.

References

AAEA. 2019. Official website. Main. http://www.aaea.org.tn (accessed March 2, 2019)

Ahmed, A., 2017. The philosophy of nuclear proliferation/non-proliferation: why states build or forgo nuclear weapons? TRAMES 21, 371–382 4.

Carranza, M., 2016. South Asian Security and International Nuclear Order: Creating a Robust Indo-Pakistani Nuclear Arms Control Regime. Routledge, London and New York, pp. 1–241.

Cavendish, M., 2003. How it Works: Science and Technology, Third Edition. Marshall Cavendish, New York, USA, Vol. 11, pp. 1551–1555.

Cimbala, S., 2012. Nuclear Weapons in the Information Age. Continuum International Publishing Group, New York, NY, pp. 216–248.

Griffiths, E., Ameri, F., 2013. Iranian Imbroglio. GM Books, CA, USA, pp. 1–283.

IAEA, 2012. Nuclear Technology for a Sustainable Future. Vienna International Centre, Vienna https://www.iaea.org/sites/default/files/rio0612.pdf.

IAEA. 2019a. Official website—about us. https://www.iaea.org/about (accessed March 4, 2019).

IAEA. 2019b. Official website—history. https://www.iaea.org/about/overview/history (accessed March 4, 2019).

International Court of Justice. 2019a. Statute of the International Court of Justice. https://www.icj-cij.org/en/statute (accessed February 25, 2019).

International Court of Justice. 2019b, International Court of Justice. http://www.icj-cij.org/en/court (accessed February 23, 2019).

International Court of Justice. 2019c. Legality of the threat or use of nuclear weapons. International Court of Justice. http://www.icj-cij.org/en/case/95 (accessed February 23, 2019).

International Court of Justice. 2019d. Nuclear tests (Australia v. France). http://www.icj-cij.org/en/case/58 (accessed February 25, 2019).

Mishra, J., 2008. The NPT and The Developing Countries. Concept Publishing Company, New Delhi, pp. 113–242.

Rooth, M., 2015. Iran and nuclear weapons: five models to explain nuclear proliferation in Iran. Department of Political Science, Lund University, Lund, Sweden, 1–52.

Rousseau, D., Walker, T., 2012. Liberalism. Routledge Handbook of Security Studies, Chapter 2, Routledge, London and New York, pp. 22–31.

Rublee, M., 2010. The nuclear threshold states challenges and opportunities Posed by Brazil and Japan. Routledge, London and New York, (17), 1, 49–70.

Sarram, M., 2015. Nuclear Lies, Deceptions and Hypocrisies. GM Books, Los Angeles - USA, pp. 1–480.

Schaffer, T., Schaffer, H., 2016. India at the Global High Table: The Quest for Regional Primacy and Strategic Autonomy. The Brookings Institution, Washington, DC, USA, pp. 1–350.

Sidel, V., Levy, B., 2007. Proliferation of nuclear weapons: opportunities for control and abolition. Am. J. Public Health 97 (9), 1589–1594.

The Commission for the Comprehensive Nuclear-Test-Ban Treaty Organization. 2018. Types of nuclear weapons. https://www.ctbto.org/nuclear-testing/types-of-nuclear-weapons/ (accessed March 5, 2019).

The National Research Council, Policy and Global Affairs, Committee on International Security and Arms Control, 2005. Appendix A physics and technology of nuclear-explosive materials. Monitoring Nuclear Weapons and Nuclear-Explosive Materials: An Assessment of Methods and Capabilities. National Academies Press, Washington, DC, USA, pp. 1–244.

Thompson, J., 2015. Universal Jurisdiction: The Sierra Leone Profile. T.M.C. Asser Press, Springer, Berlin, pp. 5–12.

United Nations. 2019a. Charter of the United Nations. UN website. http://www.un.org/en/sections/un-charter/chapter-vii/ (accessed March 7, 2019).

United Nations. 2019b. Charter of the United Nations. The Security Council. http://legal.un.org/repertory/art23.shtml http://legal.un.org/repertory/art39.shtml http://www.un.org/en/sc/about/ (accessed March 8, 2019).

United Nations Office for Disarmament Affairs. 2019. Non-proliferation of nuclear weapons. https://www.un.org/disarmament/wmd/nuclear/npt/ (accessed March 12, 2019).

Vradenburgh, A., 1991. Chapter VII Powers of the United Nations Charter: Do They Trump Human Rights Law. Loy. L.A. Int. Comp. L. Rev. 14, 1–175.

CHAPTER

3

Applications of nuclear science and radioisotopes technology in power generation, hydrogen economy, and transport

OUTLINE

Applications of Nuclear and Radioisotope Technology: The Atom for Peace and Sustainable Development
DOI: https://doi.org/10.1016/B978-0-12-821319-3.00002-6

3.1 Introduction

Nuclear science has played a pivotal role in many areas of sustainable development for several decades starting from agriculture to medicine and from power generation to industry. Many scientists have contributed to this field and have won many Nobel Prizes. In the 21st century, the role of nuclear science in power generation cannot be ignored. Nuclear science was tangible in electrical power generation for many years. With science advancement, nuclear reactors have become safer and more efficient. There is an increasing demand for energy in the world due to population growth. The United Nations (2019) reported that the world's population is likely to reach 8.5–8.6 billion in 2030, 9.4–10.1 billion in 2050, and 12.7 billion in 2100. Accordingly, different countries meet their energy needs using different methods, such as coal and fossil fuels. While developed countries are very keen on generating clean energy from clean sources, the most important of which are nuclear and renewable energy. Globally, the continuous supply of electricity is the key to economic development and environmental sustainability. Therefore, developed countries and industrialized countries consider nuclear power generation a cost-effective and environmentally friendly source of energy, unlike fossil fuel even if its price is low, it still causes greenhouse gas emissions and harms the environment. Conventional power generation facilities are among the largest sources of greenhouse gas emissions where these gases lead to global warming and climate change.

On the contrary, nuclear power plants are the only power plants that generate an enormous amount of electricity at very cost-effective in the long run and without greenhouse gas emissions. The main cost of nuclear power generation is the construction of a power plant. However, once this cost is overcome, they produce electricity at a stable and low price as compared to fossil fuel-run power plants. The electricity can be produced uninterrupted for a long time in a nuclear power plant. The fuel used in a nuclear power plant is either natural uranium or enriched uranium and the refueling of a nuclear power plant usually is done every18 to 24 months (Blazev, 2016). All these depend on the reactor type. However, strict safety measures are required for operating a nuclear power plant safely and securely. Since the 1990s, the record of the safety of nuclear power plants is outstanding. The remarkable improvement in nuclear safety and security records is based on continuous evaluation and lessons learned from rare nuclear accidents. The scarcity of nuclear accidents, in and of itself, is a landmark achievement for this industry.

The history of nuclear science is related to the discovery of proton, electron, neutron, radioactivity, and fission reaction. Many noteworthy scientists played their crucial part in developing nuclear power as an alternative to fossil fuel generation of energy. The period from 1895 to 1945 marks the flourishing period of nuclear science. Several theories have been put forward to transform nuclear science into an energy source. The era from the 1950s onwards can be viewed as an application of all nuclear research on electric power generation. Numerous designs have been developed for nuclear power plants by taking advantage of the most important nuclear accidents in the history of the nuclear industry. These scenarios of these accidents have contributed to developing the modern nuclear designs of nuclear power plants to meet the highest safety standards established by the International Atomic Energy Agency (IAEA). IAEA has made guidelines and safety measures and standards for all types of nuclear power plants starting from the construction, operation, safety, nuclear fuel transport, and used fuel management. According to Kessler (2012), about 16% of the world's electricity is supplied by nuclear energy, and the leading countries in nuclear power generation are the developed economies including the United States, Russia, Japan, United Kingdom, France, Korea, and Canada. The first nuclear power plant was inaugurated in Russia in 1954 (Basu and Miroshnik, 2019). There are different types of nuclear power plants used in the world. The design of nuclear power plants varies according to the type of nuclear reactor used. The most common types of nuclear reactors include pressure tube graphite-moderated reactors, boiling water reactors (BWRs), graphite-moderated gas-cooled reactors (GCRs), pressurized water reactors (PWRs), and pressure tube heavy water-moderated reactors (PHWRs). The design of these nuclear reactors depends on a number of factors such as nuclear fuel type, nuclear reaction type, coolant type, steam cycle, moderator type.

O'Brien et al. (2010) argue that a 1 kg fission reaction of U-235 in a nuclear power plant yields of 8.2×10^{13} J of heat energy that can be converted into electrical energy, which is sufficient to operate a 1000 MW power plant. IAEA (1999) and many other recent studies argue that the huge thermal energy produced by nuclear power plants can also be exploited to produce hydrogen, which functions as an energy carrier and not as an energy source like fossil fuels. This hydrogen is the base of the new hydrogen economy that will establish a new concept of modern energy infrastructures. Hydrogen is now researched as an effective means of storing the surplus energy being generated by renewable energy sources like wind or solar power and nuclear power. The supporters of the hydrogen economy argue that the future energy needs of the world would rely temporarily on the production of hydrogen by using fossil fuels in the short run and production by renewable and nuclear resources in the long term. The arbiters of this concept claim that hydrogen is unable to compete with electricity generated by fossil fuels or directly from other renewables. Hydrogen has several uses in transport, industry, power generation using fuel cells. Many prototypes of different vehicles and ships that have been designed to run on hydrogen proved their success. If the vehicles, marine transport, and aviation are shifted to run on hydrogen, many million tons of greenhouse gas emissions will be reduced (Kreith and West, 2004). Accordingly, this chapter will shed light on the application of nuclear science and radioisotopes technology in power generation, hydrogen economy, and transport.

3.2 Nuclear power history

The world has witnessed a rapid growth in the population accompanied by an increase in global energy demand. Meiswinkel et al., (2013) argue that the world population has increased by about 33% from six to almost eight billion between 2000 and 2020, and the energy demand is expected to increase by 62%. This growth has led to the growth of energy consumption, and a dramatic increase in greenhouse gas emissions that contribute to the global warming issue. The truth of this is due to the energy sector's reliance on fossil fuels more than ever, which created many challenges for the oil and gas era. The most important of which are the volatility of oil prices and the restrictions imposed to combat global warming, which imposes restrictions on the burning of oil, gas, and coal in power plants. Accordingly, many countries realized the urgent need to start relying on environmentally friendly alternatives such as renewable energy and peaceful nuclear energy.

According to many scholars including Meiswinkel et al. (2013) who argue that the main issues with renewable energy are the operating and maintenance cost, units' quantities, weather, and the electricity capacity they actually produce are low compared to the total electricity demand. Nuclear energy has become one of the reliable, economical, and environmentally friendly options in the global economy to meet the high demand because of its ability to produce large quantities of energy at the lowest operational cost. For instance, 1 kg of uranium from nuclear fuel can produce energy equivalent to that emitted from burning 100 tons of high-quality coal or tons of oil (Basu and Miroshnik, 2019). In addition to that, it has been reported that nuclear power plants in Europe reduce annually about 700 million tons of carbon dioxide emissions to the atmosphere (Basu and Miroshnik, 2019). Electricity generation using conventional sources is considered to be the largest contributor to global carbon dioxide (CO_2) emissions and responsible for 40% of global emissions of 10 billion tonnes of CO_2 in 2003 (Parliament of Australia, 2020).

On the other hand, it is projected that, by 2030, the cumulative carbon emissions saved due to the use of nuclear power will be more than 25 billion tonnes (Parliament of Australia, 2020). Accordingly, nuclear scientists considered "nuclear power" to be the best option for sustainable electricity supply and a carbon-free energy source. World Nuclear Organization (2020) argues that "today there are about 440 nuclear power reactors operating in 30 countries plus Taiwan, with a combined capacity of about 400 GWe. And About 55 power reactors are currently being constructed in 15 countries, notably China, India, Russia, and the United Arab Emirates." Table 3.1 illustrates the distribution of nuclear reactors types per country and the total nuclear-generating capacity.

The main power source in a nuclear reaction is the fission reaction. The enormous energy from a fission reaction comes when the neutrons cause fission in nuclei of uranium. This is a self-sustaining chain reaction providing a cheap source of power generation. Power generation by nuclear means is the peaceful use of atomic energy and may be used to promote the softer image of nuclear energy.

The history of nuclear power generation must be taken as a competition between many technologies to make a better place in the market. This competition started just after World War II and resulted in the Cold War. The technology which made a remarkable achievement was light water technology. This technology was developed in the United States and used in more than 80% of the world's nuclear power plants, which represents about 353 reactors worldwide (Watts, 2013). However, it is believed by many nuclear scientists that light water technology is not the best available now. Therefore, many new technologies are being introduced in the market, but the way light water technology has entrenched the nuclear market it would be difficult to replace.

The use of nuclear science in power generation is an outcome of the continuous and intensive research work of many scientists. Historically, the birth of nuclear science, atomic radiation, and nuclear fission began in 1895 while the flourishing of the nuclear age began in the 20th century. The first use of nuclear energy for power generation emerged in the 1950s of the 20th century and there is a long story behind this success that began in the 19th century. In 1895, ionization was discovered by Wilhelm Röntgen who discovered the X-rays whilst studying the effect of passing electric current (cathode rays) through a glass tube (Thornton and Rex, 2012). In 1896 Henri Becquerel found dark spots on a photographic plate by an ore having uranium and radium named pitchblende. These spots were due to radiation emitted by alpha particles, beta radiations, and gamma particles, and this phenomenon is called radioactivity. In 1898 Pierre and Marie Curie also isolated the first two radioactive elements "radium and polonium" from pitchblende ores (Cooper, 2001). Based on this discovery and in 1902, Rutherford revealed that radioactivity is a spontaneous reaction which emitted alpha or beta particle from the nucleus (Brescia, 2012). Reed (2009) argues that Rutherford continued his experiments until he reached a significant discovery in 1919 when he bombarded the alpha particles from radium to nitrogen and observed a rearrangement of his

TABLE 3.1 Distribution of nuclear reactors per country and the total nuclear-generating capacity.

Reactor type	Countries	Total number	GWe	Fuel type	Coolant type	Moderator type
Pressurized water reactor	United States, France, Japan, Russia, China, South Korea	299	283	Enriched UO_2	Water	Water
Boiling water reactor	United States, Japan, Sweden	65	65	Enriched UO_2	Water	Water
Pressurized heavy water reactor	Canada, India	48	24	Natural UO_2	Heavy water	Heavy water
Gas-cooled reactor (AGR)	United Kingdom	14	8	Natural U (metal), Enriched UO_2	CO_2	Graphite
Light water graphite reactor	Russia	13	9	Enriched UO_2	Water	Graphite
Fast neutron reactor	Russia	2	1.4	PuO_2 and UO_2	Liquid sodium	None
Total		441	390			

nuclei with oxygen formation as given below:

$$^{4}_{2}He + ^{14}_{7}N \rightarrow ^{1}_{1}H + ^{17}_{8}O \quad (3.1)$$

The discovery of neutrons in 1932 by James Chadwick was a crucial event in the history of nuclear science (Brown et al., 1995). Neutrons play a key role in nuclear reactions. A number of new nuclear transformations and reactions were produced through the accelerator developed by John Cockcroft and Ernest Walton in the same year and they bombarded accelerated protons on the atoms (Amaldi, 2014). Frederic Joliot and Irene Curie were successful in creating artificial radionuclides in 1934 (Jevremovic, 2005). While in 1935, Enrico Fermi used neutrons instead of protons and found a greater variety of artificial radionuclides that can be produced artificially (Hafemeister, 2016). Heavier elements were formed in the experiments of Fermi. Fritz Strassmann and Otto Hahn confirmed that fission had taken place by showing the formation of barium in 1939. At that time, everyone was curious to find an explanation about the fission reaction. Otto Frisch and Lise Meitner provided an explanation of the fission reaction process. They explained that when neutrons are captured by the nucleus, the atom becomes unstable and the atom split into two parts and produces more neutrons. They stated that about 200 MeV can be produced from a single fissional reaction (Vértes et al., 2010).

From 1939 to 1945, all the efforts were made to develop an atomic bomb through what is known as the "Manhattan project." Keeping in view the relation of mass and energy put forth by Elbert Einstein, scientists were successful in making an atomic bomb, which was supposed to be used as a weapon to deter Germans if they consider making an atomic bomb. The Hiroshima and Nagasaki bombing caused fear in the whole world and were a turning point in the history of nuclear sciences. After these tragedies, nuclear scientists devoted their efforts, scientific studies, and research to harness atomic and nuclear technology for peaceful applications. The birth of peaceful nuclear applications was announced by President Dwight D. Eisenhower on December 8, 1953, through his speech in front of the United Nations council under the title of "Atoms for Peace" (Kupp, 2005). Since then, nuclear scientists have started to develop nuclear sciences as a peaceful power generation tool instead of arming. The world's first nuclear power station was built by the Soviet Union at Obninsk in 1954 (Ojovan and Lee, 2007). This was a graphite-moderated power station that used uranium as a nuclear fuel. Due to fewer containment facilities, it was not considered a safe way to produce electricity. In the mid of the 20th century, the United Kingdom became the leader in fission R&D and the development of nuclear power technologies (Great Britain: Parliament: House of Lords: Science and Technology Committee, 2011).

During the 1950s and 1970s, a surge was seen in the design and building of nuclear power stations in the United States, Russia, and Europe to meet the energy demands of these countries in a safe way. The nuclear technologies used in the 1950s by many different countries were Magnox. Magnox reactors were developed by the United Kingdom, which are first-generation technology systems that are used to generate electric power in the United Kingdom (Great Britain: Parliament: House of Lords: Science and Technology Committee, 2011). Magnox refers to the tubes' type used in this reactor, which is made of magnesium and aluminum to contain nuclear fuel made of natural uranium and form the core of the reactor. Magnox reactors use the graphite as a moderator and carbon dioxide gases as a heat exchange coolant); RBMK in the Soviet Union (RBMK is a Russian name stands for high-power channel-type reactor and considered as an early Generation II reactor. RBMK has a unique design where each fuel assembly is enclosed in an individual channel of 8 cm diameter pipe that allows the flow of cooling water around the fuel); light water in the United States. The light water reactor is a very common type of thermal-neutron reactor that uses a normal water coolant and neutron moderator instead of the heavy water. These types of reactors come into three different categories, which are the PWR, the BWR, and the supercritical water reactor (SCWR); and CANDU in Canada (CANDU refers to Canada Deuterium Uranium, which is the common type of nuclear reactor used in Canada, and according to the Government of Canada (2020), there are about 18 reactors used to generate power in Canada. CANDU is a pressurized heavy water reactor. This type of reactor uses deuterium oxide (heavy water) as a moderator and natural uranium as a nuclear fuel).

The oil crisis of 1973 changed the dynamics of dependence on oil for power generation. The countries researched nuclear energy as its alternatives. Petit (2018) argues that the world's nuclear capacity growth has been projected from 1 GW to over 100 GW, driven by the growth of electricity consumption. In the 1980s of the last century, France built most of its nuclear plants, which produce approximately 80% of electricity from nuclear energy, and therefore France is one of the countries that lead the world in producing nuclear electric energy and has become one of the countries that export nuclear electricity to different countries in Europe (Cuff and Goudie, 2008). Various factors caused much damage to nuclear power generation in the late 1970s and 1980s. The Chernobyl incident in 1986 was one of the main setbacks to the nuclear industry. The 1990s saw a formation of the third-generation nuclear reactor designs in the United States. At the start 21st century, the developing economies like India and China invested a lot in developing and importing nuclear power generation. Furthermore, many new bills and regulations to enhance nuclear safety and security have developed. Accordingly, Safer designs of nuclear reactors and nuclear power plants have been developed. The future prospects of nuclear technology include its efficient use in other fields.

Nuclear power is now seen as a good alternative for producing electricity. This would lower the reliance on fossil fuels for power generation. Nuclear power generation has no negative impact on the environment. However, it is still viewed that nuclear waste is very damaging factor to the environment, which is not true to some extent because, in nuclear power plants, there is a proper facility specially designed to dispose of nuclear waste. Not only that, but the high-quality design of the reactor's containment building and the existence of many safety and prevention barriers that are designed to high safety standards are able to prevent any leak of radioactive materials to the environment in case of an accident. All of these technical and operational aspects are strictly checked by the IAEA, which is an autonomous body that is established to promote the peaceful use of nuclear energy.

3.3 The working principle of nuclear power plants

Nuclear power plants are usually constructed near water reserves that are used as a cooling system for the nuclear plants that transfer heat from the reactor core to the electrical generator and maintain the pressure limits. Fission is the basic working principle of nuclear power plants, which is the splitting of the uranium nucleus that releases a lot of energy (Grote and Antonsson, 2009). Albert Einstein's theory of relativity as shown by Eq. 3.2 used to estimate the amount of energy released in a fission process:

$$E = mc^2 \tag{3.2}$$

where E is the energy; m is the mass of a substance; and c is the speed of light.

In order to have a continuous source of heat, a continued and controlled chain reaction of nuclear fission is required, which depends on neutrons absorbed by a fissionable nucleus to create the fission. Fission reaction is controlled through the controlled rods that contain a fissionable nucleus, which absorbs the neutrons. The most common fissionable materials used in control rods are Cadmium and boron (Malik and Singh, 2016). Cadmium and boron are strong neutron absorbers. Eq. 3.3 shows, for example, the interaction of boron with a neutron:

$${}^{10}_{5}\mathrm{B} + {}^{1}_{0}\mathrm{n} \rightarrow {}^{7}_{3}\mathrm{Li} + {}^{4}_{2}\mathrm{He} + 2.8\,\mathrm{Mev} \tag{3.3}$$

Nuclear power plants are producing electricity based on the fission principle. The control of fission energy and its risks associated with the environment are tackled by special operational capabilities. The energy released from a single fission reaction is five or six times greater than a chemical reaction. This enormous amount of energy is a source of energy in nuclear power plants. A sustained fission reaction is a crucial step in a fission reaction, and a continuous supply of neutrons is necessary for an accelerated release of energy. Tong (2018) argues that new fissionable products are produced as a result of a fission reaction. When a fission chain reaction balances the neutron production from nonfission absorption and leaking from the boundaries of a system it is then called critical. When its production is greater than losses, it is known as supercritical. When losses exceed production, it is called subcritical. These all three forms are crucial for a working principle of any power reactor production. Therefore, the energy released by neutrons is very crucial. The main fissionable materials are isotopes of uranium and plutonium based on the reactors' type.

The design of nuclear power plants usually complex and their working principle depends on many factors such as the design, reactors types, thermal design, economics, and safety. From the technical perspective, the nuclear design considers some important factors for the optimization of nuclear energy output, which includes fertile and fissile constituents, coolants, moderators, and refueling method (IAEA, 2005). Whereas most of the nuclear power plants need to shut down the plant for nuclear refueling and maintenance because nuclear fuels are not really consumed up completely in the process. According to Argonne National Laboratory (2011), a typical reactor refueling period varies from 12 to 24 months and the shutdown period for refueling and maintenance operations is between a few weeks to a month. Modern designs of the nuclear plants investigate a new process of nuclear refueling with a minimal shutdown except for Maintenance operations. While the major economic considerations for any nuclear power plant include operating costs, maintenance costs, initial capital outlay, and nuclear fuel costs.

The main working principle of a nuclear power plant involves the conversion of thermal energy produced from the nuclear reactors into electric energy, as shown in Fig. 3.1.

In the boiling reactor, the process water is boiled in the reactor by the heat produced from the fission reaction. This reaction takes place when nuclear fuel assemblies are in place and the control rods are slowly lifted up to maintain the required sustainable chain reaction. The steam generated by the boiling water then flows to a multistage steam turbine. Pressurized steam moves the turbine blades connected to the shaft and turns mechanical energy into electrical energy. Then electricity is transferred through transformers to the national grid. The remaining steam is then fed to the condenser. The condenser works as a heat exchanger, which is mostly connected to a cooling tower. The condenser uses seawater or ponds or rivers to cool the steam. The warmed-up seawater is drained back into the sea from the condenser. While the process water is pumped back to the reactor by the feedwater pumps.

In a PWR, process water is circulated through a primary and secondary circuit. The reactor coolant pump circulates the process water into the

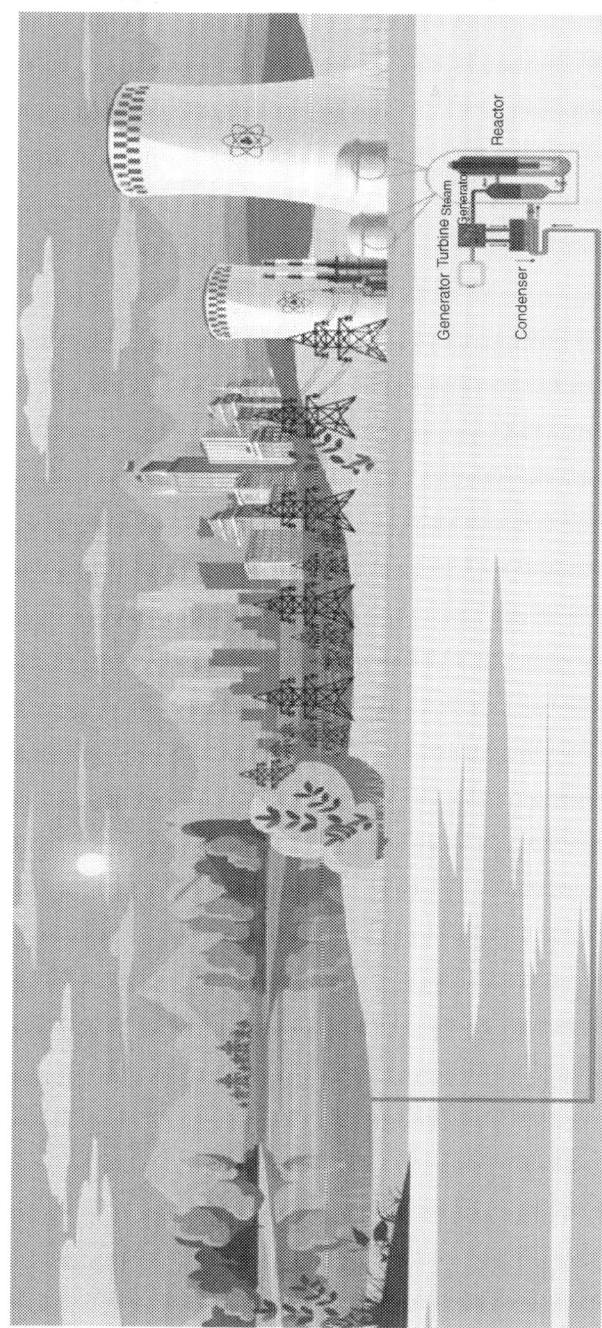

FIGURE 3.1 The power generation process from the nuclear power plant using PWR reactor. *PWR*, pressurized water reactor.

primary circuit. In the reactor, the primary circuit water is heated with the heat generated from the fission reaction. The pressure of the primary circuit water is then increased by the pressurizer to a level where water does not boil. The heat of the primary circuit water that enters the steam generates boils the process of water coming from the secondary circuit. The steam generated by the boiling water from the secondary circuit then flows to a multistage steam turbine. Pressurized steam moves the turbine blades connected to the shaft and turns mechanical energy into electrical energy. Then electricity is transferred through transformers to the national grid. Again, the remaining steam is then fed to the condenser. The condenser works as a heat exchanger, which is mostly connected to a cooling tower. The condenser uses seawater or ponds or rivers to cool the steam. The warmed-up seawater is drained back into the sea from the condenser. While the process water is pumped back to the reactor by the feedwater pumps.

The nuclear reactor is the most crucial part of the power station. The heat is produced by the core due to the fission reaction or known by nuclear fuel. Usually, nuclear fuel consists of a number of cylindrical Zircaloy rods of 1 cm in diameter that are packed with uranium oxide pellets, and these rods are bundled together into bundles. A uranium dioxide (UO_2) pellets or a mixture of uranium and plutonium oxides $(U, Pu)O_2$ are inserted into Zircaloy tubes that are bundled together (Bentaïb et al., 2015). The length, number, packing, weight of the nuclear fuel rods varies depending on the type of reactor and its design. Bentaïb et al. (2015) argue that in PWR each fuel bundle consists of about 264 fuel rods as depicted in Fig. 3.2. On the other hand, Blazev (2016) argues that the fuel bundle in BWR consists of either 91, 92, or 96 fuel rods per assembly. The fuel rods in CANDU are about 0.5 m long and typical core loading consists of 4500–6500 bundles depending on the reactor design (Fig. 3.3).

3.4 Nuclear power plant safety

The most important part of the working of any nuclear power plant is its safety. Nuclear safety and prevention systems consist of a number of safety barriers such as early detection safety barriers, safety instrumented system, process control system, emergency shut down barriers, isolation safety barriers, and many other barriers. However, the major concern in any nuclear power plant is the control of the huge heat generated from the nuclear fission reaction. Accordingly, this issue is managed by the control rods in the core is a major part of safety. The control rods are controlling the fission rate and adjusted it accordingly to predetermined parameters of flow, temperature, power levels, or pressure in the reactor to maintain a constant level of reactivity (Dennis, 1984).

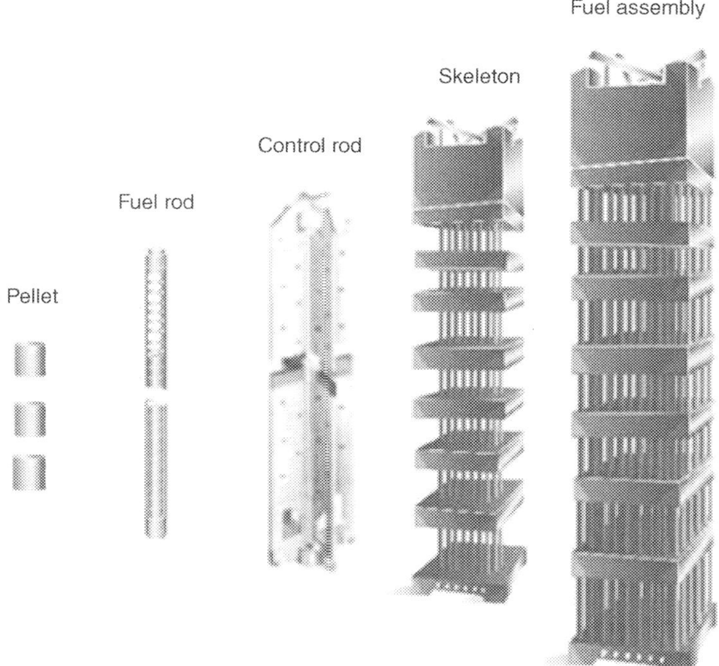

FIGURE 3.2 PWR fuel bundles/assembly. *PWR*, pressurized water reactor.

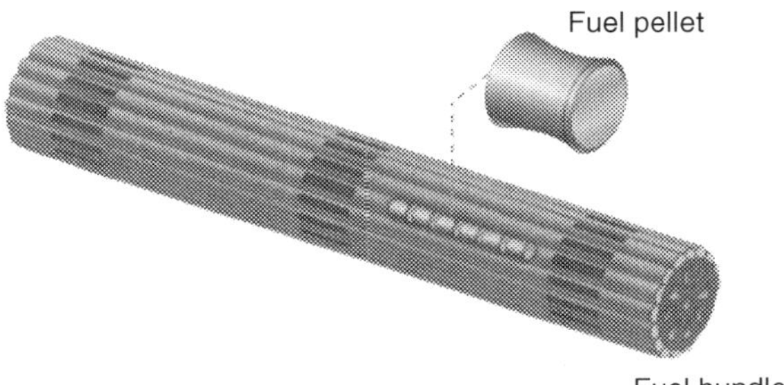

FIGURE 3.3 CANDU fuel bundles/assembly. *CANDU*, Canada Deuterium Uranium

The multiple-barrier containments in any nuclear power plant including containment of products of a fission reaction, the coolant system boundary, the fuel particles, containment structure, and surrounding cladding are the main components of an inherent nuclear safety system. The nuclear safety and security system design depends on the plant design and the type of nuclear reactors used. But the basic safety barriers are all the same in principle after the Chernobyl disaster that occurred on April 26, 1986, new technologies and safer designs have been applied in different types of nuclear power plants to enhance nuclear safety and security.

According to the Joint Committee on Atomic Energy (1967) and the Canadian Nuclear Safety Commission (2020) in the Government of Canada, the safety system of any nuclear plant must comprise of below minimum three systems in order to meet or exceeds nuclear safety standards that are as follows:

Reactor controlling safety barrier: When the reactor is operating, the power level is controlled by adjuster rods or the varying the water level in the vertical cylinders in addition to the safety instrumented systems such as sensitive detectors that monitor reactor temperature, pressure, and power level. The nuclear safety instrumented system can shut down automatically the reactor in seconds. The shut down system in a nuclear plant is an independent system (work without power or operator intervention as well as can be operated manually) and it is a fast-acting system. This system is made up of rods that drop automatically and stop the nuclear reactions in case of any emergency. The second way of shut down is through injecting a liquid or poison inside the reactor to stop immediately fission nuclear reaction. once the reactor is shut down, there is no chance of restarting on its own and must be restarted manually by the operator.

Fuel cooling safety barrier: Following the reactor shut down or even during the normal operation, still fuel continues to decay an amount of heat that must be cooled continuously. Therefore, the fuel cooling safety barrier consists of a heat transport system, steam system, and condenser cooling system. So, the heat transport system transports the heat produced from the reactor to the steam generator. This system is made of tubes that are regularly inspected for any microscopic damages. These tubes contain the coolant that can be heavy water or pressurized water or any other coolant. The steam system that uses normal water is converted into steam as a result of the heat generated from the reactor. This steam runs the turbines and generators, this steam is then cooled and condensed using a condenser cooling system that pumps the cold water in from any water source such as lakes. In case of a loss of the heavy water due to a pipe rupture, for instance, three or four emergency injection sprinkler systems, which work by pumps or pressurized tanks of nitrogen will automatically be activated to keep circulating water over the fuel assembly to cool it. These safety barriers are connected with different backups of power sources such as onsite power, at least three to five standby power generators, and emergency power generators in addition to emergency batteries.

Radiation's containment safety barrier: Any nuclear power plant is built with multiple safety barriers to contain radiation. The first layer of radiation containment starts at the reactor heart, where pallets are hardened with ceramic. These pallets are then placed in rods cladded with zirconium alloys that form the second layer of containment. these rods are loaded in

TABLE 3.2 The most widespread type of nuclear reactors.

Year	Gneration	Reactors
2030–2090	Gen.IV	Very-high-temperature reactor, gas-cooled fast reactor, supercritical-water-cooled reactor, sodium-cooled fast reactor, lead-cooled fast reactor, molten-salt reactor
2005	Gen.III/ III+	Boiling water reactor, evolutionary power reactor
1986	Gen.II	Weapon-grade plutonium reactor, graphite-gas reactor, RBMK, naval propulsion, fast neutron reactor, accelerator-driven subcritical reactor ADS, molten salt reactor
1942–1973	Gen.I	Pressurized heavy-water reactor, graphite-moderated reactor, U- enriched reactor, naval pressurized water reactors

pressure tubes that are part of the heat transport system. This is the third layer of radiation containment. The complete reactor system is contained a thick vault made of a reinforces concrete of at least 1-m thickness that housing and protect the reactor. This is the fourth safety barrier. The fifth containment safety barrier is used for fuel management. After the nuclear fuel has been used in the reactor. It is removed and stored and secured in a pool of water for a period of an average of 10 years. These pools are located in a separate building in the nuclear plant that is designed to withstand an earthquake. This water cools the nuclear fuel and provides shielding against radiation.

3.5 Common types of nuclear reactors

The first world's nuclear power plant connected to the national grid was built in the Soviet Union in 1954 at Obninsk. Since then, nuclear reactors have seen many changes and improvements. In the 1960s, American designs gained much appreciation. In the 1970s, French technology and designs were mostly used (Cuff and Goudie, 2008). The characteristics on the basis of which the nuclear power reactors are classified are neutron energy, moderator, coolant, fuel production, reactor design, and stream cycle (IAEA, 2005). Table 3.2 lists the common types of nuclear reactors based on their evolution from one generation to the next.

An overview of the common types of nuclear power generation reactors.

The following are the common types of nuclear power generation reactors according to Grote and Antonsson (2009):

1. Light water reactors:
   i. PWR
   ii. BWR

2. Pressure tube heavy water-moderated reactors (PHWRs):
   i. CANDU (CANadian Deuterium Uranium)
3. Pressure tube graphite-moderated reactors:
   i. Magnox
   ii. Advanced gas-cooled reactor (AGR)
   iii. Graphite-moderated GCRs
   iv. Chernobyle type
v. Pebble-bed reactor (PBMR)

3.5.1 Pressure water reactors

Pressure water reactors were first developed in the United States to generate electricity and were originally designed for the US Navy. They based mainly on the concept of the naval reactor program to propel nuclear submarines and naval vessels (Reynolds, 1982). The use of this type of reactor on marine ships and nuclear ships is of the utmost importance to various armies around the world. Nuclear energy has a significant advantage over fuel because it allows ships and submarines to operate for long periods without the need to refuel along with the momentum feature that allows maneuverability flexibility. PWRs became widely used in commercial nuclear power plants and are the most dominantly used reactors in the world (Fig. 3.4). In 1957 the first commercial nuclear electricity-producing plant in the United States was inaugurated at Shippingport in Pennsylvania with a production capacity of 60,000 KW (Koslowsky, 2004). PWRs are the generation II reactors. Masterson (2017) argues that the core of PWR is typically arranged in either 15×15 or 17×17 or 19×19 or 20×20 arrays of 4-5 m high assembly. The rods are usually made of zircaloy-clad UO_2, having a 1 cm diameter (Argonne National Laboratory, 2011). Its enrichment ranges from 3% to 5% (Argonne National Laboratory, 2011). These types of reactors use light water (ordinary water) as their coolant and neutron as a moderator. PWR's coolant flows in an open lattice. This allows flow mixing and is maintained under enough pressure so that no boiling takes place under normal conditions. The boric acid provides long-term reactivity control in the coolant. The control rods are made of B_4C or a mixture of Ag-In-Cd (Stacey, 2007). They are weak absorbers, so they make less flux on rod withdrawal. Approximately 190–240 fuel assemblies are having 90,000–125,000 kg of UO_2 that forms a typical PWR core (Stacey, 2007; Kingery, 2011).

3.5.2 Boiling water reactors

These types of reactors were also first established in the United States. They can be found all over the world now and considered as the second famous one after PWR. The nuclear physics of the BWRs is much like those

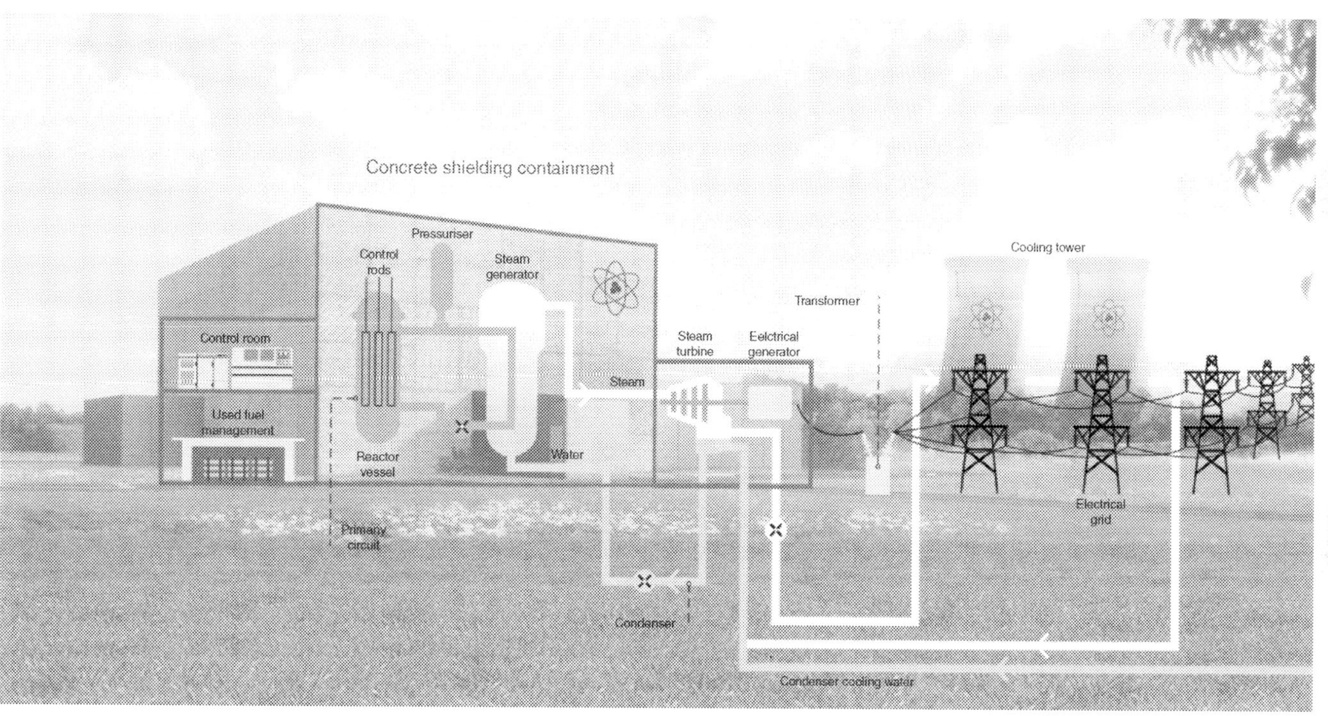

FIGURE 3.4 Schematic diagram for a nuclear power plant powered by a pressurized water reactor.

of PWRs that use a light-water nuclear reactor to generate electricity but the main difference between the two is that in the BWR, the core of the reactor heats the light water in the second circuit to the boiling point until it turns into steam, which circulates the steam turbine. BWR reactors are classified as Generation III reactors (Fig. 3.5). Stacey (2007) argues that BWRs' fuel assembly is 14 cm × 14 cm × 4m high. It comprises 8 × 8 arrays of zircaloy-clad UO_2 fuel pins with a diameter of about 1.3 cm. The enrichment of uranium is from 2% to 4%. The fuel pins array is surrounded by a zircaloy fuel channel, which prevents crossflow between assemblies. Each assembly is loaded with fuel pins of different enrichment. Control rods and recirculation flow provide the control on short-term reactivity. According to Kingery (2011), the BWR core is made up of 750 fuel assemblies having 140,000–160,000 kg of UO_2, just like the core of PWR, located in a pressure vessel. Recirculation of about 30% of the coolant flow is achieved.

3.5.3 Pressure tube heavy water-moderated reactors

CANDU reactors were developed in the 1950s in Canada and use heavy water (deuterium oxide—D_2O) as a moderator (Soysal and Soysal, 2020). This type of reactor is unique because it can use natural nonenriched uranium, enriched uranium, mixed fuel, or even thorium as nuclear fuel (Schwarz, 2017). CANDU reactors have two main advantages that distinguish them from other reactors. The first is that fuel can be supplied during the operation of the plant at full capacity without the need to stop the station, while most other designs must be stopped for refueling. Another economic advantage is that the fuel costs of CANDU reactors are very low since natural uranium does not require enrichment. Advanced CANDU reactors are classified as III+ Generation reactors, which are basically an evolutionary improvement of Gen III (Vaidyanathan, 2013). According to Brennen (2016), the fuel bundle is the basic structure of a CANDU reactor core, which consists of UO_2 sealed into 30–40 zircaloy-clad fuel pins. The cylindrical fuel bundles are typically 50 cm long and 10 cm diameter (Brennen, 2016). These are separated by spacers. In a pressure tube, 12 fuel bundles are placed end to end, which flows pressurized D_2O. Brennen (2016) argues that there are typically about 380–480 fixed Calandria in the CANDU reactor core. About 100k kg of natural UO_2 is loaded in Calandria tubes. The temperature of each pressure tube is 267°C on entry and 312°C on exit (Laughton and Say, 2013). The height of a typical CANDU core is 4 m, and its diameter is 7 m. A backup shutdown system has a gadolinium nitrate solution in the moderator. CANDU nuclear power plant uses heat to boil water until it turns into high-pressure steam. This high-pressure steam will flow through steam turbines that turn an electrical generator and generate electricity, which then connected to the national grid through transformers, as shown in Fig. 3.6.

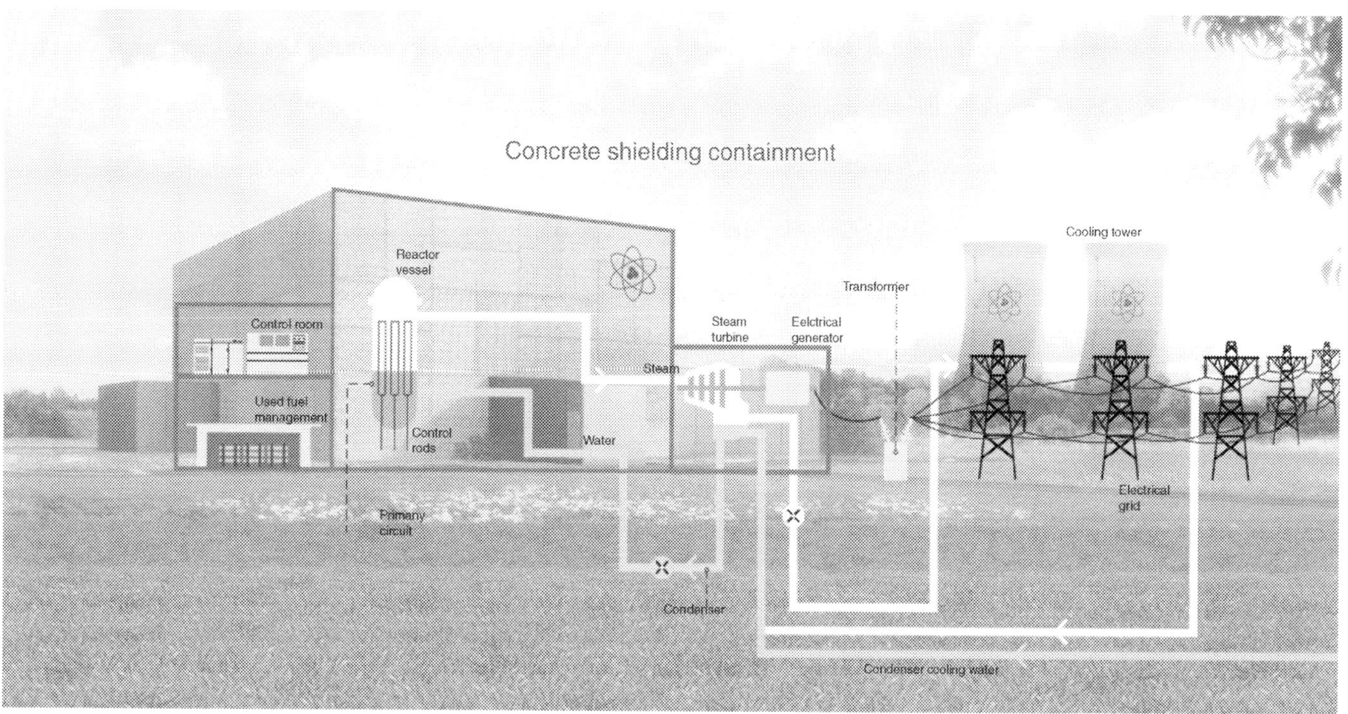

FIGURE 3.5 Schematic diagram for a nuclear power plant powered by a boiling water reactor.

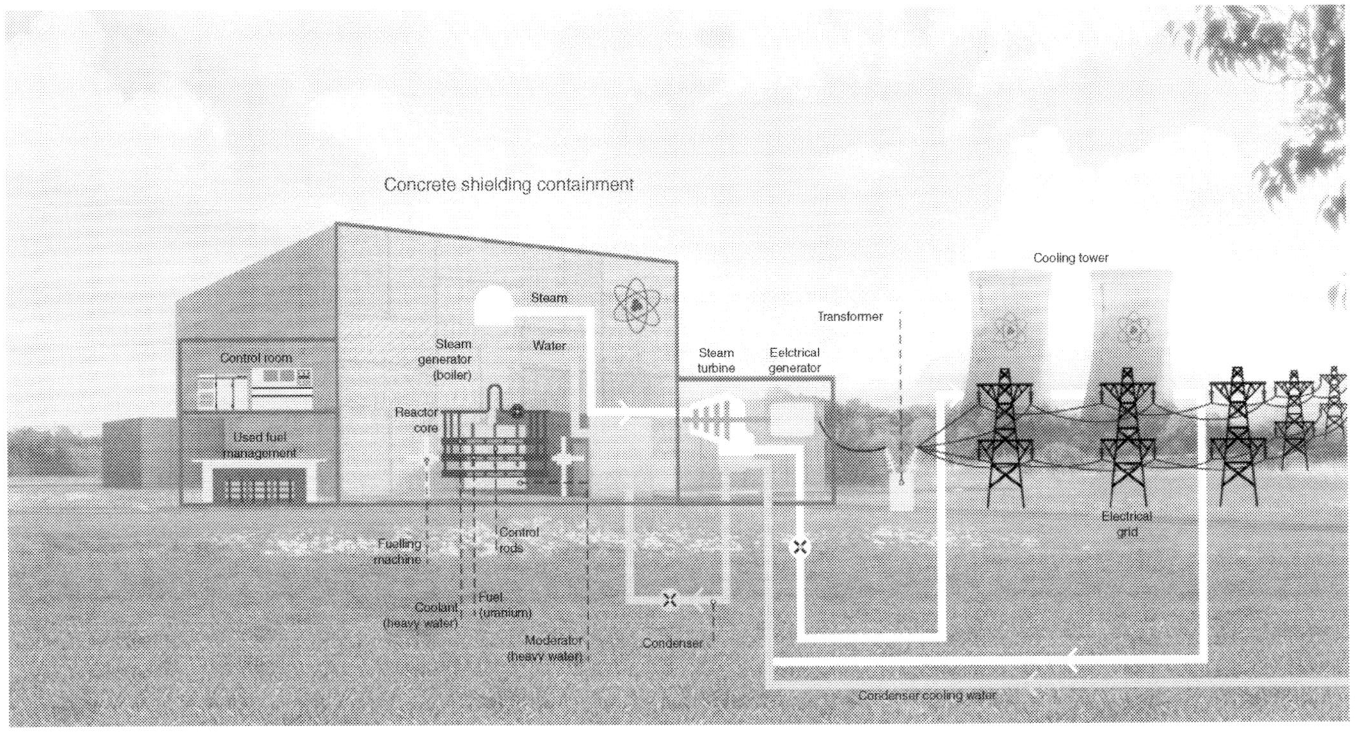

FIGURE 3.6 Schematic diagram for a nuclear power plant powered by a CANDU reactor. *CANDU*, Canada Deuterium Uranium.

3.5.4 Pressure tube graphite-moderated reactors

The world's first commercial nuclear power for electricity generation was built in the Soviet Union in 1954 and was Pressure Tube Graphite-Moderated Reactors that generate 5 MW of electricity (Onishi et al., 2007). It later evolved to the high-power pressure tube reactor. One of the pressure tube graphite-moderated reactors is RBMK (Fig. 3.7). RBMK-1000 was used in unit 4 of the Chernobyl plant, which was known to not meet international safety standards due to problems in the basic emergency cooling system, safety system, and control and control system, which all led to the Chernobyl disaster (Onishi et al., 2007). RBMKs are an early generation II power plant (Sornette et al., 2018). These types of reactors are in former Soviet countries. The core is comprised of the fuel channel tube consisting of zirconium alloyed, having 2.5% niobium. According to Stacey (2018), each channel has two fuel strings cooled separately with water at 7.2MPa. The water gets in at 270°C and gets out at 284°C. From 1.8% to 2% enriched UO_2 in about 18 fuel pins are present in each string. The length of it is 3.6 m and a diameter of 1.3 cm. An upright cylinder of 12.2 m diameter is made by 1661 graphite blocks and 222 control rods (Stacey, 2018). It has almost 200k kg of UO_2.

3.5.5 Graphite-moderated gas-cooled reactors

GCRs (Fig. 3.8) use carbon dioxide or Helium as a coolant, graphite as a neutron moderator, and natural uranium as nuclear fuel (Vaidyanathan, 2013). Magnox reactors and UNGG reactors that were built in the United Kingdom and France, respectively, are the two main types of GCR reactors. These graphite-moderated GCRs are classified as generation I reactors. In recent years, the United Kingdom has replaced its Magnox reactors with advanced gas-cooled reactors (AGR), which are a type of second-generation reactors (Vaidyanathan, 2013). According to Stacey (2018), natural uranium bars clad on a low neutron-absorbing magnesium alloy are the main component of the MAGNOX reactor. They were fitted in holes inside graphite blocks. Carbon dioxide coolant flows through it at 300 psi and leaves the core at a temperature 400°C. The height of a typical MAGNOX core is 8 m and 14 m diameter (Stacey, 2018). The advanced gas reactors are operated at 600psi. Rings of six fuel assemblies are present in a control assembly. B_4C is loaded into carbon rods to check long-term reactivity. Pairs of control rods provide control over short-term reactivity. They are inserted in special control assemblies.

As the world progresses more and more in technology, nuclear technology has also seen tangible progress in terms of designs, safety, and disposal of fuel waste. This industry had setbacks in the past due to some dangerous accidents such as the Three Mile Island accident (1979), the Chernobyl

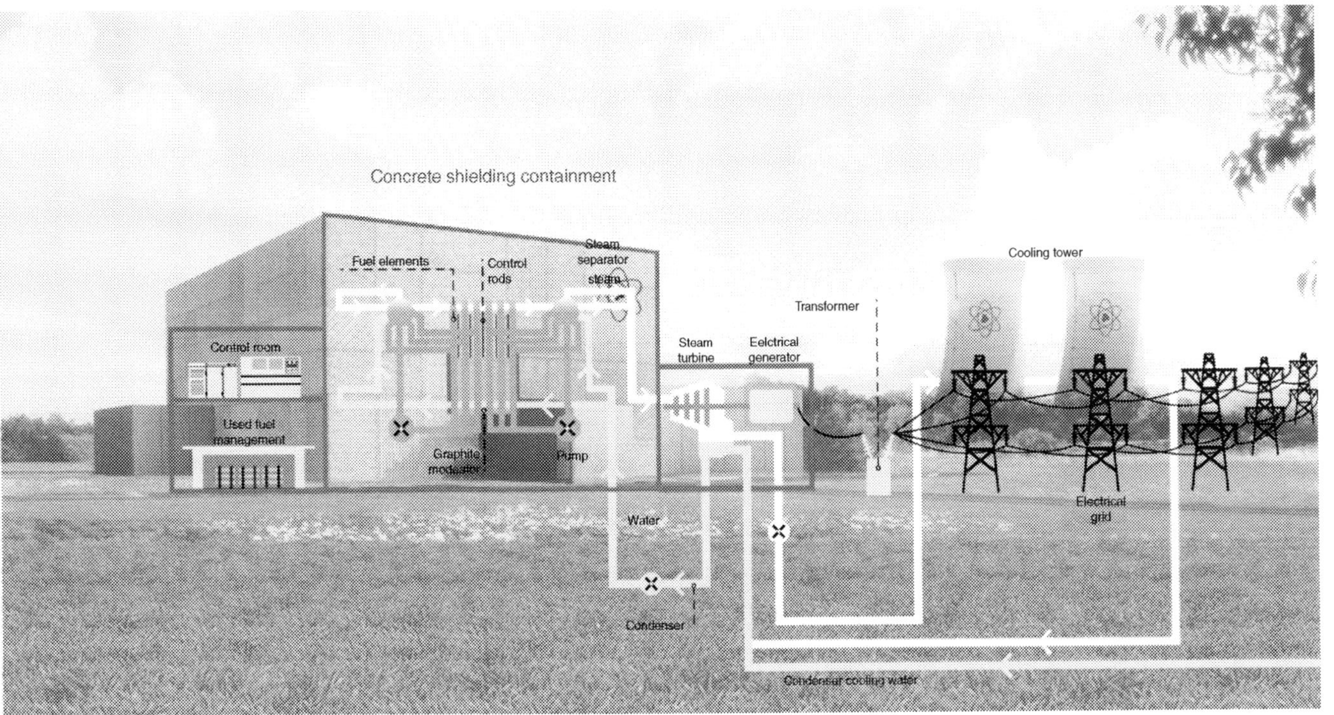

FIGURE 3.7 Schematic diagram for a nuclear power plant powered by a graphite-moderated reactor.

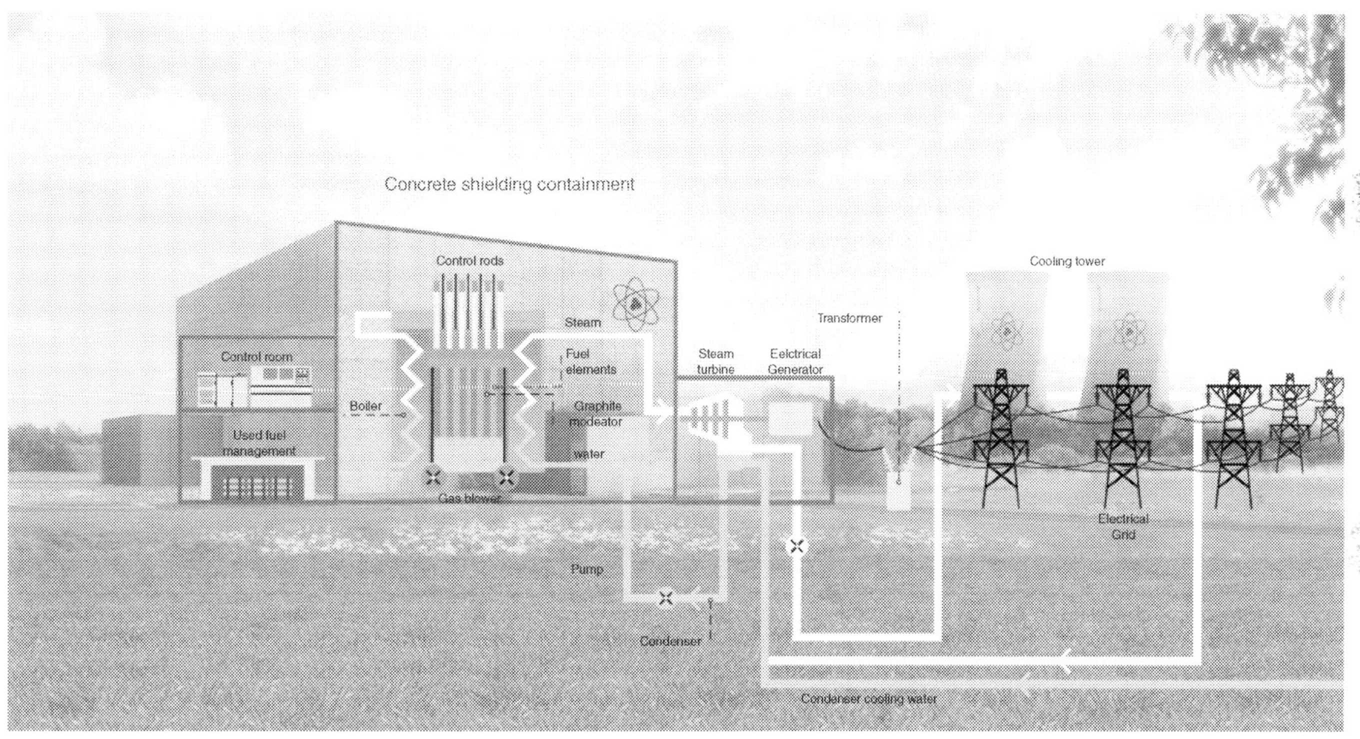

FIGURE 3.8 Schematic diagram for a nuclear power plant powered by a gas-cooled reactor (*GCR*).

disaster (1986), and the Fukushima Daiichi nuclear disaster (2011). These accidents resulted in a lot of damage. But they brought revolutionary changes in the design of nuclear power plants. As the world is going toward more eco-friendly approaches, nuclear power generation has a lot of potential for growth. More and more countries are showing interest in the peaceful use of this technology IAEA keeps a closed eye on all the significant steps being taken by countries for providing a safe environment to work in these nuclear power plants.

3.6 Sustainability of the nuclear hydrogen economy

Ryabchuk et al. (2016) argue that Jules Verne, in 1874, foresaw in his book, (The Mystery Land), that coal would be replaced by water as a source of energy in the future and stated "...that water will one day be employed as a fuel, that hydrogen and oxygen that constitute it, used singly or together, will furnish an inexhaustible source of heat and light.... Water will be the coal of the future." The Oil Embargo of 1973 made the world to search for a strategic alternative to oil, and "Hydrogen Energy" was the best alternative. It must be understood that hydrogen is not an energy producer but rather an energy carrier (IAEA, 1999). A sustainable hydrogen economy would have the least environmental impact. Stolten (2010) argues that the electrolysis of water was the first commercial technology developed for hydrogen production in the 1920s. Then, it was produced from fossil sources during the 1960s. Presently, about 97% of hydrogen is produced from fossil fuels. The share of reformation of natural gas in the world's total hydrogen production is 48 %, the partial oxidation of oil consumes about 30 %, and the gasification of coal accounts for 18% of the total production of hydrogen (Pandey et al., 2019). About 10 kg of CO_2 is produced per kg of hydrogen that is produced from a fossil fuel process of steam methane reformatting. While about 20 kg of CO_2 is produced per kg of hydrogen that is from coal gasification technology (Dincer, 2018).

As the awareness about global warming is spreading the countries are showing interest in those technologies which reduce greenhouse gases emission. Many incentives are being offered to shift toward technologies that reduce CO_2 and other greenhouse gas emissions. Elder and Allen (2009) argue that the leading technologies involved in it are high-temperature electrolysis, the SI thermochemical cycle, and the HyS hybrid thermochemical cycle. These techniques require very high temperatures to be able to produce hydrogen. According to Zohuri (2016), nuclear technology is considered an appropriate and reliable option to produce hydrogen through the heat that is fired by nuclear reactors such as the very high-temperature gas-cooled reactors Generation IV Reactors (Fig. 3.9). In principle, these reactors consist of a graphite-moderated reactor core using an

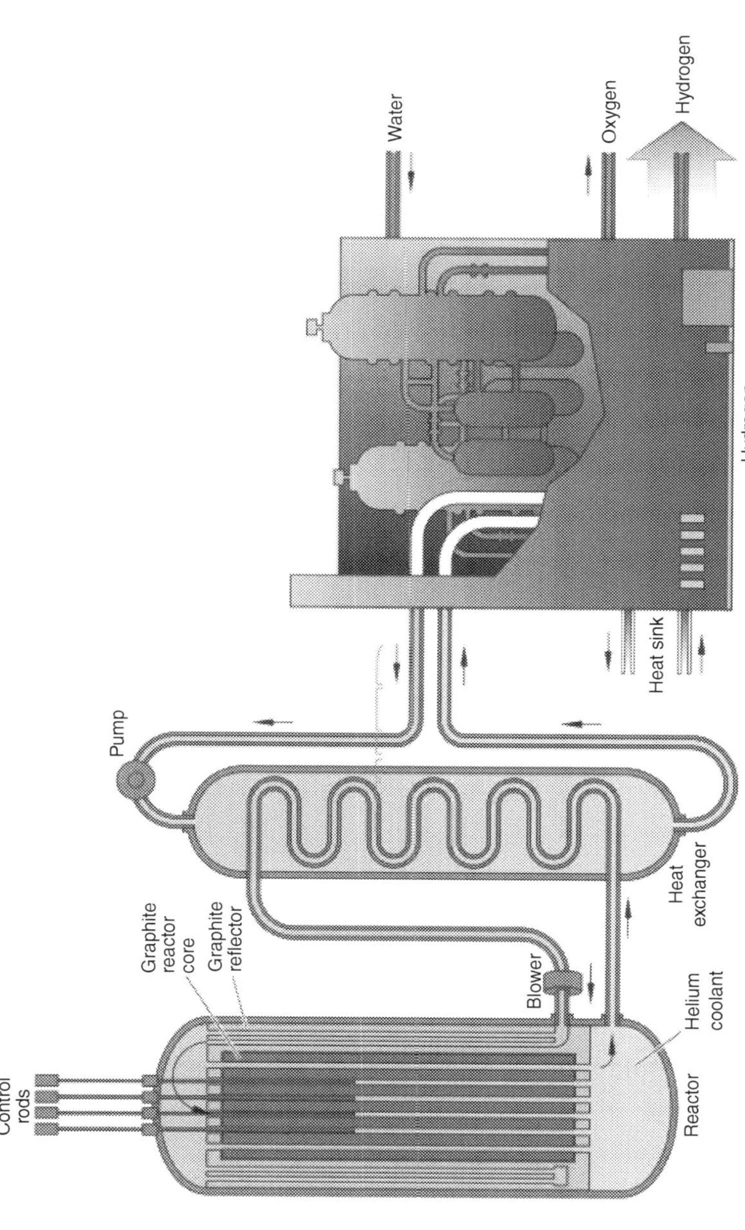

FIGURE 3.9 Schematic diagram of the very-high-temperature reactor (Available from: https://en.wikipedia.org/wiki/Generation_IV_reactor#/media/File:Very_High_Temperature_Reactor.svg).

inert gas such as helium, as a cooling fluid. Despite this, still, the size of the reactor and its type is the most crucial factor in deciding nuclear hydrogen production. Larger reactors are appropriate for cogeneration. Hydrogen generation as a single commodity may be produced by smaller or modular reactors. The integration of nuclear technology with the aforementioned techniques is an ideal option for hydrogen cogeneration.

In 1923, scientist John Burden Haldane predicted that hydrogen would be "the fuel of the future" (Rifkin, 2003). Since then, recent studies have revealed that the hydrogen economy is a futuristic choice and a wise vision for building hydrogen-based energy infrastructure. Hydrogen is a zero-carbon clean energy carrier enabling energy sustainability with reduce dependence on fossil fuels and reduce pollution and greenhouse gas emissions, which are considered as the two main significant challenges of energy production today. Using hydrogen in transportation will reduce greenhouse gas emissions because the engine run on hydrogen does not produce pollution (Staffel et al., 2019). The vision of the hydrogen economy is based on the concept of using hydrogen as an energy carrier. Hydrogen can be used in several ways as an energy carrier, such as feeding it in small quantities into the natural gas network; converting it to methane and introducing the resulting methane gas into the natural gas grid, or stored hydrogen can be converted directly into electricity via fuel cells. The hydrogen economy is an effective way to store energy generated from nuclear energy or from renewable energy sources, such as solar or wind. Hydrogen production by using nuclear energy is one of the effective and promising ways to produce hydrogen (IAEA, 2013). Elder and Allen (2009) argue that there are many research and development efforts for producing nuclear hydrogen worldwide by international energy bodies and agencies such as Japan Atomic Energy Agency, PBMR/Westinghouse, GA, and AREVA NP/CEA/EDF, which are funded by the United States DoE Next Generation Nuclear Plant program. Japan Atomic Energy Agency is the only agency working alone and is world leaders in nuclear hydrogen development.

3.7 The synergy between the nuclear plant and the hydrogen production station

One of the most promising alternatives of fossil fuels in power generation is a symbiotic association of electricity and hydrogen by keeping future demands in mind. Both hydrogen and electricity are mutually interchangeable (Sperling and Cannon, 2004). Both can be interconverted into one another during their production. These are complementary to one another as electricity should be generated at the time where it is utilized. The storage of electricity is costly while the storage of hydrogen is more feasible in this regard. Electric energy storage plays an important role in improving the integrity of the energy conservation system. Currently, electricity is only stored in the short term in batteries, therefore, storing surplus amounts of electricity in the long term requires new types of storage, such as chemical storage in the form of hydrogen. According to Shell (2017), hydrogen storage methods are divided into two main methods, which are the Physical Hydrogen Storage Methods and Chemical or Materials-Based Hydrogen Storage Methods. The physical hydrogen storage methods include pressure or cooling-based (or hybrid storage) storage methods that range from 350 to 700 bar and are widely used for transportation. In addition to storing liquefied hydrogen (LH_2), which is used in the manufacture of chips and space travel. The new hydrogen storage methods are known by materials-based H_2 storage, which is storing hydrogen in solids and liquids, and on surfaces. These storage methods are still in development. According to Hydrogen Europe Association (2017), materials-based H_2 storage methods are classified into three categories, which are: hydride storage systems (chemical bonding of hydrogen molecules with metal molecules such as palladium, magnesium, and lanthanum); liquid hydrogen carriers (chemical bonding of hydrogen molecules with high hydrogen absorption capacities such as carbazole and toluene); and surface-adsorption storage systems (adsorption of hydrogen molecules on materials with high specific surface areas such as zeolites or carbon nanotubes).

The most important challenges of the sustainable hydrogen for transportation are the production of the necessary heat to produce the hydrogen, cost, logistics, the construction of facilities for the production of hydrogen, and distribution stations to be near the users of transportation. Accordingly, the production of hydrogen and electricity at the same time by conventional methods is expensive. Gardner (2009) argues that hydrogen production with electrolysis technology requires 39 kWh of electricity to produce 1 kg of hydrogen. So, if the electricity costs are \$0.05/kWh, then the energy cost for the electrolysis process alone is \$2.40/kg of hydrogen. Therefore, the capital costs of the electrolytic facility are high, but nuclear power plants are able to overcome the problem of cost and production. So nuclear energy is crucial in this regard. The electricity and heat output from a nuclear power plant can be used to produce hydrogen continuously. Hydrogen storage is cost-effective as compared to electricity in fuel cells so that can be used in transport.

3.8 Hydrogen economy applications in the industry and transport

Energy consumption for industrial, residential, business, and transportation purposes is the world's largest energy consumer. Globally, the majority of this energy is supplied by fossil fuels. Many hundred million tons of different petroleum products are being used daily all over the globe. International Energy Association (2020) estimated that the total global final consumption for electricity for 2017 is around 21,372 terawatt-hours. It has been also estimated that over 1 billion cars around the world. In Europe itself, the estimated total energy consumption of cars is approximately 6×10^{18} J/year. While Maritime Transport Report (2017) stated that the world total commercial fleets on January 1, 2017, of 93,161 vessels. In 2012 it has been estimated that Marine Fuel demand around 6.1% of global world oil demand (Concawe, 2012).On the other hand, the average global demand for jet fuel is around 7.5 million barrels per day and the World Economic Forum (2020) estimated that about 100,000 flights take off and land every day across the world.

Nuclear reactors are able to increase their contribution to meet the growing global demand for energy. Not only that, but nuclear energy has become a powerful candidate for providing the zero-carbon energy needed for industrial and transportation applications. This can be achieved through the synergy between nuclear energy and a hydrogen production plant to provide huge amounts of hydrogen at a very low cost and without environmental emissions. For more than 70 years since the launch of the first nuclear-powered vessel, the USS Nautilus submarine in 1955, PWRs used in submarines and marine ships, as shown in Fig. 3.10, have demonstrated their efficiency and reliability in providing the required momentum for ships and submarines to remain at sea for long periods without the need t to refuel (Grant, 2010). Recent studies have also demonstrated that in addition to the impressive success of nuclear power reactors in producing green energy, it has been scientifically proven that it is possible to take advantage of the heat produced from nuclear reactors to produce hydrogen through the synergizing between nuclear plant and hydrogen production plant, as illustrated in Fig. 3.9.

The applications of hydrogen in the market can be divided into two main categories including transport applications and stationary applications in buildings, such as combined heat and power plants or fuel cell electric generators (Widera, 2020). The decentralized power generation is referred to as stationary applications. This includes commercial, public, and industrial large-scale productions, community, and residential power productions. Transport applications include vehicles, and different propulsion systems for aeronautic, marine, and defense applications. Currently, the biggest consumers are running on fossil fuel, and the transport and power industries have great potential to be shifted on hydrogen due to the fact that hydrogen generation is not a cause of any type of pollution.

According to Verfondern (2007), hydrogen is used in many industrial applications, for instance, about 51% of the total consumption of hydrogen is used in the production of ammonia as a fertilizer, 37% is utilized in refining processes of petroleum products, methanol production accounts

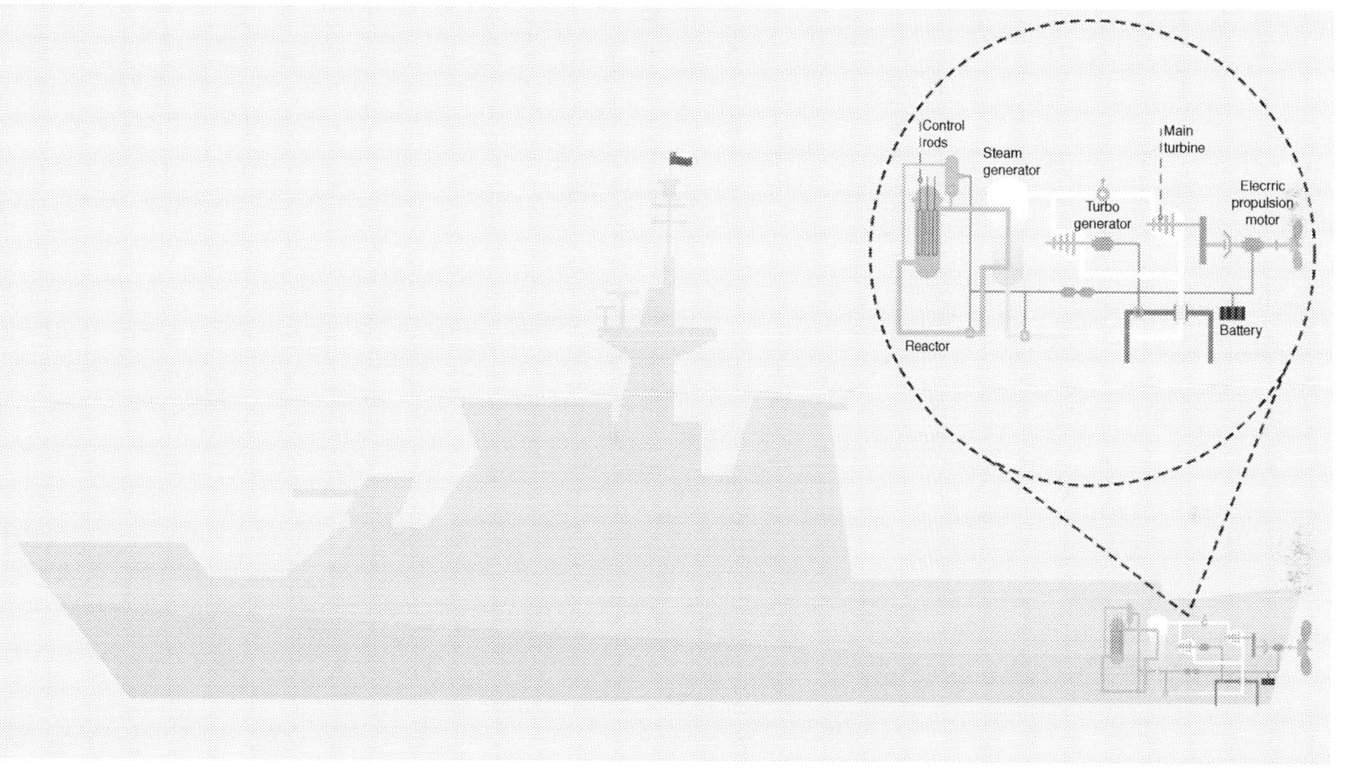

FIGURE 3.10 Pressurized water reactors that are used in submarines and marine ships.

for 8%, 1% is used as fuel space. It has been estimated that the energy used for the small-scale combustion installations within European countries is almost 12×10^{18} J/year (Verfondern, 2007). The main uses of this energy are to produce hot water and heating of buildings. This heat can be produced by the nuclear reactors or the hydrogen that can be used as a feed for the boilers and process heaters if an economic system is developed for producing hydrogen. Many fuel cell systems are being introduced in the market offering higher electrical efficiencies. Moreover, currently, many hybrid systems are being devised for an efficient generation of electricity. A combined system of a gas turbine and hydrogen fuel cell can give as much as 65% or greater efficiency (Verfondern, 2007). Accordingly, and in principle, nuclear reactors or cogeneration, or combined heat and power are able to generate electricity and heat at the same time.

In the maritime industry, vessels are considered a major source of air pollution (Kutz, 2010). In this context, the International Maritime Organization argues that the global marine industry is on a date with a fundamental shift and its future will be linked to the hydrogen future. This is due to the ongoing pressure from the marine community to reduce ship emissions to meet IMO MARPOL Annex VI regulations (DiRenzo, 2019). Many ship owners and shipbuilders have begun to think of hydrogen fuel cell technology to meet the development of emissions regulations. Several marine powers, including the European Union, the United States, and Japan, have begun pilot programs to assess the feasibility of marine hydrogen to reduce emissions while maintaining cost par with conventional propulsion technology. GGZM is the first commercial hydrogen fuel cell vessel in North America Within the Water-Go-Round project (DiRenzo, 2019). Therefore, by using hydrogen fuel cells, pollution can be controlled. Olmer et al. (2017) reported that over the period 2007–2012, ships accounted for approximately 1 billion tonnes of GHG emissions. With the tangible results achieved by the Water-Go-Round project in the United States and in addition to other ongoing projects in the field of marine hydrogen technology such as the HYSEAS III project in the United Kingdom and the HYBRIDskip project in Norway, it is clear that marine hydrogen technology is advancing rapidly from concept to the global domain (DiRenzo, 2019). Similar to the prevailing acceptance of LNG as a marine fuel, hydrogen will likely achieve similar dependence on a large scale. With hydrogen production spreading all over the world, larger ships such as container ships, ferries, and boats may begin to adopt this technology.

In the aviation industry, nuclear technology application has been examined by scientists or engineers working in the Manhattan Project during the world war to develop a new concept of nuclear aviation or nuclear-powered aircraft. This project was stopped due to the difficulty of protecting pilots from the reactor radiation at the time. The main reason behind this stop due to the fact that the reactor needs thick and heavy layers of shields to protect the crew from radioactivity while the plane must be very lightweight in order to take off. Thus, the appropriate protective shields are not compatible with flying standards at that time. In light of the current scientific and technological development, research is still ongoing to overcome this problem, and perhaps in the near future, we will witness the first generation of nuclear propulsion or hydrogen fuel cell aircraft, which will enhance planes' speed and the duration of their stay in the air without the need to land to refuel.

On the other hand, and in recent years, automotive companies have launched prototype vehicles adopting PEM fuel cells that are very energy efficient as compared to internal combustion engines (Brandon and Thompsett, 2005). Despite that, the battery electric vehicles utilize less energy than hydrogen fuel cells, but they are not as cost-effective because their batteries are costly, and the required recharging time is more. Therefore, the hydrogen-fueled PEM cell vehicles have a tough competition with petrol hybrid electric vehicles. In light of the scientific and technical progress witnessed by many developed countries, hydrogen production and vehicle distribution points became available, which enhances the use of hydrogen vehicles that will soon take their place in the market, especially in the area of public transportation such as buses. The use of hydrogen as a fuel is a renewable energy concept and reduce the emissions of nitrogen oxides.

3.9 Hydrogen fuel cycle

The search for alternatives to conventional fuel "oil" has become a necessity for developed industrial countries, especially after the noticeable fluctuation in fuel prices worldwide and a noticeable increase in the emission of greenhouse gases that contributed to the aggravation of the global warming issue. Accordingly, scientists and researchers started working on developing studies and research to obtain alternative sources of energy. Solar and wind energy have been developed to generate electrical energy. Not only that but also tidal and waterfalls energies have been used to convert the kinetic energy into electrical energy. But these alternative sources face many challenges, as they depend on sophisticated, high-cost technologies, and are not suitable for all applications as an alternative to oil because of their association with many factors that limit the efficiency of their productivity. Whereas most of these sources depend on certain climatic and geographical conditions such as the sunshine for long periods in relation to solar energy, and therefore in countries where clouds and dust abound during the year. It will be hard for these countries to rely on solar energy as a reliable and continuous source of electricity generation. Similarly, for wind energy, because it depends on many factors such as wind rate, its direction, and associated operational or maintenance cost. While it may more economically and technically feasible to benefit from the tidal phenomenon and wave movement provided that to be near the sea and this is also not available everywhere, of course. However, despite all these difficulties, the fuel cells, and after the continuous efforts of scientists, have surpassed them to be fuel for the future and an alternative to oil.

Many scholars call fuel cells "the energy source of the 21st century." Fuel cells are a form of converting chemical energy stored in hydrocarbons into direct electrical energy (Wagner, 2007). The most common fuels used in different fuel cells are hydrogen, natural gas, biogas, or methanol with the help of oxygen or air (Witzel and Seifried, 2013). According to Appleby (1990), fuel cell technology was invented in England in the mid-19th century by Sir William Grove, who did not know that his invention of fuel cells in 1839 would solve a problem facing the world in the 21st century by which electricity can be obtained from hydrogen without any combustion process. Thus, the difficult equation, which is obtaining clean energy without polluting the environment and at the lowest prices, has been solved.

The invention of the fuel cell was unworkable for more than 130 years since it was first invented, but the fuel cells emerged in the 1960s when General Electric developed cells that generate electric energy necessary to launch the two famous spaceships "Apollo" and "Jimny." In addition to providing electricity to these spaceships, it provides clean drinking water. Despite the size and the cost of the cells in those two spaceships that were large and expensive, but they performed their mission without any errors.

Fuel cells differ from conventional batteries in their dependence on combining the elements of hydrogen and oxygen to produce electricity, which the cell obtains from an external source and are not considered as part of the fuel cell itself. This gives these cells the importance in comparison with the batteries. In contrast, the situation is different in conventional batteries. The components of the battery are the basis for energy generation. The chemical reaction of the battery components must take place to produce electrical energy and this process continues until the reactive chemicals finished in the battery. At this point, the battery stops until it is recharged again. On the opposite, the fuel cells operate continuously because their hydrogen and oxygen fuel come from external sources. The fuel cells are nothing but a number of flat chips, each of which produces one electric voltage, and this means that the more chips used, the higher the voltage produced. So, the fuel cell is only an energy carrier.

Fuel cells are different depending on the fuel used but in principle, all work the same. The fuel cells are mostly fueled by hydrocarbons, alcohol, or hydrogen. The fuel cell has two electrodes a negatively charged electrode (anode) and a positively charged electrode (cathode) (Fornasiero and Graziani, 2006). These electrodes are separated by an electrolyte. The most common hydrogen fuel cell is the Proton Exchange Membrane Fuel

Cell. The coolant in this type of fuel cell is oxygen, and the electrolyte most commonly used is a membrane of polymer-like Nafion. Fig. 3.11 depicts the working principle of the fuel cell, which starts when hydrogen flows to the anode, while oxygen flows to the opposite plate, the cathode as shown by Eqs. 3.4–3.6. The catalyst causes the hydrogen molecule to split into two atoms, each of which is split into a positive ion, and a negative electron is released from each hydrogen atom that moves to the anode. The electrolyte plate only allows the passage of ions, carrying the positive charges through it while preventing the flow of electrons, the electrons move through an electrical circuit toward the cathode and the movement of electrons from the anode to the cathode represents an electric current that generates the electricity. At the cathode, the positive hydrogen ions combine with their negative electrons and with oxygen to form water that flows outside the cell.

$$\text{At Anode: } 2H_2 => 4H^+ + 4e^- \tag{3.4}$$

$$\text{At Cathode: } O_2 + 4H^+ + 4e^- => 2H_2O \tag{3.5}$$

$$\text{Overall Reaction } 2H_2 + O_2 => 2H_2O \tag{3.6}$$

According to Yürüm (2012), Proton Exchange Membrane Fuel Cells are a reliable energy source for various transportation, marine, and stationary applications, and their efficiency ranges between 50% and 65% with the operating conditions of 90°C and pressure up to 600 KPa. Hydrogen fuel cells are a feasible alternative to replacing fossil fuels as a clean energy source for electricity production.

3.10 Conclusion

Nuclear energy is the only reliable source that can generate large amounts of electricity to meet the growing global demand for energy, at low operating costs, and without emitting any harmful gases, such as greenhouse gases. Today, there are more than 400 nuclear plants in more than 30 countries around the world. Nuclear power stations generally use the energy released from nuclear fission or fusion reaction (division or fusion of atoms) as a source to produce heat that converts water into steam, then pressurized steam activates electricity generator that produces electricity. While conventional power plants produce heat from burning coal, or oil, or natural gas that releases greenhouse gases to the environment.

In directed nuclear fission reactors, the nucleus of the Uranium-235, which is used as fuel in the reactor is bombarded with a free neutron. This bombardment leads to the fission of the nucleus and the release of a huge amount of thermal energy and release of a number of neutrons, which in

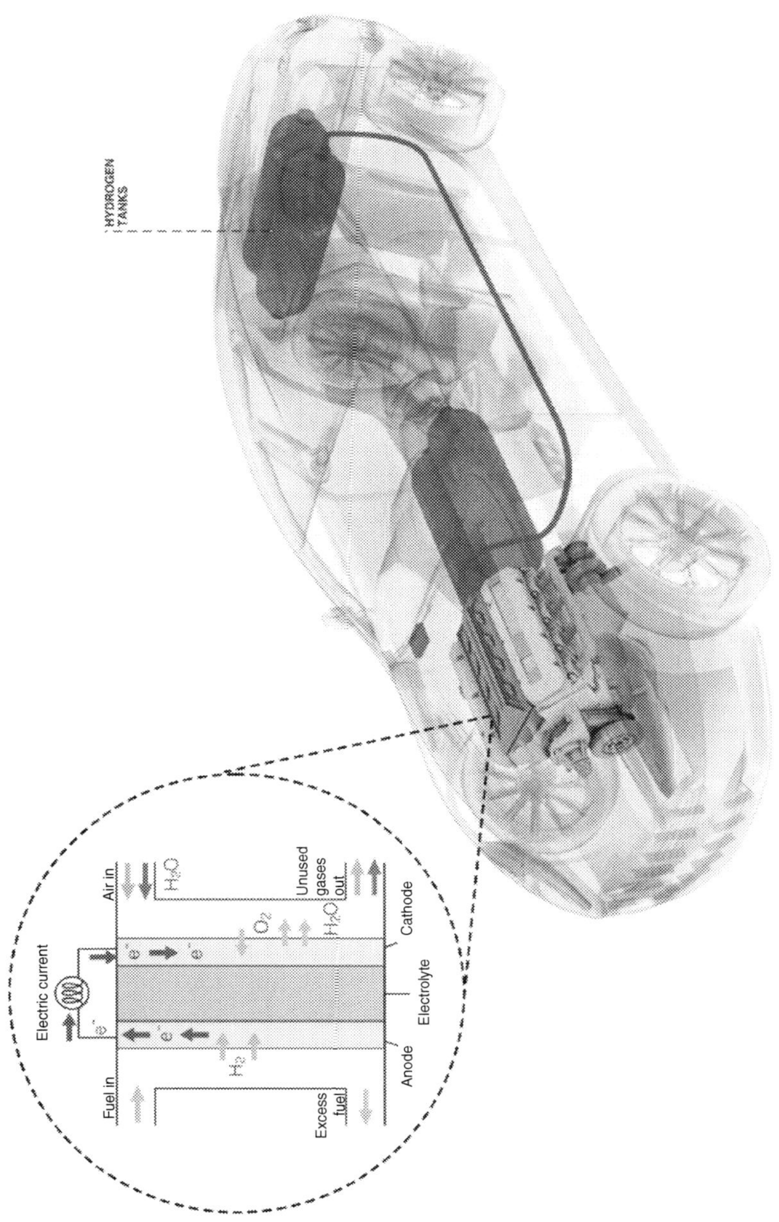

FIGURE 3.11 Hydrogen fuel cell powered car.

turn collide again with other uranium atoms. In such a fission reaction, Millions of uranium atoms will split through a millionth of a second, which creates massive thermal energy.

The energy generated by the fission process within nuclear reactors and the reaction speed is controlled through a number of safety and prevention barriers. The prime function of the safety and prevention barriers is to prevent any explosion. These include control rods, reinforced concrete containment buildings, coolants, control process systems, and safety instrumented systems. Nuclear reactors are divided into two types: the first type is research reactors and ray's production reactors, where atomic research reactors are used in the production of radioisotopes that are widely used in medicine, industry, agriculture, chemistry research, physics applications, and diagnosis and treatment of diseases, such as cancerous tumors. While the second type is energy reactors that are used for energy production. Energy reactors are classified into many types and the most common ones are PWRs, BWRs, PHWRs, pressure tube graphite-moderated reactors, graphite-moderated GCR, and many other.

In addition to producing electric energy from nuclear thermal energy, it is now possible to use nuclear thermal energy to produce hydrogen, which is used in many applications that are the basis of modern hydrogen economics. Hydrogen can be converted into energy using fuel cells and therefore can be used in multiple economic sectors such as for heating and power generation for buildings, transportation, and the industry. Hydrogen energy has the advantage of being a clean energy source. Therefore, the use or export of hydrogen energy reduces environmental impacts as well as helps in meeting global energy demands, besides its main advantage of being an energy carrier, therefore, can be easily stored. Therefore, hydrogen, which functions as an energy carrier by hydrogen fuel cells, can be stored in storage tanks that can be used to supply hydrogen-powered cars, hydrogen-powered ships, and hydrogen-powered planes, as well as build hydrogen distributors like electrical distributors to supply green cities with energy.

Hydrogen energy can be produced through the use of traditional sources of energy such as crude oil and natural gas through the process of steam reforming. Hydrogen can also be obtained from coal by gasification (i.e., conversion of coal into gaseous fuel) or through natural renewable sources where the biomass energy is converted into gas, or through electrolysis processes for solar energy and wind energy. Not only that, but it is also possible to produce hydrogen from other renewable sources such as geothermal energy and tidal energy, but at low production rates in exchange for its abundant production by nuclear energy or what is known as nuclear hydrogen production technology. Nuclear hydrogen production technology has different approaches including methane recycling using internal heat from nuclear reactors or through conventional electrolysis using electricity generated by nuclear energy, or by high-temperature electrolysis using electricity and heat from nuclear energy or by cogeneration or other modern methods.

References

Amaldi, U., 2014. Particle Accelerators: From Big Bang Physics to Hadron Therapy. Springer, Switzerland, pp. 1–284.

Appleby, A., 1990. From Sir William Grove to today: fuel cells and the future. J. Power Sources 29 (1–2), 3–11.

Argonne National Laboratory, 2011. Nuclear fuel. 1–2. Accessed 10 May 2020 and retrieved from https://www.ne.anl.gov/pdfs/nuclear/nuclear_fuel_yacout.pdf.

Basu, D., Miroshnik, V., 2019. The Political Economy of Nuclear Energy: Prospects and Retrospect. Springer Nature, Switzerland, pp. 1–275.

Bentaïb, A., et al., 2015. Nuclear Power Reactor Core Melt Accidents. Current State of Knowledge. IRSN. EDP Sciences, France, pp. 1–434.

Blazev, A., 2016. Global Energy Market Trends. The Fairmont Press, Inc, USA, pp. 1–1065.

Brandon, N., Thompsett, D., 2005. Fuel Cells Compendium. Elsevier, Oxford, UK, pp. 1–632.

Brennen, C., 2016. Thermo-Hydraulics of Nuclear Reactors. Cambridge University Press, New York, USA, pp. 11–160.

Brescia, F., 2012. Fundamentals of Chemistry: A Modern Introduction (1966). Elsevier, Academic Press, New York and London, pp. 1–832.

Brown, L., Pippard, B., Pais, A., 1995. Twentieth-Century Physics. CRC Press, Boca Raton, USA, pp. 1–644.

Canadian Nuclear Safety Commission, 2020. Nuclear power plant safety systems. Accessed 13 May 2020 and retrieved from http://nuclearsafety.gc.ca/eng/reactors/power-plants/nuclear-power-plant-safety-systems/index.cfm.

Concawe, 2012. Marine fuel facts. 1-27. Accessed 20 May 2020 and retrieved from https://www.concawe.eu/wp-content/uploads/2017/01/marine_factsheet_web.pdf.

Cooper, C., 2001. Physics. Taylor & Francis, UK., pp. 1–278.

Cuff, D., Goudie, A., 2008. The Oxford Companion to Global Change. Oxford University Press, New York, pp. 1–720.

Dennis, J., 1984. The Nuclear Almanac: Confronting the Atom in War and Peace. Addison-Wesley, USA, pp. 1–546.

Dincer, I., 2018. Comprehensive Energy Systems. Elsevier, USA, pp. 1–5540.

DiRenzo, J., 2019. IMO 2020: hydrogen's future in maritime. Accessed 21 May 2020 and retrieved from https://www.marinelink.com/news/imo-hydrogens-future-maritime-467713.

Elder, R., Allen, R., 2009. Nuclear heat for hydrogen production: coupling a very high/high-temperature reactor to a hydrogen production plant. Prog. Nucl. Energy 51 (3), 500–525.

Fornasiero, P., Graziani, M., 2006. Renewable Resources and Renewable Energy: A Global Challenge, 1st ed. CRC Press, Boca Raton, USA, pp. 1–504.

Gardner, D., 2009. Hydrogen production from renewables. Renew, Energy Focus. Accessed 18 May 2020 and retrieved from http://www.renewableenergyfocus.com/view/3157/hydrogen-production-from-renewables/9 .

Government of Canada, 2020. Uranium and nuclear power facts. Accessed 7 May 2020 and retrieved from https://www.nrcan.gc.ca/science-data/data-analysis/energy-data-analysis/energy-facts/uranium-and-nuclear-power-facts/20070.

Grant, R., 2010. Battle at Sea: 3000 Years of Naval Warfare. Dorling Kindersley Ltd. London. UK. 1–360.

Great Britain: Parliament: House of Lords: Science and Technology Committee, 2011. Nuclear research and development capabilities: 3rd report of session 2010–12. The Stationery Office, London, UK, pp. 1–116.

Grote, K., Antonsson, E., 2009. Springer Handbook of Mechanical Engineering, 10. Springer, New York, USA, pp. 1–1580.

Hafemeister, D., 2016. Nuclear Proliferation and Terrorism in the Post-9/11 World. Springer, Switzerland, pp. 1–434.

Hydrogen Europe Association, 2017. Hydrogen storage. Accessed 18 May 2020 and retrieved from https://hydrogeneurope.eu/hydrogen-storage.

IAEA, 1999. Hydrogen as an energy carrier and its production by nuclear power. IAEA-TECDOC-1085.VIENNA1-347.

IAEA, 2005. Design of the Reactor Core for Nuclear Power Plants. Safety Standards, Vienna, pp. 1–74.

IAEA, 2013. Hydrogen Production Using Nuclear Energy. IAEA, Vienna, Austria, pp. 1–379.

International Energy Association, 2020. Electricity Information 2019. Accessed 20 May 2020 and retrieved from https://www.iea.org/reports/electricity-information-2019.

Jevremovic, T., 2005. Nuclear Principles in Engineering. Springer, New York, USA, pp. 1–444.

Joint Committee on Atomic Energy 1967. Licensing and Regulation of Nuclear Reactors: Hearings, Ninetieth Congress, First Session, Parts 1–2. In: Congress. Joint Committee on Atomic Energy. U.S. Government Printing Office, Nuclear reactors, pp. 11003.

Kessler, G., 2012. Sustainable and Safe Nuclear Fission Energy: Technology and Safety of Fast and Thermal Nuclear Reactors. Springer, Berlin, Germany, pp. 1–466.

Kingery, T., 2011. Nuclear Energy Encyclopedia: Science, Technology, and Applications. John Wiley & Sons, Hoboken, New Jersey, USA, pp. 1–448.

Koslowsky, R., 2004. A World Perspective Through 21st Century Eyes: The Impact of Science on Society. Trafford Publishing, Victoria, British Columbia. Canada, pp. 1–288.

Kreith, F., West, R., 2004. Fallacies of a hydrogen economy: a critical analysis of hydrogen production and utilization. J. Energy Resour. Technol. 126 (4), 249–257.

Kupp, R., 2005. A Nuclear Engineer in the Twentieth Century. Trafford Publishing, Victoria, British Columbia. Canada, pp. 1–372.

Kutz, M., Elkamel, A., 2010. Environmentally Conscious Fossil Energy Production. John Wiley & Sons, Hoboken, New Jersey, USA, pp. 1–368.

Laughton, M., Say, M., 2013. Electrical Engineer's Reference Book. Elsevier, UK, pp. 1–992.

Malik, H., Singh, A., 2016. Engg Physics. Tata McGraw-Hill Education. New Delhi, India, pp. 1–739.

Maritime Transport Report, 2017. Review of Maritime Transport 2017—structure, ownership and registration of the World Fleet. 21–41. Accessed 20 May 2020 and retrieved from https://unctad.org/en/PublicationChapters/rmt2017ch2_en.pdf.

Masterson, R., 2017. Nuclear Engineering Fundamentals: A Practical Perspective. CRC Press, Boca Raton, USA, pp. 1–961.

Meiswinkel, R., Meyer, J, Schnell, J., 2013. Design and Construction of Nuclear Power Plants. John Wiley & Sons, Berlin, Germany, pp. 1–150.

O'Brien, G., Pearsall, N., O'Keefe, P., 2010. The Future of Energy Use. Routledge, UK and USA, pp. 1–296.

Ojovan, M., Lee, W., 2007. New Developments in Glassy Nuclear Wasteforms. Nova Publishers, New York, pp. 1–131.

Olmer, N., Comer, B., Roy, B., Mao, X., Rutherford, D., 2017. Greenhouse gas emissions from global shipping, 2013–2015. International Council on Clean Transportation, pp. 1–27.

Onishi, Y., Voitsekhovich, O., Zheleznyak, M., 2007. Chernobyl—What Have We Learned ?: The Successes and Failures to Mitigate Water Contamination Over 20 Years. Springer Science, The Netherlands, pp. 1–291.

Pandey, A., Mohan, V., Chang, J., Hallenbeck, P., Larroche, C., 2019. Biomass, Biofuels, Biochemicals: Biohydrogen. Elsevier, The Netherlands, pp. 1–540.

Parliament of Australia, 2020. Chapter 4: Greenhouse gas emissions and nuclear power. Accessed 4 May 2020 and retrieved from https://www.aph.gov.au/Parliamentary_Business/Committees/House_of_Representatives_Committees?url=isr/uranium/report/chapter4.htm#fn23.

Petit, V., 2018. The New World of Utilities: A Historical Transition Towards a New Energy System. Springer Nature, Switzerland, pp. 1–195.

Reed, B., 2009. Physics of the Manhattan Project. Trafford Publishing, Victoria, British Columbia. Canada, pp. 1–184.

Reynolds, A., 1982. Projected costs of electricity from nuclear and coal-fired power plants. Energy Information Administration, Office of Coal, Nuclear, Electric, and Alternate Fuels, U.S. Department of Energy, Washington, D.C, Vol. 2, pp. 1–185.

Rifkin, J., 2003. The Hydrogen Economy. Penguin, New York, USA, pp. 1–304.

Ryabchuk, V., Kuznetsov, V., Emeline, A., Artem'ev, Y., Kataeva, G., Horikoshi, S., Serpone, N., 2016. Water will be the coal of the future—The untamed dream of Jules Verne for a solar fuel. Molecules 21 (12), 1638.

Schwarz, P., 2017. Energy Economics. Routledge, New York, USA, pp. 1–406.

Shell, 2017. Shell Hydrogen Study Energy of the Future? Sustainable Mobility through Fuel Cells and H_2. Shell Deutschland Oil GmbH, Hamburg, pp. 1–71.

Sornette, D., Kröger, W., Wheatley, S., 2018. New Ways and Needs for Exploiting Nuclear Energy. Springer Nature, Switzerland, pp. 1–276.

Soysal, O., Soysal, H., 2020. Energy for Sustainable Society: From Resources to Users. John Wiley & Sons, Croydon, UK, pp. 1–552.

Sperling, D., Cannon, J., 2004. The Hydrogen Energy Transition: Cutting Carbon from Transportation. Elsevier, USA, pp. 1–266.

Stacey, W., 2007. Nuclear Reactor Physics. John Wiley & Sons, Weinheim, Germany, pp. 1–735.

Stacey, M., 2018. Nuclear Reactor Physics, 3rd Ed. John Wiley & Sons, Weinheim, Germany, pp. 1–723.

Staffel, I., Scamman, D., Velazquez Abad, V., Balcombe, P., Dodds, P.E., Ekins, P., Shah, N., Ward, K., 2019. The role of hydrogen and fuel cells in the global energy system. Energy Environ. Sci. (12) 463–491.

Stolten, D., 2010. Hydrogen and Fuel Cells: Fundamentals, Technologies and Applications. John Wiley & Sons, Weinheim, Germany, pp. 1–908.

Thornton, S., Rex, A., 2012. Modern Physics for Scientists and Engineers. Cengage Learning, USA, pp. 1–688.

Tong, C., 2018. Introduction to Materials for Advanced Energy Systems. Springer Nature, Switzerland, pp. 1–911.

United Nations, 2019. World Population Prospects 2019: Data Booklet. pp. 1–28. Accessed 2 May 2020 and retrieved from https://population.un.org/wpp/Publications/Files/WPP2019_DataBooklet.pdf2.

Vaidyanathan, G., 2013. Nuclear Reactor Engineering (Principle and Concepts). S. Chand Publishing, New Delhi. India, pp. 1–248.

Verfondern, K., 2007. Nuclear Energy for Hydrogen Production, 58. Forschungszentrum Jülich, Jülich . Germany, pp. 1–186.

Vértes, A., Nagy, S., Klencsár, Z., Lovas, R., Rösch, F., 2010. Handbook of Nuclear Chemistry. Springer, London and New York, pp. 1–3008.

Wagner, L., 2007. Overview of energy storage methods. Research report. Accessed 22 May 2020 and retrieved from http://www.moraassociates.com/reports/0712%20Energy%20storage.pdf .

Watts, R., 2013. Engineering Response to Climate Change, Second Edition. CRC Press, Boca Raton, USA, pp. 1–520.

Widera, B., 2020. Renewable hydrogen implementations for combined energy storage, transportation, and stationary applications. Thermal Sci. Eng. Prog. 16, 1–8.

Witzel, W., Seifried, D., 2013. Renewable Energy—The Facts. Routledge, UK, pp. 1–256.

World Economic Forum, 2020. Global Agenda, Aviation, Travel and Tourism. Accessed 20 May 2020 and retrieved from https://www.weforum.org/agenda/2016/07/this-visualization-shows-you-24-hours-of-global-air-traffic-in-just-4-seconds/.

World Nuclear Organization, 2020. Plans for New Reactors Worldwide. Accessed 5 May 2020 and retrieved from https://www.world-nuclear.org/information-library/current-and-future-generation/plans-for-new-reactors-worldwide.aspx.

Yürüm, Y., 2012. Hydrogen Energy System: Production and Utilization of Hydrogen and Future Aspects. Springer Science & Business Media, Berlin, Germany, pp. 1–341.

Zohuri, B., 2016. Nuclear Energy for Hydrogen Generation through Intermediate Heat Exchangers: A Renewable Source of Energy. Springer, pp. 1–403.

CHAPTER

4

Applications of nuclear science and radioisotope technology in the industries, and in environmental sustainability

OUTLINE

Applications of Nuclear and Radioisotope Technology: The Atom for Peace and Sustainable Development

DOI: https://doi.org/10.1016/B978-0-12-821319-3.00007-5

4.1 Introduction

After the tragedies caused by the nuclear bombs in World War II, nuclear scientists devoted their efforts, scientific studies, and research to harness atomic and nuclear technology for peaceful applications and advancing sustainable development. Since then, nuclear and radioisotopic technologies have contributed to various sectors of development, and the industry sector is one of many sectors that benefited a lot from these technologies. This chapter provides an overview of a number of nuclear uses in the industrial sector.

The industry sector has exploited the phenomenon of radiation penetration of the material in developing sensitive measures to measuring kits that are able to measure the thickness, density, and level of many liquids and materials that exist in very complex or in dangerous environments.

Furthermore, the nuclear imaging technology provides an opportunity for factories to improve production quality, inspect their products to verify that they are free of defects, and not only that but also used to examining jet engines, detecting cracks in construction and structures, and testing the strength and quality of welding. Radioactive tracers also played an important role in understanding the geological formations of oil, mineral, and water reservoirs, which determines their abundance and the optimal extraction methods to enhance their production. These radiotracers also play an important role in preserving the environment by detecting and sensing leakages from flowlines or feeding dams and monitoring greenhouse gas (GHG) emissions in the environment that cause many environmental issues including but not limited to the issues of global warming and ocean acidification.

Nuclear energy has also proved to be a suitable option in generating clean energy without emissions of greenhouse gases, thus protecting the environment. Nuclear energy ensures the continuity and the sustainability of energy in the light of the global increasing demand for energy as a result of the increase in population density in various parts of the world. Not only that but nuclear energy can be used to produce hydrogen and thus support the transport sector, hydrogen economy, and storage of surplus electrical energy. In parallel to nuclear power generation, it can be used for efficient and reliable water desalination and therefore, supply sustainable and reliable water to meet the global water demand. The increase in global population is not limited to an increase in global demand for energy and drinkable water but has resulted in an increase in waste, the most important of which is wastewater, where nuclear technology was able to treat and turn this wastewater into a new source of drinkable water.

There are many industrial and commercial applications that could not be covered in this chapter such as nuclear technologies used by cosmetic companies to cleanse cosmetic products and another example of industrial and commercial nuclear applications that are credited in saving the lives of tens of thousands of lives and are used widely in millions of homes, offices, buildings, aircraft, are the smoke and fire detectors that rely on a radioactive isotope called americium-241, in addition to many other applications where the list goes on.

Nevertheless, this chapter will attempt to shed light on the largest possible number of applications of nuclear technology in the industry. For example, it will discuss both electrical and nonelectrical applications powered by nuclear energy that can provide sustainable solutions to a number of energy and natural resource sustainability challenges. Whereas nuclear technology can produce energy and water together through the production of electrical energy using nuclear reactors, and the use of surplus energy and heat in desalinating seawater and storing the surplus electrical energy in the form of hydrogen, nuclear desalination has also proven to be a viable option to meet the increasing demand for drinking water around the world as a result of the steady growth in population densities and the scarcity of natural resources, giving hope to arid and semiarid regions facing acute water shortages. Therefore, the industrial nuclear applications in the cogeneration of nuclear energy and water desalination contribute to reducing gas emissions and thus the sustainability of energy resources and environmental protection.

This chapter will also shed light on the exponential increment in the pace of development in nuclear technology, atomic energy scientists have developed several nuclear technologies that have contributed to the reliable promotion and production of minerals. These technologies include but are not limited to natural radiation approach; nuclear radiation approach; nucleonic analysis; nuclear imaging; and scanning tools. These techniques are used to evaluates and characterizes the targeted sediment around and including the prospect using the natural radioactivity of the material, relying on the fact that different substances exhibit different spectra of radioactivity, which can be used to identify the type makeup of the sediment and available minerals, hence determine to how best exploit and extract the prospect. These nuclear approaches are also very useful to determine the concentration of various elements using the nuclear footprint to identify factors present either by tracing or otherwise radioactivity content present in the sediment and ore. Not only that but, the nuclear on-stream analysis combined with modeling techniques make it possible to extract information from the data generated during the mining process. Adding to their merit is the fact that these technologies are nondestructive.

Since most natural resources' reservoirs are isotopically distinctive, isotopic compositions may be used to fingerprint an element or compound. Elements are known to have variations in their isotopic compositions due to mass and bonding environmental differences. Isotopes including hydrogen, carbon, nitrogen, sulphur, calcium, iron, copper, zinc, and uranium are key indicators in many exploratory and environmental science. Accordingly, this chapter will also address how radiotracers are intensively used in a wide range of industrial applications such as exploration and extraction activities in mining, and oil and gas, leaks detection such as flowlines leaks or dams or reservoirs leaks, conserve resources, and reduce the effects of greenhouse gases on the atmosphere.

Furthermore, this chapter will reveal how nuclear and radioisotope engineering has led to the development of a new generation of safe and accurate measurement devices that are called the smart nucleon gauges. Nuclear technology has led to the development of a new generation of measuring devices that able to work in a harsh environment to assess various static or continuous measurement parameters, including the level, flow, weight, density of components, whether liquid or solid or gas, stored in closed tanks or in flow lines. They can also be used in manufacturing lines to calculate the volume of liquids in opaque packages and to verify the efficiency and remove noncompliant products.

It is worth mentioning that the discoveries of X-rays in 1895 and Gamma Rays in 1900, respectively, have led to a new advancement of nuclear science by introducing nuclear imaging and radiography that are widely used in the industry. Nuclear imaging and radiography are nondestructive imaging techniques that are used to examine the insides of objects in minute details using radiation imaging like X-rays, gamma rays and/or neutron or electron beams (EBs) which is not possible with the conventional testing methods. In industrial capacity, this technique has a wide variety of practical applications. From finding cracks in concrete and a wide variety of welds to detecting flaws in structures that would not be detectable otherwise. On the other hand, radiography can also be used for structural analyses of jet engines, pressurized vessels and pipelines, etc. Furthermore, nuclear radiography is also able to detect contamination of materials which makes it a viable quality control tool in industries. All these will be discussed in more details in this chapter.

Another application of nuclear technology will be discussed in more detail in this chapter in relation to combating food fraud. Food authenticity investigation techniques include vibrational spectroscopy, nuclear magnetic resonance, and fluorescence spectroscopy, ion mobility mass spectrometry to verify the authenticity of a material rather than relying on labels and paperwork that can be easily forged. At the beginning of the 21st century, stable isotope analysis has been used to establish the geological sources of foodstuffs. The composition of stable isotopes found in the food serves as a "record" of the environmental factors at the point of origin of the food as well as of the farming and feeding methods used during the manufacturing process. The basic principle of the isotope analysis approach relies on comparing the ratio of stable isotopes in the country of origin such as hydrogen, oxygen, and carbon and their abundance in a product sample. This approach is considered as a special fingerprint connecting any crop to the location where it is grown in that specific area depending on the isotopes of the water. Thus, isotopes of matter that the food product absorbs from the atmosphere reveal where they originate in contrast to DNA that only reveals the lineage.

Finally, this chapter will review sustainable nuclear solutions to face the challenges of global population growth and increasing demand for energy and water. The world has witnessed an exponential growth in population that has resulted in exponential growth with a plethora of related problems, such as pollution, global warming, and increased demand for food, energy, medicine, and other facilities, and the list goes on. But the fundamental challenge is the increase in the demand for energy to operate all these facilities, without negative impacts on the environment. Among renewables sources, at the moment, none have been able to match up with the high and sustainable demand for energy except nuclear technology. Because of its potential to generate vast amounts of energy at the lowest operational cost, nuclear energy has become one of the most economical and environmentally sustainable alternatives in the world to meet the rising demand. The nuclear technology has now been a strong contender to provide the zero-carbon power generation needed for industrial and domestic applications, water desalination, wastewater treatment, hydrogen economy for transportation applications. This can be done by the synergy of a nuclear power plant with a water desalination plant, and a hydrogen production facility to produce large volumes at extremely low rates without having adverse emissions that are harmful to the environment. Nuclear desalination, for instance, began to gain momentum back in the 1960s and it involves transforming seawater into drinkable water using a nuclear reactor plant as a source of electricity. Another problem that became glaringly difficult with the rise in population is the handling of waste, as the waste produced also rose exponentially. Massive amounts of untreated wastewater due to inadequate disposal practices have contributed to the incorporation of toxic and harmful chemicals from industrial, agricultural, domestic activities into surface water and groundwater. Water-borne diseases commonly found and distributed by these wastewaters, such as cholera, dysentery, typhoid, and shigellosis are major killers of millions of people worldwide even with the traditional treatment means. The latest developments in nuclear wastewater treatment systems using irradiation approaches have helped solve crucial problems in providing clean and healthy water supply, pollution control, water conservation, and environmental safety.

In general, nuclear technology has proven itself as a competent tool with enormous benefits in a variety of applications ranging from power generation to improving industrial efficiency and optimization of natural resource exploration and extraction including oil, gas, minerals, and water, to the environment protection from greenhouse gases and marine life from ocean acidification. Not only that but, nuclear technology is credited for saving the life of tens of thousands of people from landmines, combating food fraud, and improving the quality of materials for safe and reliable use in addition to many other industrial applications. The power of nuclear technology has transformed many industries in the past decades. Although some challenges remain, workers in this field remain optimistic that they will be overcome in the near future to provide better solutions to a large number of problems for the benefit of humankind and advancing the wheel of progress and prosperity through sustainable development for the resources and the environment. All these issues will be discussed in more detail in the following sections.

4.2 The role of nuclear science in mineral exploration

With the exponential increment in the pace of development in technology, atomic energy scientists have developed several nuclear technologies that have contributed to the reliable promotion and production of minerals. These technologies include but are not limited to natural radiation approach, nuclear radiation approach, nucleonic analysis, nuclear imaging, and scanning tools. Not only that but, nuclear on-stream analysis combined with modeling techniques have made it possible to extract information from the data generated during the mining process. Adding to their merit is the fact that these technologies are nondestructive.

Many of these technologies are employed today throughout the mining process. Nuclear technologies applications start from the early first stage of mineral exploration, including identification and elemental analysis and its abundance estimation. The next step is procuring safe access to the mine by assessing the mine's caving, slope stability, water and gas penetration, leakage, and in-situ leaching (if any). Once the deposit is accessible safely, the next important stage starts where nuclear technologies play significant roles in extracting and processing the minerals. This process starts with determining the grade and sorting the ores and creating a material balance system (accounting of the input, output, and streams of material between streams in a process), and dynamic monitoring of the operations. In the mining process, there are three main processes of extracting a mineral, which are underground mining, open-pit mining, and placer mining. The nature of the mine, its location, the market price of the targeted minerals all determine what is the most suitable extraction process to be employed. However, environmental factors also need to be considered during the extraction, processing, and transportation of the minerals. Water and waste management, monitoring of contamination of natural water sources (underground or otherwise), energy conservation, efficient use of the resource, and soil evaluation are essential factors in the mining process. The conventional extraction process usually employed supercritical fluid extraction, solid-phase extraction coupled with mechanochemical-assisted extraction or ultrasound-assisted extraction, and many other, which all depends on the ground and minerals conditions and all associated with having a high level of uncertainty compared to the application of more reliable and accurate nuclear techniques.

According to IAEA (2014), the most common nuclear approaches that are implemented in mineral exploration and extraction are (1) natural radiation approach (gamma-ray logging), this approach is a wireline logging method used to evaluates and characterizes the targeted sediment around and including the prospect using the natural radioactivity of the material, relying on the fact that different substances exhibit different spectra of radioactivity, which can be used to the identify the type makeup of the sediment and available minerals, hence determine to how best exploit and extract the prospect; (2) nuclear analytical approach. This approach includes neutron activation analysis or charged particle activation analysis or prompt gamma neutron activation analysis or gamma–gamma backscattering, or gamma-induced X-ray fluorescence, which are used to determine the concentration of various elements using the nuclear footprint to identify factors present either by tracing or otherwise radioactivity content present in the sediment and ore; (3) nuclear radiation approach. This approach is used to detect the physical traits (by the emission or absorption of radiation by nuclei) of the sediment around and including the prospect using one or more of the following methods (gamma transmission and backscattering or neutron moderation, or transmission and backscattering (back-diffusion), or beta and alpha transmission and backscattering, or radioactive tracers); and (4) radioactive tracers' approach. It is a form of isotopic labeling technique, this approach uses labeled radioactive isotope(s) that have been planted in place of a typical molecule in a compound to track the path and reaction that compound undergoes, which comes in handy to determine the flow and dispensation of components in a dynamic system.

Given below are short descriptions of some of the most commonly used nuclear techniques of mineral exploration and processing (Fig. 4.1).

4.2.1 Natural radiation-based analysis (radiometric surveys)

This approach detects and maps the natural radioactive emissions from rocks and soil. Naturally occurring radioactive materials are usually belonging to one of the main three radioactive chains, which are Uranium (U238) chain, or Thorium (Th232) chain, and Potassium (K40) chain, which is traceable by the radiometric survey from the ground. This approach uses the correlation between the densities of naturally occurring radioactive elements among the preselected elements (IAEA, 2014). Initially used for Uranium mining, this approach is now being also applied to gauge ore grade and zone thickness of other radioactive elements emitting gamma-rays and even ores that contain trace amounts of radioactive elements with gamma-ray radiation.

According to Morse (2019), the three main radioactive chains Uranium (U238), Thorium (Th232), Potassium (K40) are commonly found in earth's crust, which therefore indicates the potential minerals in ore sediments and mineral productions. These elements undergo radioactive decay, which produces a decay product called the daughter isotope. The half-life of uranium is much longer than its long list of daughter isotopes, which are also radioactive, forming an equilibrium in an ore/rock between the decayed element and the quantity of isotopes present, given that the ore does not tamper in any way. Thorium and its daughters are also radioactive, hence

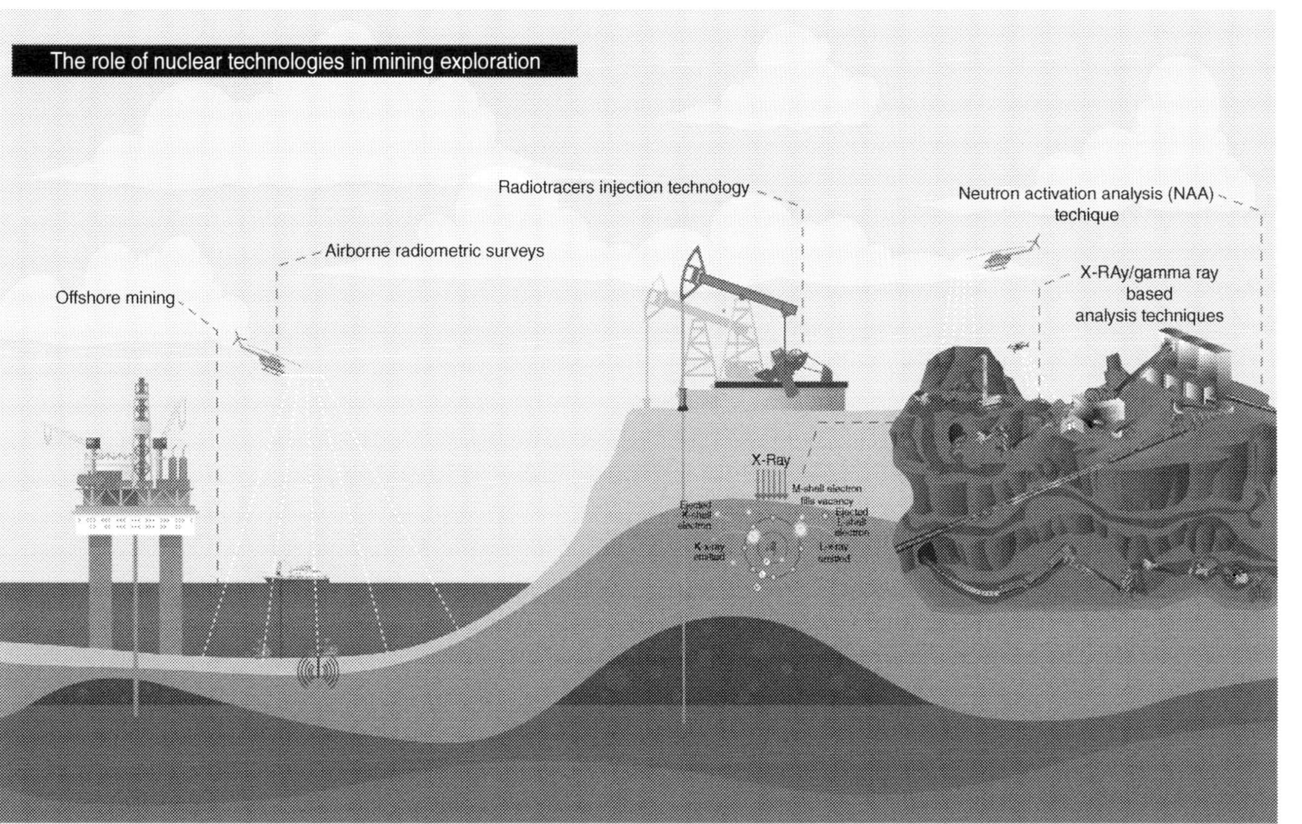

FIGURE 4.1 The role of nuclear technologies in mining exploration.

forming the same trail of decay as uranium, which makes it possible to identify the type of ore by studying the decay chain. For instance, quartz, micas, calcite, and many other rock-forming minerals contain potassium that emits gamma rays at 1.46 MeV and is used in gamma-ray logging to gauge the grade of the deposit using spectrometric analysis. Natural radioactivity is a property of the nucleus and therefore, geophysicists search for areas where gamma-radiation is nonuniform. Spectroscopy is one of the tools used to measure and distinguish elements and its belonging chain by measuring the energy level discriminations emitted from each particular radiation. Radiometric surveys can be an airborne survey, which provides details of gamma counting and spectrometric data, or can be a ground survey that able to provide geological and radiometric data to locate the host rocks (Morse, 2019).

4.2.2 Gamma ray-based analysis

This approach comprises two main categories, namely: gamma transition and gamma activation (IAEA, 2014). Gamma transition, which includes both isobaric and isometric transitions, is used for densitometry to determine ore density in industrial flows. In contrast, gamma activation is used to determine the presence of many elements simultaneously using high-energy photons. Gamma-based densitometry has been widely accepted for decades in the mineral industry. Density profiles and distributions are calculated using gamma transmission and backscattering coupled with gamma tomography. Gamma transmission can be used in laboratories or in situ to obtain a relevant understanding of the operating conditions and analysis of precious metals in mining and processing for industrial applications.

4.2.3 Neutron activation analysis

This approach is used to identify the concentration of elements in the vast quantities of ore (Silberman, 2012). This approach relies on the excitation of the sample with neutrons. As a result, the sample emits gamma radiation, which allows for the precise identification and quantification of elements in the sample. Neutron-based analysis is best suited for bulk and in situ analysis. As the approach uses neutrons that require a steady source of neutrons to function undisturbed (Silberman, 2012). With the ready availability of compact and long-life neutron generators such as ^{137}Cs has made it possible for this approach to be used for minerals exploration. The foundation of neutron analysis includes inelastic neutron scattering, gamma activation analysis in conjunction with radio spectrometry that measures the emitted particles or gamma rays, which characterize the available elements. Furthermore, the amount of energies emitted indicates the abundance of that element in the ore (Silberman, 2012). Neutron techniques can be used for off-line analysis of ores in bulk or in-situ mineral resource characterization utilizing vertical, horizontal borehole logging, and core drillings such as the case with uranium in situ mining that uses neutron prompt fission techniques, which is more cost-efficiently. The analytical method for neutron activation produces an elemental concentration depending on the volume, therefore, address problems with sampling involving other classical methods. Furthermore, Al Nabhani and Khan (2019) argue that prompt fission neutrons (PFNs) technique can be used to explore minerals and this technique is widely used for in-situ Uranium-235 recovery. The PFN logging technique comprises a pulsed source of neutrons flux. In the pulsed source of neutrons flux, a high voltage of 14 MeV pulses at 1000 cycles per second and emits about 108 neutrons per second that strike and excite atoms, transferring their energy into these atoms. The excited atom(s) will release the excess energy to attain a stable low-energy state. This expelled energy is detected by the spectrometric probe, which is then analyzed and identify the type of element.

For instance, in the in-situ Uranium-235 recovery process, Al Nabhani and Khan (2019) explain that the PFNs logging tool emits neutrons flux to the targeted geological formations, which collide with only ^{235}U isotopes and lead to slow-neutron-induced fission of ^{235}U in the formation. Epithermal neutrons and thermal neutrons returning from the formation after the collision following fission of natural ^{235}U isotopes in the rock formations are counted separately in detectors in the logging tool called thermal/epithermal neutron detector. The thermal/epithermal neutron detector gives the percentage ratio of ^{235}U where the ratio of epithermal to thermal neutrons is directly proportional to ^{235}U isotopes. The time-gated ratio of epithermal to thermal neutron counts provides a measure of uranium content. Uranium content measurements obtained from PFNs logs have shown good agreement with core measurements, so this technique provides a major data source for delineation and exploitation of uranium isotopes mineralization. The PFN technique provides a precise direct measurement of in-situ uranium recovery, in particular the ^{235}U isotopes over even very narrow intervals. Not only that, but the PFN technique is also able to identify other fissionable materials such as plutonium isotopes by detecting their thermal neutrons. Once the drilled well is classified as economically feasible and ready for production, a recovery project starts. Fig. 4.2 below depicts a diagram of in-situ Uranium-235 recovery process flow.

4.2.4 X-ray analysis techniques

This includes X-ray fluorescence, which refers to the emission of fluorescent X-rays from particles that have been pushed into an excited state

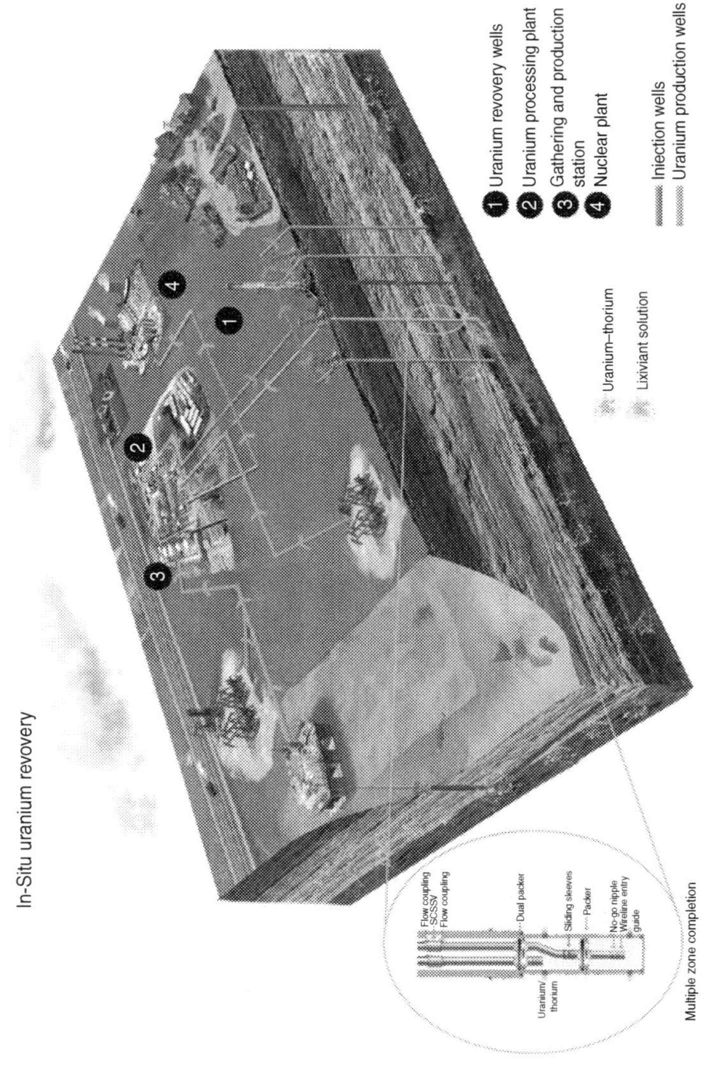

FIGURE 4.2 In-situ Uranium-235 recovery process.

by being bombarded with X-rays or gamma rays. X-ray analysis is at the base of nucleonic analysis and control in most processing units in the mining process. Using energy-dispersive X-ray diffraction, which employs polychromatic photons as the source, X-ray analysis is usefully applied to the mineralogical analysis, where it yields substantial optimization of the grinding and floatation processes as well as augments recoveries from the mining chain. X-ray Fluorescence (XRF) technology is ideal for qualitative and quantitative analysis of mineral composition because the X-ray fluorescence of every element in the ore is distinctive. A solid or liquid sample is irradiated from a controlled radiation tube emitting high-energy X-rays. Most atoms have multiple orbitals of electrons (K shell, L shell, M shell). IAEA (2014) argues that when an atom absorbs X-rays, it causes the electrons in the atom to jump to higher energy orbitals or shells and become excited as they contain more energy than they would naturally have. The excited atom tries to become stable and therefore these excited electrons with more energy attempt to jump back to their low energy orbits and therefore release their excess energy in form of XRF peaks with variable intensities, which are detected in the XRF spectrum. The peak intensity points out the element, and peak height characterizes the concentration. The energy of this X-ray corresponds to the specific energy difference of the two quantum electron states. The XRF analysis functions by monitoring and measuring this energy. A set of characteristic fluorescent X-rays is emitted by all the elements present in a sample. Each of the X-rays is unique to a particular element in that sample, making XRF spectroscopy particularly useful for elemental Analysis (IAEA, 2014). This "Elementary Fingerprint" is best demonstrated by studying the spectrum of X-ray radiation and its scattering peaks. Furthermore, Silberman (2012) argues that particle-induced X-ray emission or proton-induced X-ray emission spectrometry is another form of X-ray analysis that uses a beam of accelerated protons (1–3 Mev) as a source instead of electron-based X-ray analytical techniques. Proton-induced X-ray emission is more sensitivities technique than electron-based X-ray analytical technique than that offers a higher depth of penetrations with more peak ratios to trace elements.

4.2.5 Radiotracers approach

Radiotracers have been intensively used in a wide range of industrial applications to refine and track operations, conserve resources, and reduce the effects of greenhouse gases on the atmosphere. In the mining industry, for instance, Kyser (2017) argues that, since most reservoirs in natural environments are isotopically distinctive, isotopic compositions may be used to fingerprint an element or compound. Elements are known to have variations in their isotopic compositions due to mass and bonding environmental differences. They have been used as key indicators in exploratory and environmental science. These elements include hydrogen, carbon, nitrogen, sulphur, calcium, iron, copper, zinc, uranium, and many other elements (Kyser, 2017). Given that the element has at least two isotopes or more then it is possible to use any element for isotopic tracing. Isotopes, hence, add importance to mineral discovery and will improve deposit footprint and more efficient detection of false-positive element concentrations. Scientifically, mineral deposits consist of high redox-active element concentrations. They are regions of variable electron density that are habitats for microbial species whose metabolism relies on oxidation–reduction reactions. Furthermore, Kyser (2017) argues that the elements produced by microbes as by-products of metabolic responses may be mobilized as aqueous or gaseous deposits with special isotopic compositions in their waste products or from the microbes' decay. During the secondary scattering processes, these compounds migrate to the surface, especially along with fractures and defects, are variably attenuated through soil and Fe-Mn oxide, and reach the biosphere through the surface (Kyser, 2017). There may be unique elemental and isotopic signatures of these complexes that represent the deposit at depth. The actual power of isotopic compositions is to illuminate unique processes relevant to the mineralization mechanism, which can maximize the deposit footprint and to more accurately detect false positives in element concentrations that represent the deposit at depth.

4.3 The role of radiotracers in enhancing oil and gas recovery

Bjørnstad and Somaruga (2012) argue that tracer applications can be used both offshore and onshore in virtually every step of oil field production. Inter-well tracer technology is an essential method for the secondary and tertiary recovery of oil from the reservoir. In conventional and geothermal reservoirs, inter-well tracer monitoring technique is used to achieve a greater understanding of reservoir geology and to improve production and reinjection systems. The key aim of performing inter-well tracer tests in conventional oil and geothermal reservoirs is to track the injected fluid flows between injection wells and production wells, and to qualitatively or quantitatively evaluate the flow dynamics including inhomogeneity of reservoirs and volumetric sweep between wells. For water-based tracers, small concentrations of radioactive particles are blended into the water or gas. The tracer is then injected through an injection well and then measured in the production wells surrounding it (Fig. 4.3). The tracer response proves to be extremely useful in explaining the flow pattern and thus obtain a

FIGURE 4.3 Radiotracers injection into the reservoir to enhance oil and gas recovery.

deeper understanding of the reservoir. To enhance the recovery of oil, this knowledge is essential. The structures of the reservoir are typically layered and often involve significant heterogeneities that add to spatial challenges in the degree of flow. Therefore, it can be hard to foresee the successful flow of fluids; this is where the tracer technology becomes useful, given, the tracer moves with injected fluid.

Accordingly, radiotracers help to analyze the flow patterns of injected water during secondary oil recovery, precisely the course of injected water flow, residence time, breakout time, flowrate, and many other important parameters. A suitable quantity of radiotracers are inserted in the injection well, then samples are collected from the nearby production wells within the same reservoir. An auto-liquid scintillation counter analyzes these samples after pretreatment. To ascertain the reservoir properties and the reservoir's flow pattern, the effects are measured, replicated, and interpreted. For instance, if the injected tracer is found in numerous samples in the nearby production wells, it means that these wells are interconnected to the same reservoir and therefore extracting oil from the same reservoir and have similar flow patterns. While wells whose samples do not contain any radiotracer that means, these wells are isolated due to a fault in the geological structure.

For optimal and economical oil recovery, it is vital to know the scale of different oil fields, their geological formations, and their interconnectivity. Radiotracers are able to dynamically answer all relevant questions and provide clear evidence of what is happening at a depth of thousands of meters. Radiotracers can be either natural or chemically developed for such purposes. Radiolabeled tracers including ^{14}C, ^{35}S, ^{125}I, ^{3}H, ^{14}C, ^{35}S, ^{131}I, and many others can be inserted independently or in combination. Since the water molecules consist of ^{2}H and ^{18}O isotopes, therefore, can be used as efficient tracking tools to locate the source of water coproduced, both in geothermal and oilfield applications, the relative contribution of various water sources to the water co-produced with oil can be measured based on stable isotope indices. However, the ^{18}O quality of pumped water is likely to be changed in geothermal reservoirs and high-temperature oil reservoirs due to the ^{18}O contact between water and host soil. While ^{2}H is considered cautious and can be used safely to quantify relative contributions (Bjørnstad and Somaruga, 2012).

4.4 Leak detection in industrial applications using radiotracers

Jadiyappa (2018) argues that the discovery of radiation, radioactivity, and isotopes goes back nearly a century ago and is credited to Roentgen - 1895, Becquerel - 1896, and Soddy - 1913 respectively. For several decades, they have been applied in a wide range of industries including agriculture, medicine, hydrology, sedimentology, and industry. In the industry, radiation and radioisotope are used in many industrial applications and leaks detection is one of these applications especially at a very complex environment that is difficult to reach easily. There are three fundamental leak detection methods using radiotracers techniques, which are the calculation of flow frequency or residence time distribution or the "direct" tracer technique.

Radiotracer technology is frequently used in the oil and gas industry, for instance, to detect leakage in the underground pipelines. This is achieved by injecting radioactive isotopes that are a beta emitter with a half-life of

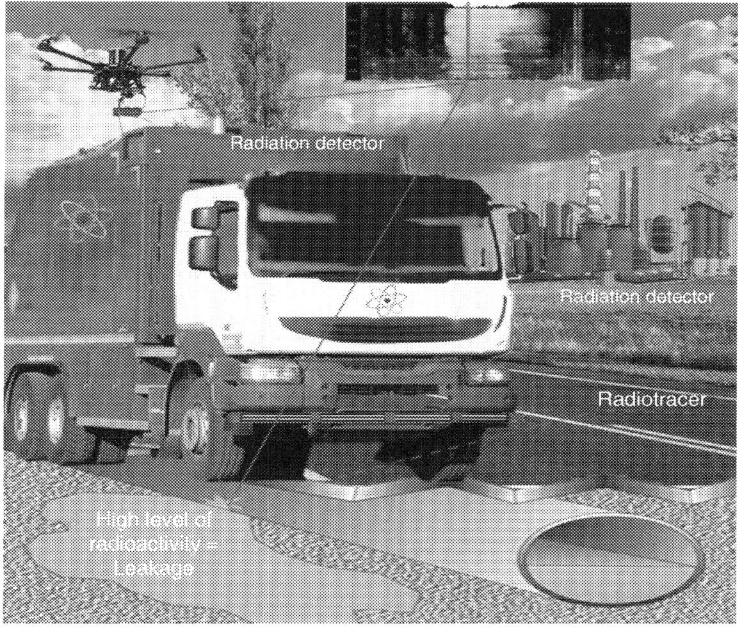

FIGURE 4.4 The use of radiotracer technology for leak detection of flowline in the oil and gas industry.

a couple of hours or a few days that last long enough to locate the spill. According to IAEA (2019), Iodine-131, Sodium-24, Bromine-82, and Krypton-85 are all radioisotopes widely used to conduct leak checks in flowlines, based on the chemical and physical properties of fluids circulating within the production line under examination. Once the suitable radiotracer is selected and injected into the pipeline, the leak source in the pipeline will be known at the surface, where a high level of beta radiation is measured and detected by the detectors (Fig. 4.4). Therefore, radiotracer technology helps to determine the exact location for the leak for repair action and saves time, cost, and effort. It is vital to use a radioisotope that emits beta radiation because beta radiation has a suitable penetrating capacity to pass through the soil to give a detectable above-ground reading. The level of the beta radiation would be high at the source of the leak since no energy is lost to the pipe wall. While an alpha source would be of no benefit to use in for leaks detection because the soil would trap alpha radiation. On the other side, gamma rays provide high energy that able to penetrate through the pipe's walls, and therefore hard to distinguish the leak besides that it may pose a risk to safety or health.

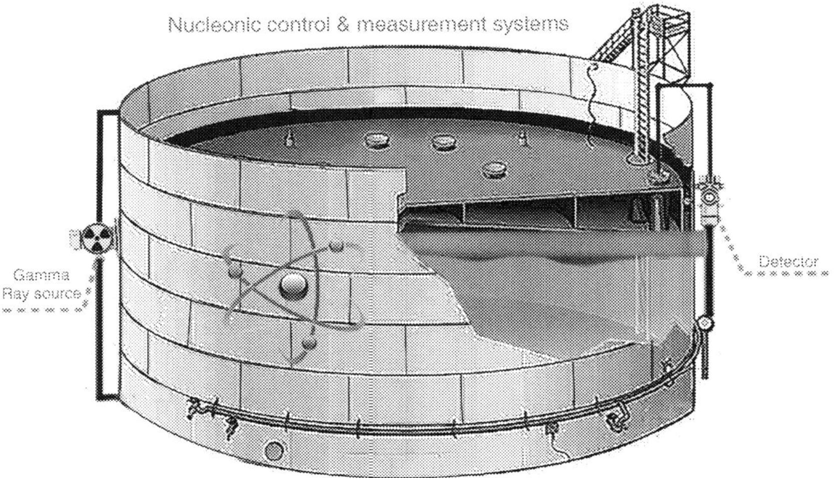

FIGURE 4.5 A gamma-ray-based nucleonic gauge to measure the level, density, and weight.

4.5 Nucleonic gauges

Nucleonic gauges or nucleonic control systems and also referred to as ionizing radiation gauging devices are instruments used for measuring and assessing the interaction between ionizing radiation and matter (IAEA, 2005). Nucleonic control systems are able to measure different parameters including density, weight, levels of any matter whether in a static or variable situation (Fig. 4.5). Accordingly, industry worldwide started to employ this technology for refining and revamping procedures to monitor and enhance production efficiency in many industries including but not limited to mining and mineral ore processing, environmental monitoring, the paper and plastics industries, the cement industries, civil engineering, petrochemical industries, the oil and gas industries, and many other industries. Globally, some hundred thousand nucleonic control systems work in different industries. Since nucleonic gauges have the essential advantage of not requiring physical material interaction and can work in difficult environments. Therefore, the technology of nucleonic control systems competes well with traditional control systems and became ideal for high-speed manufacturing lines with multiple fluids or processes running at intense temperatures and harsh environments.

Nucleonic control systems usually make use of beta or gamma or neutron or X-ray. IAEA (2005) argues that Promethium-147, Thallium-204, Krypton-85, Strontium-90, Yttrium-90, Americium-241, Caesium-137,

Cobalt-60, Americium-241, Californium-252, Iron-55, Cadmium-109 are the most common radioisotopes sources of beta, gamma, neutron, X-ray production for nucleonic control systems in the industry. The isotopes emitting beta or gamma, or neutron or X-ray are housed in a sealed tank placed inside another shielded container to ascertain safety standards. The radiation is only released in the desired direction of the targeted specimen. A small transmitter for measuring radiation is mounted on the opposite end of the specimen required to be measured. The emitted gamma radiation, for instance, is then attenuated by the composition of the medium and material thickness as it moves through the tank or flow pipe. The emitted gamma radiation is registered by the transmitter that converts energy from gamma photons into pulses of lights that enter the photomultiplier. They are amplified and converted into electronic signals, then interpreted based on the assumption that as the mass rises, less radiation can pass and detected by the detector; the electronics of the detector use gamma reading to determine measurements and producing values as a result. On the other hand, there are some nucleonic gauges that do not use sources of radiation and instead are reliant on measuring natural radioactivity. Nucleonic measuring methods can also visualize the internal configurations and motions of objects through various technologies, such as computed tomography (CT). Transmission and backscatter are the two most widely used measurement methods. Without damaging or modifying the characteristics, these devices, therefore, can quantify the material. The high-energy gamma radiation penetrates the sealed container walls, making it possible to test the substance inside without removing the containers.

4.6 Nondestructive industrial radiography

Nuclear radiography is a nondestructive imaging technique used to examine objects from inside in minute details. Nuclear Radiography relies on X-rays, gamma rays, or neutron beams. These techniques enable to capture the very important details inside an object due to the fact that the radiation is able to penetrate multiple layers, which is not possible with the conventional inspection techniques that can only detect imperfections on the surface or near the surface. In the industry, many scholars including Wang (2015) and Beck and Williams (1995) argue that this technique has a wide variety of practical applications to detecting flaws and cracks that would not be detectable otherwise such as finding cracks, locating the presence of materials that are embedded inside other materials, such as the presence of water or electric wires present in concrete or structures as well as the cracks found in different types welds or structures. It is a very useful technique for structural analyses of jet and car engines, pressurized vessels, and pipelines. Not only that but nuclear radiography is also able to detect contamination, materials, and defected parts which make it a viable quality control tool in industries and widely used in the manufacturing industry to assess the consistency of manufacturing. Therefore, nuclear radiography helps to maintain the stringent quality control standards for safe operation and production in many industrial installations.

Many scholars including Blitz (2012) and Chaplin (2017) agree that the most common conventional nondestructive testing methods used in the industry to evaluate the integrity and properties of the materials are:

- Ultrasonic testing, employing sound waves to detect cracks and defects.
- Liquid or dye penetrant inspection where low surface tension fluid allowed to penetrate into the specimen that is detected by ultraviolet or fluorescent or nonfluorescent depending on the type of dye used.
- Magnetic particle inspection used the detect inconsistencies on the surface and/or subsurface of ferromagnetic materials through introducing a magnetic field into the ferromagnetic materials and the discontinuity in the material allows the magnetic flux to leak and easily detected.

These methods have a lot of limitations from quality control, safety and reliability perspectives compared to advanced nuclear nondestructive testing methods, which are as follows.

4.6.1 X-ray computed tomography

Wilhelm Röntgen's invention of X-rays in 1895 when he took the first radiography image of his wife's left hand contributed in the emergence of the current scientific revolution of radiographic technology that is entering many medical and industrial applications (LA Porta et al., 2020). X-ray capable of penetrating a number of materials and stimulating light emission in inorganic crystals. The emitted radiations are then collected and detected by thin-layer chromatography plates and then produce an image on the film about the scanned object. Scientifically, X-rays are produced when electrons are accelerated by an electrostatic field over a short distance of a few centimeters in X-ray tubes and then fired through the static or rotating anode (Wang, 2015). The higher the voltage, the more intense the electron stream is, and the deeper the penetration. In linear accelerator and betatron equipment, commercial X-ray tubes use voltages that range from 20kV up to 30MV depending on the required application. The electron stream impacts an anode, emitting around 1% of the energy as a pulse of X-rays, and the rest is released as heat. Röntgen's discovery has led to the development of the first X-ray imaging devices in the 1960s that was later transformed into CT, which has become the primary tool for imaging in multiple areas from medical diagnostics to process engineering where gases, liquids, and solids, are examined in chemical reactors, power plant coolant systems, heat exchangers and hydrodynamic devices, pipelines,

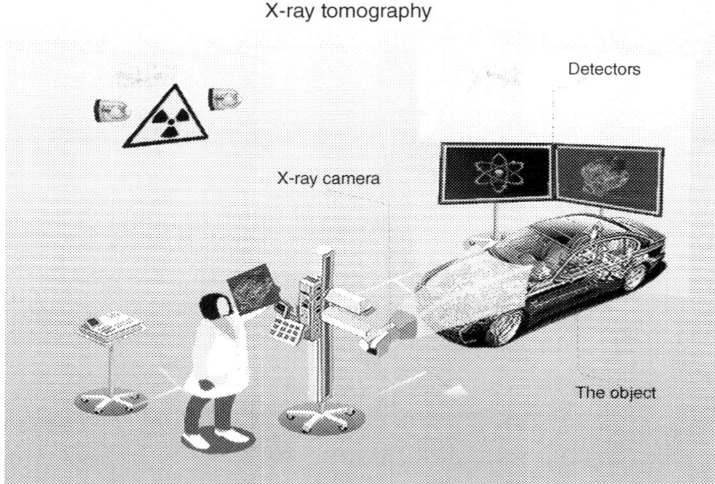

FIGURE 4.6 The use of X-ray CT technology imaging technology in checking automobile engine. *CT*, computed tomography.

petroleum refining equipment, transport engines (Fig. 4.6), and for many other applications (Heindel, 2011). X-ray CT technology is based on the attenuation of X-rays which is basically a reduction in the intensity of X-rays as they pass through an object. Building on this principle, X-ray attenuation can be used to identify and differentiate elements/compounds present inside an object using the difference in densities as a key indicator (Wang, 2015). Not only that but, recently scientists recognized and fully understood the amount of transmitted radiation that depends on the object's mass, thickness, and atomic number to pass through the living and nonliving objects, and these results culminated in the development of industrial radiography.

4.6.2 Gamma-ray-tomography

Gamma-ray tomography imaging works on similar principles as X-ray imaging but uses a different kind of radiation source, which is gamma rays. Gamma-ray was first detected by Paul Villard in 1900 (Joesten et al., 2007). When a nucleus becomes excited, it becomes unstable and jumps from its lower energy level to a higher level, then excited nucleus will try to become stable by going back to its lower-energy state (Fox, 2014). At this stage, it will try to get rid of the excess energy by releasing gamma rays or photons. Energy differences in nuclei usually fall in the range of 1-keV (thousand electron volts) to 10-MeV (million electron volts) range that is common for most tomography applications (Wang, 2015). Therefore, when a nucleus moves from a high-energy level to a lower energy level, a photon is emitted and carry off the excess energy; nuclear energy-level differences correspond to photon wavelengths in the gamma-ray region. These gamma-ray photons have the energy to ionize atoms, as electrons are knocked off their orbits. The photons produced as a result of gamma-radiation are different from those produced by X-rays. Gamma-ray photons carry much more energy than those of X-rays, making this radiation capable of penetrating much deeper than X-rays. Gamma-ray tomography works by placing the object between a radioisotope generator and a detector. Multiple observations are then collected from different angles covering the whole surface area of the object. The amplitude of the gamma beam (photons per second) is calculated by the detector as it received the radiation that passed through the object under study, essentially creating an image of the object. This is effective in imaging very static objects or structures. While instant image generation is useful for the imaging of dynamic structures and processes. Wang (2015) argues that ^{241}Am, ^{137}Cs, ^{133}Ba, and ^{60}Co are the most common gamma-ray emitters for gamma-ray CT.

4.6.3 Neutron radiography

Neutron radiography has become more advanced technology at the commercial scale (Garbe, 2020). The process of neutron radiography is based on the neutron attenuation properties of the imaged object. This is also a nondestructive technique, but it is different from X-rays. Since neutrons only interact with the nuclei of atoms, some items which might be difficult or impossible to observe with X-ray imaging can be observed easily with neutron radiography. Both gamma and X-rays are often more helpful in the study of denser materials, including concrete and structures. However, neutron radiography, however, deals well with low atomic number materials by-products of hydrogen, such as water, oil, concrete, rubber bombs and corrosion by-products (Berger and Iddings, 1998). Consequently, this form of radiography technique is often used for the inspection of condensers, adhesives, toxic materials such as radiation sources and their shielding, airbags, metal artefacts for corrosion, or for the identification of foreign hydrogen in sealed devices.

4.7 Radiotracers applications in detecting dams and hydroengineering systems leaks

Water leakage from dams or hydroengineering systems used for power or the industrial tanks is considered as a serious safety and economical issue. It is therefore very important to precisely identify the location of the leakage as well as to estimate the amount of the leakage that occurred to be cured immediately. The available literature on such leakage studies is relatively limited with regard to the effective and reliable techniques used to identify the location leakage precisely. However, In the last few decades, the radiotracer technique has been developed that helps in providing a precise assessment of any leak.

According to Owczarczyk et al. (1992), the basic principle behind the radiotracer technique for dam's leak detection is through the introduction of certain radiotracers with the suitable sorption and half-life characteristics such as Indium-113m or Gold-198 In the water held in the reservoir or the dam under examination. Water and radiotracers flow toward the leak, then radiotracers and their particles will accumulate on surfaces of the soil matrix surrounding the leak location. Radioactive spots formed thereby in areas where leakages take place are then located precisely using an underwater radiation detector (Owczarczyk et al.,1992).

4.8 Nuclear technology role in fraud food control

To verify the authenticity of imported food products, it has been very common practice for most of the border police and customs agencies to only look at labels and papers containing food products, which often—can be easily forged. but with the rapid development of nuclear science-based technology, it became very easy to verify the quality of food items and react more quickly at vital checkpoints to prevent the flow of contaminated products or adulterated. Food protection programs are of considerable significance and must have the ability to track the origin of food products in the global supply chain. According to IAEA (2020), to determine the provenance and authenticity of the food, this can be achieved by using stable isotope tracing analyses (the naturally occurring stable isotope "fingerprints" found in the food chain), complementary fingerprinting (metabolomics), and profiling techniques. The nuclear approach is an integral part of a systematic approach to food safety to be able to identify the sources of food products through rigorous analytical techniques. Knowing the origin of food helps to establish food credibility, fight unethical practices, and prevent food adulteration. The related monitoring mechanisms strengthen consumer interest in supply chains and thereby help to eliminate foreign trade restrictions and concerns.

Typically, there are two ways to counterfeit any product, either by manipulating labels and papers wrapped outside food containers or by manipulating with the product content. However, nuclear energy, which is a nondestructive technology that is capable to detect any tampering or forgery of the wrapping content as well as the component itself. IAEA

(2020) argues that the first use of the stable isotope analysis method for determining food authenticity was started in the early 1970s. While the adulterated food tends to be chemically and functionally similar to the real commodity; therefore, the use of the stable isotope analysis method is an efficient mechanism because it presents unequivocal proof of food adulteration or substitution via tracing and comparing isotopes of the adulterated food with the original product's isotopes. Much of the efforts have been directed by counterfeiters to find alternatives to manipulating food ingredients for economic motives, such as increasing ingredients with affordable alternatives such as water and sugar syrups in fruit juice, sugar and sugar syrups in honey, water and sugar in the wine, cheaper oils in maize oil, vanillin added to natural vanilla extract, and so on (IAEA, 2020). Furthermore, IAEA (2019), reported that Manuka honey derived from the nectar of the New Zealand Manuka tree flower, for example, has natural anti-microbial properties and can get up to 1000 New Zealand dollars per kilogram. Truffles from Slovenia, a prosperous enterprise. White truffles can sell for up to EUR 2300 per kilogram; Jamaican Blue Mountain coffee, coveted for its scent and low acidity, is one of the most expensive in the world. All these expensive products are vulnerable to counterfeiting; Not only that but, Thai Hom Mali rice, a luxury fragrant long-grain variety that accounts for an average of 15% of Thai rice exports, is a vast market, and hence has room for a lot of theft and counterfeiting (IAEA, 2019). This is not only limited to expensive products but also includes highly consumed products such as fruits and vegetables.

At the beginning of the 21st century, stable isotope analysis has been used to establish the geological sources of foodstuffs. The composition of stable isotopes found in the food serves as a "record" of the environmental factors at the point of origin of the food as well as of the farming and feeding methods used during the manufacturing process. The basic principle of the isotope analysis approach relies on comparing the ratio of stable isotopes in the country of origin such as hydrogen, oxygen, and carbon and their abundance in a product sample. This approach considered as a special fingerprint connecting any crop to the location where it is grown in that specific area depending on the isotopes of the water. For instance, milk is obtained from grass-dependent cows, these grasses rely on rainwater for irrigation (Fig. 4.7). This rainwater carries the certain fingerprint of certain isotopes, and so the imprint of the water isotope is transferred to the milk that is made up of 88% of this water. Accordingly, Chesson and Valenzuela (2010) argue that using stable isotope analysis of delta values of hydrogen and oxygen for the milk associated with cow drinking water or the water stored in the grass eaten by cows helps to allocate the geographic origin of the milk. Stable isotope concentrations and the metabolic processes of digestion are not specific to milk but apply to all agricultural and animal origin food products. Thus, Isotopes of matter that

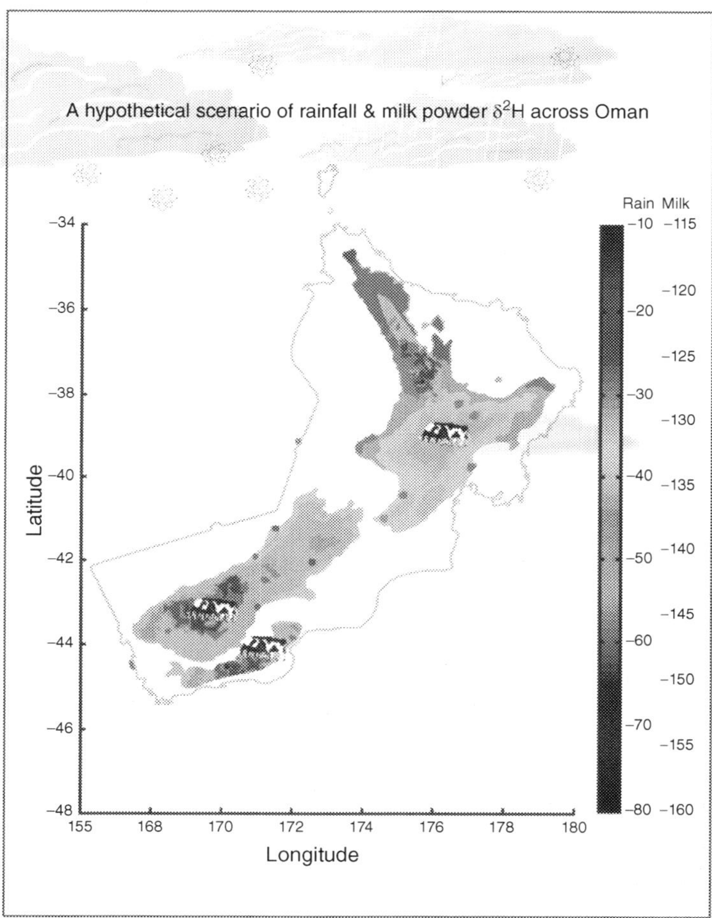

FIGURE 4.7 The use of stable isotope analysis to establish the geological sources of foodstuffs (Hypothetical scenario).

the food product absorbs from the atmosphere reveal where they originate in contrast to DNA that only reveals the lineage.

Since there is substantial natural variance and overlap in "signatures" in isotopes throughout foods from all over the world, therefore, isotope analysis approach of food source is a comparative method based on the collection of measured values from real goods and related environments. Accordingly, this approach is to leverage the correlation between the atmosphere and the isotopic food composition and use variables such as climate data to "predict" isotope signatures for any given place as the origin point of food. In this regard, the Global Network of Precipitation Isotopes database developed by FAO and the International Atomic Energy Agency (IAEA) help to manage global food component isotope databases.

Besides the conventional methods used to verify the authenticity of imported food products and combat food fraud such as chemical, chromatographic, molecular, and protein-based techniques, nuclear techniques including vibrational spectroscopy, nuclear magnetic resonance, and fluorescence spectroscopy, ion mobility mass spectrometry, which have been used widely over the past few years due their high accuracy and reliability (Hassoun et al. 2020). Ion mobility spectrometry, for instance, is considered to be a quick and sensitive tool that is able to detect trace substances. It is not only used for explosives and chemical warfare detection but also used in modern medical diagnostic and process control applications and also to verify the authenticity of food products. According to Vautz and Baumbach (2006), a gas sample or any other sample type should be first converted into gaseous phase so that it can be easily ionized using either ultraviolet light, ss-radiation, or partial discharges. Ionized particles will travel toward the detector in a weak electrical field. They collide with drifting gas moving in the opposite direction during their flow and are thus slowed down based on their size, shape, and charge. Consequently, different ions reach the detector at different drift times. Finally, the detector will analysis and measure the number and the concentration of ions reaching the detector.

4.9 Industrial applications of ionized radiation in cross-linking polymerization

Soon after the invention of the crosslinking of olefin polymers using ionizing radiation in the mid of 20th century, radiation crosslinking application has emerged and begin to be implemented in a wide range of processing industries started (Sun et al., 2017). This technology is now used for the manufacturing of a wide range of products, such as heat shrink tubing and tapes, consumer product encapsulations, polyolefin foams. Radiation technique is also used to cross-link insulations and coatings in the wire and cable industry to make them capable of withstanding high temperature, suppressing flame propagation and exhibiting improved abrasion tolerance and fluid resistance. Moreover, radiation crosslinking of polymeric was found to be very useful for the reinforcement of water treatment pipelines, oil and gas pipelines, and in crosslinking of the vehicle tires plies.

Not only that but radiation technique is also used in the medical industry to produce hydrogels and to adjust ultra-high polyethylene molecular weight (UHMWPE) reliable for medical implantation purposes. Radiation for crosslinking purposes can be generated using the EB accelerators and based on the required applications, the spectrum of radiation is broadened by several EB parameters, such as energy and strength. For instance, high-power accelerators able to boost the production rate with the required irradiation treatment. Radiation crosslinking process is a green technology that does not require any external chemicals or thermal treatment. Processing radiation improves the intermolecular bonds between polymeric chains, such as mechanical properties, resistance to corrosive liquids, thermal stability, ease of application. As a result, the use of processing radiation in the industry leads to improving the quality and reliability of products.

4.9.1 Radiation-based cross-linking polymerization of electrical wires and cables

The inclusion of irradiation techniques in the wire and cable industry using radiation-based cross-linking approach has made it possible to manufacture products with special characteristics that are required in certain sensitive applications (Makuuchi and Cheng, 2012). Radiation-based cross-linking polymerization approach has been embraced by wires and cables industries to increase their products' efficiency and reliability so that they can work in harsh conditions such as environments with very high or very low-temperature, where standard wires and cables are unsafe to be used in some critical applications such as aerospace, medicine, aircraft, and electronics. The scientific fact is that radiation-based cross-linked three-dimensional polymer structure prevents the cables from melting at elevated temperatures and therefore causing a fire. Accordingly, the radiation-based cross-linked cables and wires are safer and more reliable.

4.9.2 Radiation-based cross-linking polymerization of polymeric foams

Polymeric foams technology emerged before the Second World War. The basic idea of the polymeric foam is the creation of a polymeric matrix with a huge number of cells that work as a reinforcement agent. Many scholars including (Sun et al., 2017) argue that polymeric foams are widely used in the industry because they are ideal for absorption of energy and lightweight insulation as the matrix is designed to transfer loads between the fibers and the air molecules trapped inside to provide good insulation capabilities and therefore are used as noise suppression shields. They are characterized also by thermal insulation properties, low density, low thermal conductivity, and chemical resistance. Based on these characteristics, they are widely used in construction, automotive, packaging, sports, and shoe industries

(Głuszewski et al., 2018). Although polymeric foams are made using a wide variety of polymerization agents, some of the commonly used ones include polyurethanes, polystyrene, and polyolefin. Due to the importance of polymeric foams in many industrial applications, the producers of this material have been keen to enhance its physical and chemical properties through the use of radiation-based cross-linking polymerization approach. The process starts with creating sheets of the polymer compound which are treated with gamma radiation or EB to enhance cross-linking of the polymers in the sheets. In the case of cross-linking polymerization using gamma radiation, the dose rate approximately 4 kGy/h, while in case of EB irradiation the dose rate about 14,000 kGy/h (Głuszewski et al., 2018). This treatment provides added advantages to polymer foams to be more heat resistant, withstand wider temperature ranges, become less weight and highly reliable and durable so that can be used for critical applications such as medical, space applications.

4.9.3 Radiation-based cross-linking polymerization of medical devices

Medical instruments designed for implantation purposes are subject to rigorous testing taking into account multiple aspects and circumstances such as the materials used, health concerns, structural specifications, and even the manufacturing process among others. Sun et al. (2017) argue that acetabular cup (for hip replacement) or the patellar structure (for knee replacement), for instance, may be subjected to forces and loads that are about eight times the bodyweight. But the implant not only has to adhere to the physical pressures but also to the chemical atmosphere inside the body. Studies have revealed that implants start to deteriorate within about seven years of implantation due to the oxidative properties of body fluids. According to many scholars such as Hussain et al. (2020), implants are usually manufactured using ultra-high molecular weight polyethylene (UHMWPE) due to its high wear-resistance, ductility, and biocompatibility. UHMWPE materials still suffer from the issue of deterioration with the time as a result of the internal oxidative stress reactions and to overcome this issue, radiation-based cross-linking approach is used to improve its mechanical and tribological performances (Hussain et al., 2020). This fosters the crosslinking of UHMWPE and becomes more durable and wear-resistant material with a long-life span inside the body. The liners of the UHMWPE are irradiated using EB at about 2 MeV. The small infiltration of the accelerated electrons triggers a significant and gradual decrease in cross-linked density over depth. This makes, the inner and outer parts of the implant protected by the radiation-based cross-linking technology and therefore, become capable of enduring heavy tribological loads.

4.9.4 Radiation-based cross-linking polymerization of tires

Radial tires consist of the so-called radial structure of cords that are placed at 90° to the direction of travel to provides superior and reliable performance (Carbone, 2011). Since the angle is important to the performance of the tire, manufactures make sure the plies are set in place, so they do not get displaced at any point during the molding process. To ensure the orientation of plies, manufacturers employ EBs to cross-link the plies. This is much better than the orthodox vulcanization process that usually results in the displacement of the plies during the final forming process, which may result in serious safety issues. Manufacturers use thicker plies in order to achieve an even distribution of material throughout the tire to prevent any deformation during the thermal curing. However, Sun et al. (2017) argue that radiation-based cross-linking technique is required to have safer tiers and usually done at a range of 30–50 kGy. Radiation-based cross-linking for the rubber provides increased stability, balance, safety, cost-saving, and therefore, there will be no need to use thicker plies anymore, the finished product uses even less rubber with less chance of cracks, puncture, or explosion.

4.9.5 Radiation-based cross-linking polymerization of food packaging

Advances in plastics sciences have revolutionized the food processing industry more than anything else. The use of plastics packaging in the food processing industry first emerged in the 1930s and since then the use of plastics packaging has been playing important role in the transformation of this industry. Food packaging is usually made from petroleum-based polymers that offer a plethora of benefits, which are not available in any single other material. Silvestre et al. (2017) argue that plastic packaging provides protection against contamination from microorganisms such as bacteria, molds, yeasts or parasites or poisonous substances. Moreover, plastics used in packaging are often nonporous and nonpermeable which prevents absorption of moisture and therefore, increasing the shelf life further.

Majority of these polymers, derived from the feedstock, are low-cost readily available materials and usually have high tensile strength and durability. According to Silvestre et al. (2017), the most common plastics in use in the food packaging industry are polyethylene, distinguished by low vapor permeability; polypropylene, oxygen permeable; and polystyrene, partially water-permeable, although often polystyrene may be distinguished by flexible and low heat resistance.

Plastics have a lot of advantages as well as some severe disadvantages. Such materials usually take a long time to degrade which makes them the primary pollutant to the environment and may become toxic due to degradation from elements and/or chemicals making the food inside unhealthy to consume. However, irradiated food packaging can solve these problems through the use of radiation-based cross-linking technology (Ebnesajjad, 2012). This makes them odorless, hygienic, nontoxic, low- and high-temperature tolerance, strong gas barrier properties, sealing ability to render food-safe, and long-lasting fresh.

4.9.5.1 Other industrial applications of radiation-based cross-linking polymerization

In addition to the above, there are wide-ranging potential uses for ionizing radiation in cross-linked polymerization in the market, including high-performance carbon fiber composites, nano-, and micro-composite sensors, medically focused uses, and many other products.

4.10 Nuclear technology offers a feasible option for power cogeneration and production of fresh water

Molden (2013) argues that about 2.8 billion people around the world are living in water-scarce zones. Accordingly, scientists, politicians, exploring innovative alternatives for supplying fresh water, considering the overwhelming need for water and prevalent access constraints. The availability of fresh water from seawater is one of these alternatives and the possibility of using nuclear technology to desalinate water also part of this approach and began to gain momentum back to the mid-last century. The method of nuclear desalination involves transforming seawater into drinkable water using the surplus heat and electricity generated by the nuclear reactor plant.

In principle, the desalination process includes either distillation, membrane processes (multistage flash reverse osmosis (RO), or a multi-effect distillation a combination of RO and distillation) (IAEA, 2007). From the two primary categories of desalination processes used worldwide, one can be identified as a thermal process in which feed water is heated. The resulting vapor is processed as pure water (distillate) (IAEA, 2007). The other can be recognized membrane desalination processes in which feed water is pumped through a selectively permeable membrane to flush out the dissolved solids. Multi-stage flash distillation, multi-effect distillation, and thermal and mechanical (TVC, MVC) vapor compression are the primary thermal methods (IAEA, 2007). While the RO is the principal membrane operation. Most of the energy and associated financial expenditures are spent on pushing seawater under pressure into a selectively permeable membrane, a filter so that on the other side of the membrane, the ions that make it "salty" are left behind, and filtered water is produced.

Energy prices, however, are growing as environmental measures to minimize carbon emissions demand new and renewable sources of energy for manufacturing processes. Much of today's modern desalination plants use fossil fuels, leading to a rise in greenhouse gas emissions (IAEA, 2007). Many studies have been inspired by concern about global warming and water shortages and found that the combination of a desalination plant with a nuclear power plant to provide drinkable water is an optimal solution. On the other hand, and compared to intermittent renewable energy sources such as wind and solar are still facing many challenges to prove their reliability. According to many techno-economic analyses, IAEA (2007) argues that nuclear technology to be more feasible and there are several advantages of combining a desalination facility with a nuclear power plant (nuclear desalination)including but not limited to carbon-free electricity that reduces GHG emissions and therefore protecting the environment, low costs compared to the fossil plants, and a reliable continues power supply beside the advantage of nuclear power plants locations near to the sea. Thus, nuclear desalination is an appealing choice for areas with existing nuclear power plants.

In nuclear desalination, the surplus heat from a nuclear reactor is used to evaporate seawater and condense pure water. According to World Nuclear Association (2020), Small to medium-sized nuclear reactors are suitable for producing desalinized seawater at high quantities, which can produce an average of 250,000 m^3/day at a very low cost with zero environmental pollution. In this context, it is estimated that the US Navy nuclear-powered aircraft carriers desalinate about 1500 m^3/day for their onboard use (World Nuclear Association, 2020). Finally, we can conclude that nuclear energy can be exploited for a dual generation of electricity and water desalination simultaneously and has the ability to respond to the changing demand for electricity and water. For example, the surplus of nuclear energy can be fed to the desalination plant using RO technology to produce more water. Where potable water can be stored easily. Or the surplus energy can be used to produce hydrogen that can be stored and converted into electrical energy in the long run.

4.11 The role of nuclear power in environmental remediation for industry & pollution management

The increase in energy demand has followed exponential global population growth. Meiswinkel et al. (2013) estimate the global energy demand to increase by 62% as the population swelled from six billion to almost eight billion during the past 20 years of an increase of almost 33%. This escalation in energy demand has also accompanied an exponential growth in GHG emissions into the atmosphere. Greenhouse gases are gases that function as insulators trapping the heat radiated from the sun onto earth by not letting it radiate out of the atmosphere, which consequently has worsened the global warming issue, increasing the average overall temperature of the planet. The accruement of the global warming effect mandates long-lasting changes to the global weather, which has a butterfly effect with other more profound changes affecting the lives in the earth planet. The main driver of the greenhouse effect is carbon dioxide (CO_2), a strong greenhouse agent (Upadhyay et al., 2019). Hore-Lacy (2010) argues that global CO_2 emissions total out at approx. 25 billion tons per year, from fossil fuel usage. The primary reason for these high emission rates is the heavy dependence of most of the energy sector on fossil fuels as the source of energy. The heavy dependence on only one source of energy, such as fossil fuels has created several challenges. The most important of them is the uncertainty of oil prices and the constraints put on to fight against global warming, which places limits on the consumption of oil, gas, and coal in power plants. The majority of countries have thus acknowledged the immediate need to start focusing on environmentally sustainable solutions such as sustainable energies and safe nuclear plants. In this context, Abdallah and El-Shennawy (2013), reported that the generation of electricity through the burning of fossils fuel is the primary culprit of the increase in CO_2 emissions, responsible for contributing 40% of the global carbon dioxide.

Many scholars in the field of energy, including Meiswinkel et al. (2013) altercate on the problems attached to renewable energy which comprise of but are not limited to, operating and maintaining costs of the equipment, weather patterns, noncontinuity of production, as well as the capacity to generate electricity sustainably in an available area of operation and budget, which fall short to the ever-increasing energy demand. On the other hand, nuclear energy has become one of the most economical and environmentally sustainable alternatives in the world to meet the rising demand for energy. This is due to its potential to generate vast amounts of energy at the lowest operational cost and according to Basu and Miroshnik (2019), the energy capacity of one kilogram of uranium, for example, will create energy equivalent to that produced from 100 tons of high-quality coal or 60 tons of oil. Moreover, nuclear power plants in Europe have been estimated to eliminate about 700 million tons of carbon dioxide from the atmosphere annually (Basu and Miroshnik, 2019), since no carbon dioxide generated as a byproduct when producing electricity through nuclear means. Liptak (2008) argues that the total world's nuclear plants save 2.5 billion tons in CO_2 emissions yearly. Therefore, it is anticipated that by 2030, the total greenhouse emissions will be avoided due to the use of nuclear technology to reach 25 billion tons. Nuclear scientists recognize "nuclear power" as the best choice for a renewable supply of electricity and a carbon-free method of producing energy. The World Nuclear Organization (2020) reported that "today there are about 440 nuclear power reactors with a total capacity of about 400 GW operating in 30 countries plus Taiwan… and about 50 power reactors are currently being constructed in 15 countries notably China, India, Russia, and the United Arab Emirates." As a whole, nuclear power generation is considered a viable solution to the problem of increasing global energy demand by providing a renewable source of energy. As the world increases its nuclear capacity for generating power, the dependence on fossil fuels as the source energy started to lower progressively. It is still assumed by many, however, that radioactive waste can be very dangerous for the environment in the case of any breach of containment of the radioactive waste produced as a byproduct of nuclear energy generation, which is not valid to some degree as there are appropriate facilities in nuclear plants planned to manage safely such radioactive waste. Besides, the high-quality structure of the containment buildings of the reactors and the presence of several protection measures, all of which are constructed in compliance with high safety requirements, are capable of avoiding the release of radioactive materials to the ground, air, or water in the case of an accident. The IAEA directly regulates all of these technical and operational aspects to ensure a high level of nuclear safety and security.

Energy consumption for industrial, residential, business, and transportation purposes is the world's largest energy consumer. Globally, the majority of this energy is supplied by fossil fuels that release millions of tons of greenhouse gases into the environment every day. Many hundred million tons of different petroleum products are being used daily all over the globe for this purpose. International Energy Association (2020) estimated that the total global final consumption for electricity for 2017 is around 21,372 terawatt-hours. It has also been estimated that over 1 billion cars around the world. In Europe itself, the estimated total energy consumption of cars is approximately 6×10^{18} J/year. While Maritime Transport Report (2017) stated that the world's total commercial fleets on January 1, 2017, of 93,161 vessels. In 2012, It was estimated that Marine Fuel demand around 6.1% of global world oil demand (Concawe, 2012).On the other hand, the average global demand for jet fuel is around 7.5 million barrels per day, and the World Economic Forum (2020) estimated that about 100,000 flights take off and land every day across the world.

A symbiotic arrangement of nuclear energy and hydrogen is one of the most viable alternatives to fossil fuels in electric power generation, taking into account expected demands in the future. Electricity and hydrogen are both mutually synonymous (Sperling and Cannon, 2004). During their creation, both can be interconverted into one another. As energy should be produced at the time that it is used, these are complementary to each other. Storing electricity is expensive, while hydrogen storage is more viable in this respect. In enhancing the credibility of energy management, electrical energy storage plays an essential role. Energy is usually contained in batteries only in the short term, the long-term storage of excess volumes of energy requires different forms of storage, such as chemical storage in the form of hydrogen. According to Shell (2017), hydrogen storage methods are split into two primary approaches, which are, solid hydrogen storage and chemical or material-based hydrogen storage.

While on the other side, hydrogen is one of the most viable alternatives to fossil fuels for transport. The most important obstacles for renewable hydrogen for transport seem to be the generation of the required heat for the manufacture of hydrogen, expenses, transportation, and the installation of hydrogen generation facilities and distribution stations close to consumers. The production of hydrogen and electricity at the same time by traditional methods is thus costly and adds to the release of greenhouse gases into the atmosphere. However, nuclear power plants can address these issues of cost and manufacturing. In this sense, nuclear technology is essential. It is possible to use the surplus energy and heat generated by a nuclear power plant, as a byproduct of nuclear fission reaction, to continually generate hydrogen. In addition, hydrogen stored as fuel cells is a very cost-effective alternative, meaning that it can be used in transport.

Vessels are considered a significant cause of air pollution and contribute to ocean acidification issues in the marine shipping industry (Kutz, 2010). Olmer et al. (2017) argue that ships accounted for approximately 1 billion tons of GHG emissions over the period 2007–2012. In this sense, the International Maritime Organization believes that the global maritime sector is at a critical turning point and that its success will be connected to the success of hydrogen as a renewable fuel source. This is due to the maritime industry's efforts to reduce pollution to comply with IMO MARPOL Annex VI requirements (DiRenzo, 2019). In order to comply with the standards of emissions regulations, many ship owners and shipbuilders have started to consider hydrogen fuel cell technologies seriously. Several maritime entities, including the European Union, the United States, and Japan, have initiated pilot projects to explore the reliability of using marine hydrogen to reduce emissions while still retaining the efficiency of conventional propulsion technology. GGZM is North America's first successful commercial vessel built for the Water-Go-Round project that is operating using hydrogen fuel cells (DiRenzo, 2019). Thus, one can safely claim that the use of hydrogen as a renewable fuel source for a reduction in pollution is practically viable. With substantial results from the

Water-Go-Round project in the United States and other, in progress, marine hydrogen technology ventures such as the HYSEAS III project in the United Kingdom and the HYBRID skip project in Norway, it is clear that marine hydrogen technology is fast progressing from a concept to practical applicability on the global domain (DiRenzo, 2019). Hydrogen would likely reach significant dependency on a broad scale, close to the widespread recognition of LNG as a marine fuel. With hydrogen production expanding all over the world, this technology will continue to be embraced by larger ships such as container ships, ferries, and vessels.

Accordingly, and in order to accommodate the increasing global demand for energy, nuclear technology is likely to grow its output. The nuclear generation has now been a strong contender to provide the zero-carbon generation needed for industrial and transportation applications. This can be done by the synergy of a nuclear power plant and a hydrogen manufacturing facility to produce large volumes of hydrogen at extremely low rates without having adverse emissions that are harmful to the environment.

4.11.1 The role of nuclear technology in reducing greenhouse gas from agriculture activities

The world has seen an unprecedented rise in population growth, followed by a drastic uptick in the demand for food. As a result, the demand for fertilizers increased to reach optimum crop production to provide adequate food for a rising world population. Byrnes and Bumb (1998) argued that by 2020, the world population is expected to reach 8 billion, which would require an increase in food grains production to more than 3 billion tons that would require more than 300 million tons of fertilizers. Therefore, the improper use of such huge quantities of fertilizers and nutrients represents a major environmental challenge that contributes to global warming and ocean acidification issues, due to the release of greenhouse gases such as carbon dioxide and nitrous oxide from livestock waste and other synthetic fertilizers.

In order to address and mitigate GHG emissions from agricultural practices, nuclear energy, and radioisotopes have proved very useful. Accordingly, nuclear technologies have played an important role in the calculation of absorption rates, the evaluation of soil quality and humidity, soil degradation, and, ultimately, the rationalization of fertilizer use. The use of the labeled nitrogen probe technique, for example, helps to quantify soil moisture and thereby decide the amount of water needed, extenuate irrigation, and preserve water. This approach is paired with the method of calculating absorption rates that allow us to know precisely the amounts of water and fertilizers needed. Fertilizers used by farmers contain the very expensive nitrogen or phosphorous, often seen in large and random quantities, amounting to millions of tons and dollars annually. According to the FAO (2017), the growing world demand for nonfertilizer nutrients, including nitrogen, is forecast at 45 million tons annually. The global demand for nitrogen itself is about 35 million tons of the world's overall demand. While nitrogen is an integral part of all agricultural processes, it should be used properly, as excess use of the substance can lead to contamination of the soil, air, and water, resulting in global warming and climate change issues including ocean acidification.

In this context, nuclear scientists have been able to leverage atomic energy and radioisotopes technology by using labeled phosphorous or labeled nitrogen radioisotopes, to calculate the exact amount of fertilizer required for the crop, which helped in mitigating the effect of the global warming problem as well as contributed in saving a lot of resources and fertilizers.

4.11.2 The role of nuclear and radioisotopes technology in the assessment of ocean acidification and climate change impacts

Greenhouse gases between the Earth and its protective ozone layer could induce changes in weather patterns on a global scale (IAEA, 2010). Often the tiniest temperature fluctuations can affect weather systems, disturb the production of food, and cause extreme droughts or floods. The greenhouse gases of the most profound importance are CO_2, CH_4, N_2O, which are produced from both natural and human activity, and up to 80% of all ocean pollution as a result of human-based activities (Ray and McCormick-Ray, 2013). According to IAEA (2010), the oceans are huge carbon-absorbing basins and its acidification is exacerbated by the accumulation of rising concentrations of CO_2 where ocean microscopic phytoplankton consumes large amounts of atmospheric carbon produced by the combustion of fossil fuels from vehicles, power plants, and houses. Therefore, by consuming high concentrations of greenhouse gases, the seas help prevent global warming to some extent. But contaminated oceans are much less capable of moderating the carbon balance, thereby eventually contribute to accelerated climate change. Moreover, IAEA (2010) argues that Ocean acidification poses significant threats to endangered species of birds and aquatic life, seas, fisheries, and coral reefs. Not only that but it also adds adversely to the melting of ice caps; dwindling fish stocks; aquatic habitat loss and the recurrent outbreak of harmful algal blooms (HAB) or commonly known by the "Red Tide," as shown in Fig. 4.8 that contributes to the poisonous pollution of shellfish and fish and the lethal paralytic shellfish poisoning (PSP) or ciguatera fish

FIGURE 4.8 Nuclear and radioisotopes technologies help in predicting and assessing ocean acidification such as harmful algal blooms (*HAB*) as a result of the emission of greenhouse gases.

poisoning that can actually kill a lot of people; changing the chemistry of seas and coastlines, which are important for human survival as well as are of immense economic value, with 12 million fishermen working in more than 3 million vessels landing nearly 90 million tons of fish annually, providing jobs for over 200 million people worldwide. Therefore, IAEA (2010) affirms that the oceans are crucial to the atmosphere, capturing approximately 2 billion tons of carbon dioxide (CO_2) per year, making them one of the main protections of the Planet against global warming.

Nuclear and radioisotopes technologies such as radiotracer and radioassay tools are used in many areas to increase knowledge of marine environments and to enhance their conservation and security including seafood safety assessment (IAEA, 2010). Radioactive and stable isotope tracers boost aquatic biological process awareness and help track the transport of different forms of heavy metals and industrial contaminants. Radiotracer and radioassay are very useful instruments for obtaining data on bio-kinetics, bioaccumulation, and food chain transmission of metals and harmful algal blooms toxins in aquatic species, including those valued as aquatic materials. The findings of these types of studies can be used in evaluations that help risk-based management decisions concerning the safety of the human intake of seafood. According to IAEA (2013), receptor binding assay, for instance, is a key tool that calculates efficiently, and precisely the levels of "red tide" toxins that may be found in shellfish. In this process, samples are first washed and purified to classify and measure essential aquatic pollutants, such as heavy metals, petroleum hydrocarbons, organochlorine pesticides, PCBs, dioxins/furans, and organotin, radionuclides, chemical compounds or trace metals, and other molecules under examination. During the process of extraction, purification, a number of chemical processes are used followed by the use of either ultra-low-background high-resolution gamma spectrometry, alpha spectrometry, atomic absorption spectrophotometry, gas chromatography, or mass spectrometry to examine the sample for more precise details (IAEA, 2010). In addition, the low plutonium levels of natural nuclear tracers (such as uranium and thorium) in surface waters are used to trace the flow of CO_2 as it is transferred between the atmosphere and the ocean. These motions are monitored to consider the balance of carbon isotopes in open ocean waters by sampling suspended particles and sediments. This helps to build models of climate change in the ocean world since changes in this equilibrium can be influenced by temperature. In this regard, IAEA has established a web-based website called Marine Radioactivity Information System. This system provides more than half-million marine radioactivity measurements of more than 100 different radionuclides or radionuclide ratios in seawater, sediments, biota, suspended matter, and marine species (IAEA, 2020).

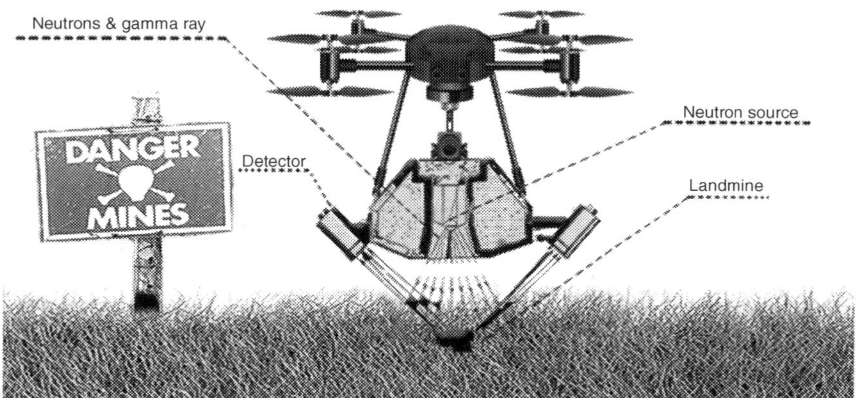

FIGURE 4.9 The role of nuclear technology in the detection of landmines.

4.11.3 The role of nuclear technology in protecting the environment by detecting landmines

Abd El-Samie et al. (2012) argue that there are more than 100 million landmines buried in more than 70 countries that cover more than 200,000 km^2 of the world's surface. Wars are the main reason behind laying such a huge number of these mines, the most important of which are the First and Second World Wars, civil wars, and terrorism. These mines pose a serious danger to human life, they kill and maim thousands of innocent civilians in addition to destroying the environment and the land. The removal of existing landmines is a complex, expensive, and risky task due to the high level of uncertainty and unreliability of available technologies. The available demining technologies are sometimes unable to distinguish landmines from mineral debris.

Nuclear technology such as nuclear detection methods that use Neutron Thermal Analysis or Fast Neutron Analysis is appropriate and effective options in detecting landmines from the perspectives of safety, reliability, costs, and speed (Fig. 4.9). Nuclear detection methods have the ability to detect landmines at great depths and also to identify sensitive elements of explosives in the landmines such as hydrogen, carbon, nitrogen, and oxygen (Megahid, 1999). Nuclear technology is the only technology that is able to distinguish between elements of explosives in the landmines and similar elements that are naturally available in the soil and thus can distinguish between metals, soil, or mineral debris. Not only that, but the use of nuclear quadrupole resonance (quadrupole moment of nitrogen) helps to detect the presence of explosive compounds in the landmines, such as Trinitrotoluene (TNT) (Megahid, 1999).

Moreover, neutron thermal analysis depends specifically on the activation of the nitrogen nucleus, which is abundant in most explosives using the neutrons emitted from a radioisotope source. When the nitrogen nuclei are activated in the landmine, Nitrogen nuclei become excited and jump to a higher energy level. then Nitrogen nuclei try to become stable by jumping back to the low energy level and get rid of the excess energy in form of gamma rays, as shown by Eq. 4.1:

$$^{14}_{7}N + ^{1}_{0}n \rightarrow ^{15}_{7}N \rightarrow ^{15}N_{-} + \gamma \tag{4.1}$$

The released energies from gamma rays are detected and analyzed using gamma spectrometry, where the absolute proportions and measurements determine the abundance of elements and determine the material in the landmine.

On the other hand, fast neutron analysis relies in the principle of measuring the energy spectrum and the arrival time of gamma rays emitted from the induced fast neutrons that interact with the neutrons of the elements are used in the landmine, such as hydrogen, carbon, nitrogen, and oxygen. Each of these elements interacts differently with the induced neutrons and therefore, can be distinguished through the characteristic gamma spectra. Time measurement is useful to determine the location of different scattering. While the observed energy determines the type of all the elements present in the landmine. Since induced neutrons are able to penetrate deep into the ground up to half a meter, therefore, this helps to determine the depth and location of the landmine.

4.11.4 The role of nuclear technology in wastewater treatment using ionized radiation approach

Owing to the rise in population and urbanization, the pollution from domestic and industrial waste has increased substantially and has now become an imminent danger to the environment (Schmidt-Thomé et al., 2014). Wastewater, as defined by UNEP/UN-Habitat, is a mixture of one or more of the following effluents: residential, agricultural, manufacturing, horticultural, aquaculture, and stormwater. Massive amounts of untreated wastewater due to inadequate disposal practices have contributed to the incorporation of toxic chemicals into agricultural and groundwater outlets (UN-Water, 2017). The environmental risks associated with pollutant discharges are usually local but may have repercussions that may be transboundary and continents.

Wastewater pollution may have a significant impact on public health, the environment, society, the economy, and food security. Unfortunately, most of this waste is discharged into the surface waters without proper treatment (Gupta and Gupta, 2020). Such pollution contributes to changes in the levels of nutrients in the biomass and diversity of organisms, in the bioaccumulation of organic and inorganic compounds, and changes in the trophic interactions between species. In the developed countries, the issue is further compounded by rapid population development which is usually not accompanied by an increase in water and wastewater treatment facilities, and therefore, these communities sufferer from water-borne diseases such as cholera, dysentery, typhoid, and shigellosis, which kill millions of people worldwide globally. Commonly these diseases are generated by the untreated wastewater or wastewaters that are treated by traditional means. Conventional Treatment of wastewater is usually made up of three phases, namely primary, secondary, and tertiary. According to FAO (1992) and Abdel Rahman and Hung (2019), in the conventional wastewater treatment, the pretreatment stage extracts rough and thick solids from the waste stream, using physical filtering systems such as screens and grit chambers. Then, suspended solids are extracted by leaving them to settle using gravity sedimentation with or without coagulation and flocculation, key treatment methods are then added. The sewage from primary treatment is led to the secondary treatment that removes any residual or suspended solids or organic materials through biological treatment methods. Any heavy metals, synthetic bio-refractory organic contaminants, and soluble microbial products derived through biological treatment may still be present in effluents from the secondary stage. Effluents from the secondary stage are further refined during the tertiary stage of treatment by eliminating the persistence of organic materials and heavy metals using modern wastewater treatment technology, which includes filtration, sorption, gas stripping, ion-exchange, automated oxidation processes, and distillation. Finally, based on the possible usage and properties of the processed effluent, disinfection is administered.

Despite this, traditional wastewater treatment plants are only capable of lowering the level of pollution to acceptable levels. However, they are not entirely capable of extracting harmful contaminants, and therefore their final products are certainly not suitable for human usage. The latest developments in nuclear wastewater treatment systems have, however, helped solve crucial problems in water supply, pollution control, water conservation, and environmental safety. With the assistance of ionizing radiation in the handling of wastewater, it is possible to use nuclear-treated wastewater as a new source that can fulfill the needs of commercial, farming, even can be used as potable water (Fig. 4.10).

The working theory behind ionized radiation in wastewater treatment is that ionizing radiation transfer such energy, typically greater than 5 MeV

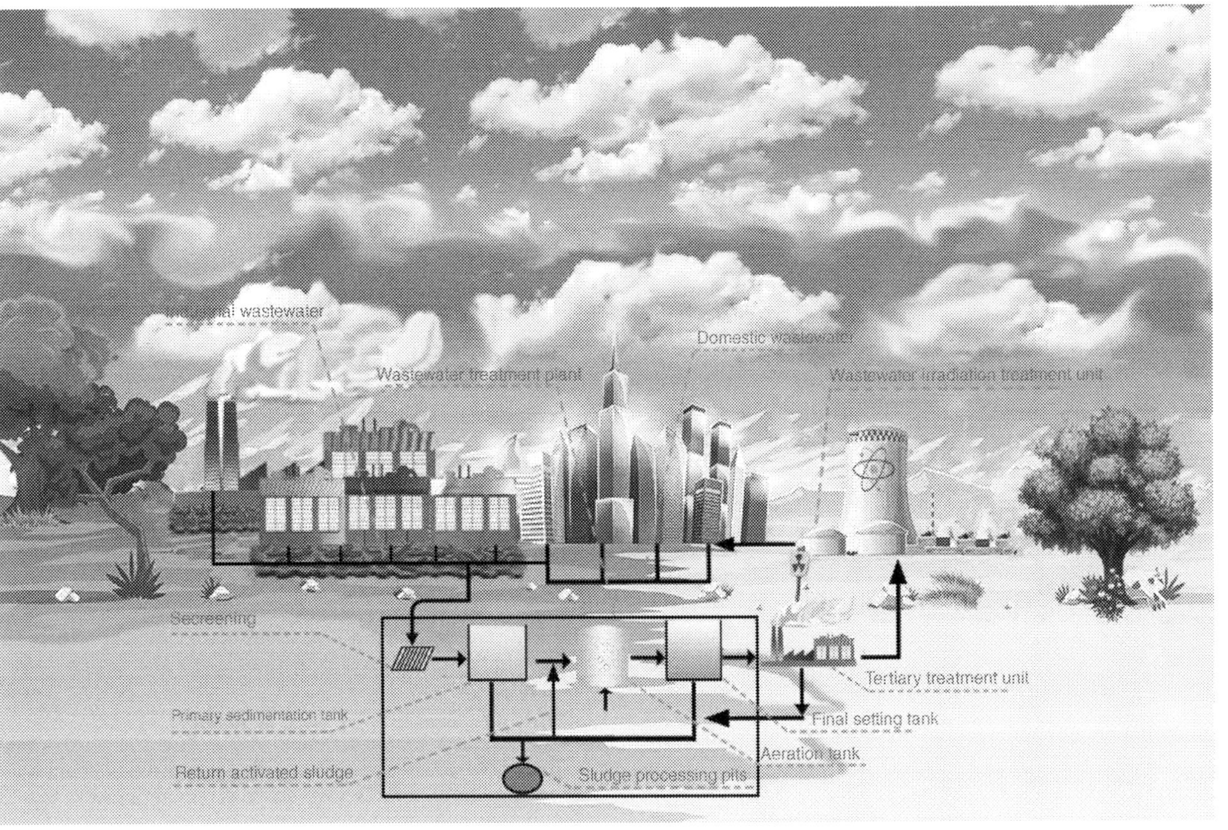

FIGURE 4.10 Wastewater treatment using ionized radiation by coupling conventional wastewater treatment plant with irradiation unit.

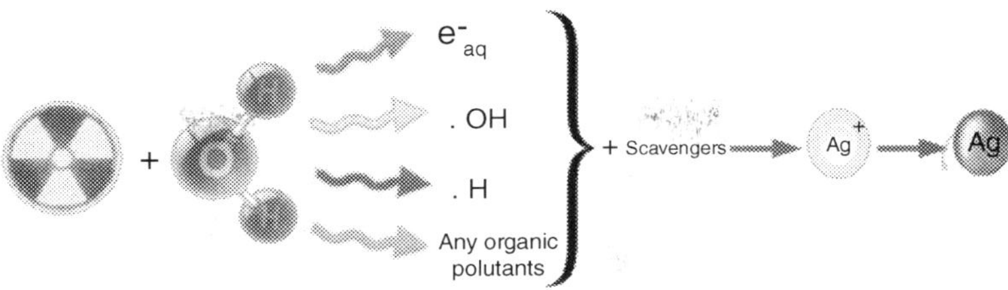

FIGURE 4.11 The hydro radiolysis process generates free water radicals and breaks the ions of organic and metal contaminants.

that excites and ionizes atoms it hits by removing their electrons or splitting the chemical bonds between their molecules and thus imparts some of its energy in the radiolysis (Abdel Rahman and Hung, 2019). Ionizing radiation may take the form of particles such as alpha, beta, and neutron particles, or electromagnetic waves, which are accelerator-generated, such as gamma rays or X-rays or EBs.

Jiang and Iwahashi (2019) argue that the process of ionization of atoms has substantial redox capabilities. When used with hydro radiolysis, the process generates free radicals, due to the high energy radiation causing ionization of water molecules, which results in the production of both reducing and oxidizing intermediaries (H atoms [·H] and hydrated electrons [e^{-}_{aq}], and hydroxyl radicals [·OH] (Fig. 4.11). Ionizing radiation is also a powerful tool for the treatment of contaminated water by directly breaking or reacting with organic contaminants through free radicals (oxidizing and reducing pathways). However, by adding oxidizing scavengers to ·OH, this decreases the quality of the metal ions to a lower degree and therefore eliminates heavy metal ions.

Agricultural wastewater, for instance, is generally tainted with chemical pesticides such as diazinon, dimethoate, procloraz, metiocarb, imidacloprid, and carbofuran, herbicides, and fungicide may migrate to surface and groundwater. However, Abdel Rahman and Hung (2019) argue that using ionizing radiotherapy technology either electron-beam or gamma rays has shown its efficacy in disintegrating a variety of these contaminants using the tertiary treatment process and the radiation dose used varies from 0.1 to 10 KGy, which has shown near-total destruction of these contaminants.

Moreover, industrial effluent wastewater is another excellent example of how the nuclear ionization technique will perfectly handle this wastewater. Industrial effluents wastewater poses significant health and environmental issues that may arise from the existence of organic dyes these effluents. To eliminate these toxins, ionizing radiation treatment using either electron-beam or gamma rays is recommended. Abdel Rahman and Hung (2019) reported that by irradiating aqueous solutions of polluted Reactive Blue 15 and Reactive Black 5 dyes with doses ranging from 0.1

to 15 KG, a complete decoloration was observed. In addition, several research studies have shown that de-coloration due to e^- and H reactions increased linearly with an increase in the dosage of irradiation, whereas decoloration increased logarithmically due to OH^- radical reaction (Abdel Rahman and Hung, 2019). Such a solution may include toxins and anions, such as Cl^-, CO^{32-}, HCO^{3-}, SO^{42-}, Mercaptobenzothiazole, and stable chemical compounds that pose a major danger to the atmosphere and health. For example, mercaptobenzothiazole has low biodegradability which can be difficult to decompose by traditional biological processes. Jiang and Iwahashi (2019) argue that mercaptobenzothiazole can be degradable by gamma irradiation through free radicals attacking its sulphur atoms, and a gamma irradiation dose of 1.2 KGy may induce 82% degradation. Up to 97% of Endosulfan will effectively eliminate with a dosage of 1.02 KGy of gamma irradiation.

4.12 Conclusion

Nuclear applications encompass many fields in our daily life. The industry is considered one of the most common fields that use nuclear and radioisotopic technologies. There is currently no developed industry that does not use nuclear and radioisotopic technologies in their production steps, technical inspecting process, quality, and defects monitoring process. This chapter provided a brief overview of some nuclear applications in industry, which are just the tip of the iceberg.

Industrial nondestructive radiography such as X-ray or gamma-ray CT or neutron radiography, for instance, utilizes ionizing rays, to detect faults in the flowlines, examine weld strength, and detect any deposits, therefore this technology widely used in examining oil and gas pipelines, as well as examining water pipes in major projects. Moreover, it can be used in examining and detection of cracks in the buildings, structures and the bodies of aircraft, cars, and ships.

Nuclear and radioisotopic technologies are also considered as one of the most sophisticated technologies through which petroleum and minerals are explored on the ground and optimized their production, nuclear reactive analysis using neutron activation approach, for instance, reveals the type of elements, and the proportion of their presence in the ground. Interwell radiotracer technology is another nuclear application used in the oil and gas industry to enhance the secondary and tertiary recovery of oil from the reservoirs. Not only that but, we can calculate the velocity of oil flow inside the pipelines by injecting the radioactive isotope into the petroleum pipeline and tracking the passage of the isotope in the pipeline, and also detecting if there is a leak in the oil pipelines.

Nucleonic gauges are another vital nuclear application used widely in many industries due to their ability to measure accurately different parameters including density, weight, levels of any matter in any harsh, or risky environment depending on the interaction between ionizing radiation and matter. Accordingly, Industry worldwide started to employ this technology for refining and revamping procedures and materials to monitor and enhance production efficiency in many industries including but not limited to mining and mineral ore processing, environmental monitoring; the paper and plastics industries; the cement and civil engineering industries, and petrochemical industries, the oil and gas industries, and many other industries.

Soon after the invention of the crosslinking of olefin polymers using ionizing radiation in the mid of 20th century, radiation crosslinking polymerization technique has begun to be implemented in a wide range of processing industries such as oil and gas pipelines, vehicle tires plies, electrical cables and wire, polymeric foams, food packaging, medical devices. Radiation crosslinking polymerization technique is a green technology that does not require any external chemicals or thermal treatment and instead, it relies on gamma radiation or EB as a source of the irradiation process. Processing radiation improves the generation of intermolecular bonds between polymeric chains, such as mechanical properties, resistance to corrosive liquids, thermal stability, withstanding high temperature, suppressing of flame propagation, and exhibiting improved abrasion tolerance, fluid resistance. As a result, the use of processing radiation in the industry improved the quality and reliability of products.

Nuclear and radioisotopic technologies play a significant role to verify the authenticity of imported food products, which often—can be easily forged. However, it became very easy to verify the quality of food items and react more quickly at vital checkpoints to prevent the flow of contaminated products or adulterated using Isotope analysis that helps to establish the geological sources of foodstuffs and the authenticity of their composites.

Nuclear and radioisotopic technologies have proven their capabilities in eliminating greenhouse gases and climate change as a result of the increase in energy demand. This escalation in energy demand due to the exponential global population growth has also accompanied an exponential growth in GHG emissions into the atmosphere, which impacts adversely on our ecosystem. Oceans' acidification is one of these adverse consequences. The heavy dependence on fossil fuels as the main source of energy has contributed to the global warming issue. The generation of electricity through the burning of fossils fuel is the primary culprit of the increase in CO_2 emissions, responsible for contributing 40% of the global carbon dioxide. On the opposite nuclear energy has become the most economical and environmentally sustainable alternatives in the world to meet the rising demand for energy because of its potential to generate vast amounts of energy at the lowest operational and with zero GHG emissions into the environment. Not only that but, nuclear energy can solve many problems in the communities at the same time. It can be used to generate electricity, desalinate water, treat wastewater, and protecting the environment at the same time.

In sum, the use of nuclear and radioisotopic technologies in the industry is very numerous and these technologies play vital roles in the industry in order to safe, high-quality products with no industrial defects. Therefore, nuclear and radioisotope technologies serve human beings by providing safe and reliable products and play a significant role in the sustainability of industrial, economic development, and environmental protection.

References

Abd El-Samie, F., Hadhoud, M., El-Khamy, S., 2012. Image Super-Resolution and Applications. CRC Press, New York, USA, pp. 1–502.

Abdallah, L., El-Shennawy, T., 2013. Reducing carbon dioxide emissions from electricity sector using smart electric grid applications. Hindawi Publishing Corporation. J. Eng. 2013, 1–8.

Abdel Rahman, R., Hung, Y, 2019. Application of ionizing radiation in wastewater treatment: an overview. Water 12 (1), 19.

Al Nabhani, K., Khan, F., 2019. Nuclear Radioactive Materials (TENORM) in the Oil and Gas Industry: Safety, Risk Assessment and Management. Elsevier, Cambridge, USA, pp. 1–328.

Basu, D., Miroshnik, V., 2019. The Political Economy of Nuclear Energy: Prospects and Retrospect. Springer Nature, Switzerland, pp. 1–275.

Beck, M., Williams, R., 1995. Process Tomography—Principles, Techniques and Applications. Butterworth-Heinemann, Oxford, UK, pp. 1–384.

Berger, H., Iddings, F., 1998. Neutron Radiography. A State-of-the-Art Report-NTIAC-SR-98-01. Non-destructive Testing Information Analysis Center, Austin, TX, USA, 1–80.

Bjørnstad, T., Somaruga, C., 2012. Application of Radiotracer Techniques for Interwell Studies. IAEA Radiation Technology Series No. 3, IAEA, Vienna, Austria, pp. 1–231.

Blitz, J., 2012. Electrical and Magnetic Methods of Non-destructive Testing. Springer Science & Business Media, Berlin, Germany, pp. 1–261.

Byrnes, B., Bumb, B., 1998. Population growth, food production and nutrient requirements. J. Crop Prod. 1 (2), 1–27.

Carbone, M., 2011. Investigating Mechanical Behavior of Cord-rubber Composites by Multi-Scale Experimental and Theoretical Approach. Ph.D. Thesis. Università degli Studi di Napoli Federico II. Naples, Italy, pp. 1–201.

Chaplin, R., 2017. Industrial Ultrasonic Inspection: Levels 1 and 2. Friesen Press, Victoria, Canada, pp. 1–292.

Chesson, L, Valenzuela, L, 2010. Hydrogen and oxygen stable isotope ratios of milk in the United States. J. Agric. Food Chem. 58 (4), 2358–2363.

Concawe, 2012. Marine fuel facts. 1–27. Accessed 19 Oct. 2020 and retrieved from https://www.concawe.eu/wp-content/uploads/2017/01/marine_factsheet_web.pdf.

DiRenzo, J., 2019. IMO 2020: hydrogen's future in maritime. Accessed 21 Oct. 2020 and retrieved from https://www.marinelink.com/news/imo-hydrogens-future-maritime-467713.

Ebnesajjad, S., 2012. Plastic Films in Food Packaging: Materials, Technology and Applications. William Andrew. Oxford, UK, pp. 1–384.

FAO, 1992. Wastewater Treatment and Use in Agricultural-FAO Irrigation and Drainage Paper 47. FAO: Rome http://www.fao.org/3/t0551e/t0551e00.htm#Contents.

FAO, 2017. World Fertilizer Trends and Outlook to 2020. Food and Agriculture Organization of the United Nations, Rome, pp. 1–28.

Fox, M., 2014. Why We Need Nuclear Power: The Environmental Case. Oxford University Press, New York, USA, pp. 1–306.

Garbe, U., 2020. Neutron radiography: WCNR-11. Materials Research Forum LLC, Millersville, USA, pp. 1–316.

Głuszewski, W., Stasiek, A., Raszkowska-Kaczor, A., Kaczor, D., 2018. Effect of polyethylene cross-linking on properties of foams. NUKLEONIKA 63 (3), 81–85.

Gupta, S., Gupta, I., 2020. Drinking Water Quality Assessment and Management. Scientific Publishers, Jodhpur, India, pp. 1–273.

Hassoun, A., et al., 2020. Fraud in animal origin food products: advances in emerging spectroscopic detection methods over the past five years. Foods 9 (8), 1069.

Heindel, T., 2011. A review of X-ray flow visualization with applications to multiphase flows. J. Fluids Eng. 133 (7), 1–16.

Hore-Lacy, I., 2010. Nuclear Energy in the 21st Century: World Nuclear University Press. Elsevier, USA, pp. 1–168.

Hussain, M., et al., 2020. Ultra-high-molecular-weight-polyethylene (UHMWPE) as a promising polymer material for biomedical applications: a concise review. Polymers 12 (2), 323.

IAEA, 2005. Technical data on nucleonic gauges. IAEA-TECDOC-1459. 1–118.

IAEA, 2007. Economics of nuclear desalination: new developments and site specific studies. IAEA-TECDOC-1561, 1–226.

IAEA, 2010. Isotopic tools for protecting the seas. IAEA in Austria, 1–34.

IAEA, 2013. Detection of harmful algal toxins using the radioligand receptor binding assay. a manual of methods. IAEA-TECDOC-1729, Vienna, 1–55.

IAEA, 2014. Development of radiometric methods for exploration and process optimization in mining and mineral industries. In: Report of the Consultant meeting, Vienna, pp. 1–36.

IAEA, 2019. IAEA launches project to help countries fight food fraud. Accessed 12 Oct. 2020 and retrieved from https://www.iaea.org/newscenter/pressreleases/iaea-launches-project-to-help-countries-fight-food-fraud.

IAEA, 2019. Radiotracer techniques for leak detection – Brochure, April 2004. Training in Radiation Technologies Industrial Applications – Leak detection – Intercomparison test (RER1020). IAEA, Vienna, Austria, pp. 1–69.

IAEA, 2020. Marine radioactivity information system (MARIS). Accessed 26 Feb. 2020 and retrieved from https://www.iaea.org/resources/databases/marine-radioactivity-information-system-maris.

IAEA, 2020. Traceability and authenticity. Accessed 11 Feb. 2020 and retrieved from https://www.iaea.org/topics/traceability-and-authenticity.

International Energy Association, 2020. Electricity information 2019. Accessed 19 Oct. 2020 and retrieved from https://www.iea.org/reports/electricity-information-2019.

Jadiyappa, S. (2018). Principles and applications in nuclear engineering—radiation effects, thermal hydraulics, radionuclide migration in the environment. doi: 10.5772/intechopen.79161.

Jiang, L., Iwahashi, H., 2019. Current research on high-energy ionizing radiation for wastewater treatment and material synthesis. Environmental Progress & Sustainable Energy. American Institute of Chemical Engineers, USA, pp. 1–9, https://aiche.onlinelibrary.wiley.com/doi/pdf/10.1002/ep.13294.

Joesten, M., Hogg, J., Castellion, M., 2007. The World of Chemistry: Essentials. Cengage Learning, CA, USA, pp. 1–608.

Kutz, M., 2010. Environmentally Conscious Fossil Energy Production. John Wiley & Sons, Hoboken, NJ, USA, & Canada, pp. 1–368.

Kyser, K., 2017. Advances in the use of isotopes in geochemical exploration: instrumentation and applications in understanding geochemical processes. In: Proceedings of Exploration 17: Sixth Decennial International Conference on Mineral Exploration, pp. 521–526.

LA Porta, C., Zapperi, S., Pilotti, L., 2020. Understanding Innovation Through Exaptation. Springer Nature, Switzerland, pp. 1–194.

Liptak, B., 2008. Post-Oil Energy Technology: The World's First Solar-Hydrogen Demonstration Power Plant. CRC Press, Boca Raton, FL, USA, pp. 1–536.

Makuuchi, K., Cheng, S., 2012. Radiation Processing of Polymer Materials and Its Industrial Applications. John Wiley & Sons, Hoboken, NJ, USA, & Canada, pp. 1–415.

Maritime Transport Report, 2017. Review of maritime transport 2017—structure, ownership and registration of the World Fleet. 21–41. Accessed 19 Oct. 2020 and retrieved from https://unctad.org/en/PublicationChapters/rmt2017ch2_en.pdf.

Megahid, R., 1999. Landmine Detection by Nuclear Techniques. In: 2nd Conference on Nuclear and Particle Physics, Cairo, pp. 564–570.

Meiswinkel, R., Meyer, J., Schnell, J., 2013. Design and Construction of Nuclear Power Plants. John Wiley & Sons, Hoboken, NJ, USA, pp. 1–150.

Molden, D., 2013. Water for Food Water for Life: A Comprehensive Assessment of Water Management in Agriculture. Routledge, London and New York, pp. 1–688.

Morse, J., 2019. Energy Resources In Colo/h. Routledge, London and New York, pp. 1–431.

Olmer, N., Comer, B., Roy, B., Mao, X., Rutherford, D., 2017. Greenhouse gas emissions from global shipping, 2013–2015. International Council on Clean Transportation. 1–27. Accessed 21 Oct. 2020 and retrieved from https://theicct.org/sites/default/files/publications/Global-shipping-GHG-emissions-2013-2015_ICCT-Report_17102017_vF.pdf.

Owczarczyk, A., Wierzchnickl, R., Urbartski, T., Chaielewski, A.G., and Szpilowski, S. (1992). The detection of leakages in open reservoirs by the radioisotope sorption method. Department of Nuclear Methods of Process Engineering, Institute of Nuclear Chemistry and Technology. 1–20.

Ray, G., McCormick-Ray, J., 2013. Marine Conservation: Science, Policy, and Management. John Wiley & Sons, Hoboken, NJ, USA, pp. 1–384.

Schmidt-Thomé, P., Nguyen, T., Pham, T., Jarva, J., Nuottimäki, K., 2014. Climate Change Adaptation Measures in Vietnam: Development and Implementation. Springer, London and New York, pp. 1–100.

Shell, 2017. Shell hydrogen study. energy of the future? Sustainable Mobility through Fuel Cells and H_2. Shell Deutschland Oil, Hamburg, pp. 1–71.

Silberman, N., 2012. The Oxford Companion to Archaeology. Oxford University Press, New York, USA, Vol. 1, pp. 1–1507.

Silvestre, C., Cimmino, S., Stoleru, E., Vasile, C., 2017. Chapter 20: Application of radiation technology to food packaging Sun, Y., Chmielewski, A., Eds. Applications of Ionizing Radiation in Materials Processing. Institute of Nuclear Chemistry and Technology, Warsaw, pp. 461–485.

Sperling, D., Cannon, J., 2004. The Hydrogen Energy Transition: Cutting Carbon from Transportation. Elsevier, USA, pp. 1–266.

Sun, Y., Chmielewski, A., Przybytniak, G., 2017. Chapter 11: Applications of ionizing radiation in materials processing. Crosslinking of Polymers in Radiation Processing. Institute of Nuclear Chemistry and Technology, Warszawa, pp. 249–268.

UN-Water. (2017). Wastewater the untapped resource. The United Nations World Water Development Report 2017, France, 1-180.

Upadhyay, A., Singh, R., Singh, D., 2019. Restoration of Wetland Ecosystem: A Trajectory Towards a Sustainable Environment. Springer Nature, Singapore, pp. 1–247.

Vautz, W., Baumbach, J., 2006. Ion mobility spectrometry for food quality and safety. Food Addit. Contam. 23 (11), 1064–1073.

Wang, M., 2015. Industrial Tomography: Systems and Applications. Woodhead Publishing, Cambridge, UK, pp. 1–773.

World Economic Forum, 2020. Global Agenda, Aviation, Travel and Tourism. Accessed 19 Oct. 2020 and retrieved from https://www.weforum.org/agenda/2016/07/this-visualization-shows-you-24-hours-of-global-air-traffic-in-just-4-seconds/.

World Nuclear Association, 2020. Nuclear desalination studies. Accessed 15 Oct. 2020 and retrieved from https://www.world-nuclear.org/information-library/non-power-nuclear-applications/industry/nuclear-desalination.aspx.

World Nuclear Organization, 2020. Plans for new reactors worldwide. Accessed 18 Oct. 2020 and retrieved from https://www.world-nuclear.org/information-library/current-and-future-generation/plans-for-new-reactors-worldwide.aspx.

CHAPTER

5

Application of nuclear science and radioisotopes technology in the sustainability of agriculture and water resources, and food safety

OUTLINE

5.1 Introduction

As the population of the world is increasing at an alarming rate, there is great pressure on agriculture to provide food, fibre, and livestock. For sustainable agriculture, there is a dire need to find new ways that would increase the production of crops as well as livestock. Another major

Applications of Nuclear and Radioisotope Technology: The Atom for Peace and Sustainable Development
DOI: https://doi.org/10.1016/B978-0-12-821319-3.00006-3

problem faced by agriculture is climate change. The climate patterns are changing, which are affecting the sowing and harvesting time of many crops, ultimately affecting their production. Sustainable agriculture is now being mostly seen in the view of using nuclear and radioisotopes technology. For instance, the radioisotopes technology assisting in building new models for tackling changing weather patterns that have a great impact in agriculture sustainability.

The utilization of nuclear science in agriculture provides new ways of ensuring food availability to the world population. The history of using nuclear science in agriculture dates to the Cold War era in the United States. In the beginning, it was the mitigation process in response to any nuclear attack. Atomic Energy Commission started to use nuclear energy for peaceful purposes by using it in agriculture. They were of the view that nuclear power would improve the agriculture of the United States. The concept of nuclear energy changed after August 6, 1945, when an atomic bomb was dropped on Hiroshima. It was considered as an evil power, but soon research was started to use it peacefully. After World War II, the world needed a lot of mouths to be fed. The Green Revolution of using pesticides and chemical fertilizer was not the only answer to the question of food security. Atomic energy in the form of radioisotopes was first started in ecosystems and other biological processes. They were used as the trace elements at the start. Radio-phosphorus was first used in the physiological processes of the plant. Photosynthesis and crop diseases were the main research areas for studying the effects of radioisotopes on these processes. The processing of foods by using radioisotopes also showed positive results. In this way, the concept of nuclear energy improved in the eyes of the public as a result of the increase in the level of nuclear awareness. The agriculture of that period was called atomic agriculture (Oatsvall, 2014).

The use of radioisotopes is declared safe by all nuclear agencies of the world. The labeled and shorter lifespan of radioisotopes makes them a significant factor in increasing yields of crops. Radioactive elements release a different amount of radiation while decaying depending on their radioactivity level and their half-lives, which can be used in many fields including health care, agriculture and, in the field of physical sciences in order to improve the living standard of life and also for the primary research purposes and in the immense dimension of its applications. The shelf life of many fresh and canned foods can be increased by using the food irradiation technique. Many seeds that have shown many positive gains in yield are produced by using nuclear mutation. The high-yielding varieties of rice, groundnut, oilseeds, jute, and pulses have been successfully used in fields by farmers. Plants absorb different types of nutrients from the soil. Radioisotopes tracers are useful in measuring the absorption level in the plants, livestock, which results in production increase and saving water, nutrients, and fertilizers. The effect of a single element on plants is tough to study. By using the tracking technology of nuclear science, it is now being investigated. A trace radioisotope of that element is provided in the soil. When the plant absorbs it, the plant is studied by using photographic paper. This makes an autoradiograph. The dark areas show where the element has been affected. It gives valuable information about fertilizing procedures (Sahoo and Sahoo, 2006).

Livestock is one of the most critical sectors of agriculture. This sector alone can complete the hunger demands of billions of people. The developed countries have achieved remarkable improvements in the productivity of livestock by using modern nuclear techniques in the health, fertility, and feeding of animals. Radioisotopes are playing a crucial role in the biological processes of disease resistance, reproduction, and growth in many kinds of farm animals. The isotopes like ^{15}N, ^{14}C, ^{51}Cr, and ^{125}I are used as tracers in tagging constituents of animal feed. The modern nutritional concepts are developed by isotopic processes, which have shown a significant increase in the quality and production of meat, milk, and egg (Khanal and Munankarmy, 2009).

The preservation of food is one of the most factors essential factors influencing food security. Millions of tons food are wasted per year in the world due to inferior preservation methods. Nuclear magnetic resonance (NMR) is one of the most powerful analytical techniques that can be used in food science. This has gained much importance in food technology in the analysis and evaluation of many foods. In NMR spectroscopy, some nuclei of isotopes of ^{31}P, ^{1}H, and ^{13}C are used due to their magnetic strength. NMR preserves food by enabling the analysis of constituents of food. The food products are composed of very complex ingredients. The evaluation of them helps in forming food preserving techniques (Hatzakis, 2019). Not only that but this technique also helps in the detection of ingredients for adulterated food that are imported or locally made, which will be discussed in more detail in the section on peaceful nuclear applications in Industries.

The insect pests are the major destructive reason for the loss of millions of tons of food such as valuable grain products, vegetables, fruits. They infect crops either directly or by transferring pathogens like bacteria and fungi. However, sterile insect technology (SIT) is the most sophisticated nuclear technology to mitigate and control pest hazards in humans, plants, livestock, and crops. Remarkably, SIT can facilitate the loss of fertilization by exposing millions of larvae to specific radiation doses and after that, releasing sterile insect males in the affected areas where they naturally and successfully mate with the wild females but rather inhibiting egg production. Technical experts and scientists assert that such a technique is eco-friendly technology and consider it as "the birth of an ecological technology to prevent reproduction." This is because the insects released in the environment are sterile and cannot be settled in the ecosystem unlike other methods of insect control (e.g., pesticide) that do not take the safety of the fauna, crops, and human health into consideration thereby causing many diseases and ecological imbalances.

On the other hand, the applications of nuclear science and radioisotope technology in the sustainability of agriculture, water resources, and food safety have played a significant role in solving soil erosion dilemmas. Soil erosion caused by water and wind plays a crucial role in removing top fertile soil from land affecting agricultural production. Globally about 2000 million hectares of arable and grazing lands are eroded annually. Accordingly, nuclear science and radioisotope technology managed to evaluate soil erosion using the evaluation of radioisotopes available in the atmosphere that are deposited in the soil by the rain. This method relates to the comparison of the reference site with the studied site based on the mathematical quantitative analysis models that track the change in the differentiated ^{137}Cs inventories into sedimentation rates and soil erosion.

This chapter will cast light on some important applications of nuclear science and radioisotope technology in the sustainability of agriculture, water resources, and food safety.

5.2 Peaceful nuclear application in agriculture

5.2.1 Sterile insect technique

SIT is the most sophisticated nuclear technology to mitigate and control pest hazards in humans, plants, livestock, and crops. Sterile insect technique was first developed in the early 1930s by the entomologists Edward Knipling and Raymond Bushland (Gillman, 1992). This technology was acknowledged as "the greatest entomological achievement of the twentieth century" by Orville Freeman who was the then US Secretary of Agriculture (Jennings, 2014). This acknowledgment was based on its huge success toward the eradication of screwworms that had infested Venezuela in the mid-19th century to the late 1970s as well as Mexico and Belize in the later 1980s. Additionally, this technique has massively contributed to suppressing insects that are a great threat to vegetables, livestock, fiber crops, and as well in saving herds of cows and domestic goats which are essential sources of milk and meat protein. Remarkably, SIT can facilitate the loss of fertilization by exposing millions of larvae to specific radiation doses and after that, releasing sterile insect males in the affected areas where they naturally and successfully mate with the wild females but rather inhibiting egg production. The fact that the proportion of sterile males exceeds those of nonsterile males is the biggest secret behind the success of this technology especially considering that the targeted insect population is reduced to the desired level and effectively resulting in their extinction. Technical experts and scientists assert that such a technique is eco-friendly technology and consider it as "the birth of an ecological technology to prevent reproduction." This is because the insects released in the environment are sterile and cannot be settled in the ecosystem unlike other methods of insect control (e.g., pesticide) that do not take the safety of the fauna, crops, and human health into consideration thereby causing many diseases and ecological imbalances.

The essential parameter behind the creation of sterile insects is the nuclear radiation, which is defined as the absorbed energy in the mass unit of specified material and is expressed in Système International d'Unités (SI) units as gray (Gy), where 1 Gy is equivalent to 1 joule (J) of absorbed energy in 1 kg of a specified material (1 Gy = 100 rad) and known by the absorbed dose of radiation. Through Ionizing radiation (IR), molecules break down resulting in several effects in the irradiated material and accurate delivery of the ionizing dose, the efficiency of the irradiation process is expected. Scientifically, the energy to be used for the SIT was estimated at 10 MeV for electrons and less than 5MeV for X-rays or gamma rays according to the epidemiological and experimental studies (Bakri et al., 2005a).

In the sterile insect technique, the use of a protective shield of lead and other suitable materials with a high atomic number enables this technique to sterile millions of larvae by exposing them to the right doses of gamma rays from the self-contained irradiations device. The radiations that are used for sterilization process are generated from either radioisotope Cesium-137 that is able to generate the energy of up to 0.66 MeV or Cobalt-60 that is able to generate gamma rays with photon energies of up to 1.33 MeV or electron beam that are generated by accelerators with less than 10MeV without the direct input of any radioactive material. While, for the X-rays that are generated from electron beams are similar to gamma radiations and emits energy below 5 MeV (Bakri et al., 2005a).

The canisters holding insects are made from aluminium or steel or plastic that will be lowered automatically from the loading position to the irradiation position. The canister then rotates around the gamma radiation source to facilitate the uniform distribution of the dose radiations (that was mathematically calculated based on quantitative distributions models) into all targeted insects. Moreover, the canister is strategically located in a shielded position in the irradiation chamber, and preselected sterility dose is provided by setting time and specific effective absorbed doses that are usually ranging from an average of 5–450 Gy depending on the type of insects, relative biological effectiveness, penetrability, safety, cost, and time. For instance, in the case for Acari insects 30–280 Gy is applied, 10–180 Gy for Hemiptera insects; from 130 to 400 Gy for Lepidoptera insects (Bakri et al., 2005b). In most cases during SIT research, specific doses are accurately given to the insects, and then the effect is measured. Notably, not always high doses are recommended to increase sterility because it consequently reducing insect quality and competitiveness. Therefore, it essential to select a specific dose that gives the required effect.

After exposing the insects to specific radiation doses, they become sterile due to the fact that their organisms have differentiated and undifferentiated cells that are radiation-sensitive stem cells and mitotically active cells causing mutation and damages. After the process, the insects become reproductively sterile as a result of lethal mutations that occur due to the damage caused by their genetic cells. The sterile insects are later released into the atmosphere of the infected area through the use of drones and aircraft where they mate with the females leading to the inhibition of mitosis and later the death of the fertilized eggs (Fig. 5.1). It is worth noting that this technique is friendly to the environment and has not potential effect on plants and humans as opposed to the mutagenic chemicals that were earlier tested in the mid-19th century as substitutes of radiation to facilitate the process of inducing sterility in the insects. The important fact is that insects have become resistant to chemosterilants. Scientifically and laboratory, the chemical method is proved to be ineffective since most of the chemosterilants are carcinogens and thereby posing a serious risk to workers' safety, the integrity of the ecological food chain, human health, and the environment.

5.3 Soil erosion tracing technology

The slow and persistent erosion of the topsoil is among the significant issues that cause global alarm concerning the future of the agriculture economy. Flooding, rain, wind, and continuous inappropriate tilling are the leading causes of soil erosion and the depletion of its productive soil resources. According to FAO (2019), globally about 2000 million hectares of arable and grazing lands are eroded annually. More than 55% of this damage is caused by rainwater erosion cause while 33% of damage is caused by wind erosion. In this sense, it raises the urgent need for developing soil conservation strategies and monitoring programs that can aid in the assessment of their spatial redistribution and redistribution rates to determine its agricultural feasibility for farmers, which is integral in assisting them in saving their effort, time, and money.

The use of nuclear and isotopic technology has contributed significantly to the outperformance of conventional erosion measurement strategies such as the use of hydrological methods, erosion plots as well as geodetic and volumetric methods that often takes too much of the time and capital and labor-consuming. Using Caesium-137 isotope erosion tracer (^{137} Cs) makes it quicker and more comfortable for farmers to accurately evaluate soil erosion. The ^{137}Cs is a human-induced environmental radionuclide that is usually emitted into the atmosphere due to nuclear power plants'

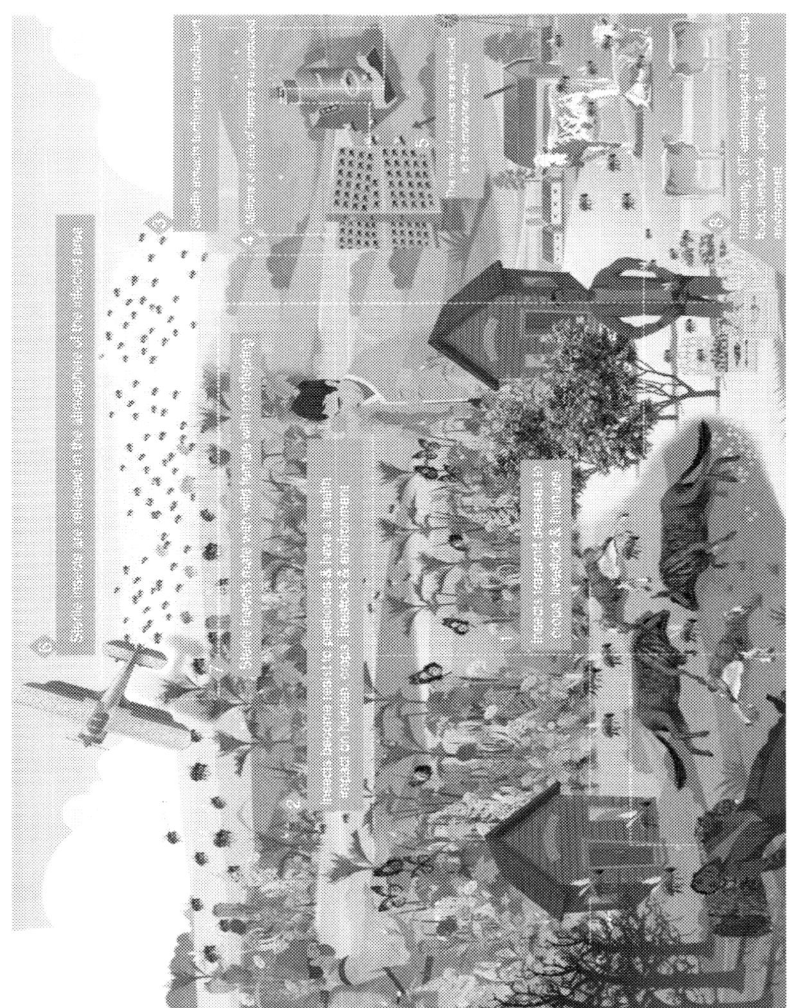

FIGURE 5.1 Sterile insect technique.

accidents and from the explosion of nuclear bombs, as well as the spread of nuclear weapons tests (Clark, 2018). Gradually, the ^{137}Cs are deposited into the land surface by rains and precipitated up to depths of 10–20 cm into the soil according to FAO and IAEA (2017) but it may extend to 40 cm (Fig. 5.2). ^{137}Cs later is coagulated with the colloids in the soil and transferred to other areas through rains, plant uptake, or leaching. Therefore, soil erosion is considered as the main responsible agent for the redistribution of the ^{137}Cs inventories that move alongside the particles of the soil.

During soil erosion evaluation, a reference site of a stable and undisturbed with Caesium-137 is used. A reference site referred to a regular site in which neither sedimentation nor erosion occurs to ensure that ^{137}Cs represent the initial fallout initially reduced through radioactive decay. The principle integrated into this method relates to the comparison of the reference site with the studied site based on the mathematical quantitative analysis models that track the change in the differentiated ^{137}Cs inventories into sedimentation rates and soil erosion. Thus, the studied site can only be considered eroded if the Caesium-137 content in the site under study is lower than the reference site. However, sedimentation can affect the studied site in case the content of the Caesium-137 in the site under study is higher compared to the reference site. Caesium-137 content can be measured either in situ or in the lab through sample collection, which must be dried, weighed, and crushed. Laboratory gamma spectroscopy comprising of a higher purity germanium semiconductor detector (HPGe detector) connected to a multichannel analyzer and equipped with an amplifier and computer with a data assessment software, which is used in the analysis process of the ^{137}Cs content in the soil.

Gamma spectroscopy can easily measure and detects ^{137}Cs from another isotope because it gives firm peaks at 662 KeV energy that is distinguishable in the gamma-ray energy spectrum (FAO and IAEA, 2017). Data processing and interpretation occur following the collection of data regarding the calculated soil erosion rates and ^{137}Cs inventories. The resulting data can be utilized in the estimation of the impact of crop rotation and land management in soil erosion; characterization of erosion over a range of land uses and environments; and in the assessment of the efficiency of various soil conservation measures. In short, this technique has proven that it is a practical approach to interpret the soil samples to aid the assessment of soil redistribution processes.

5.4 Study of plant nutrition through radioisotopic technology

The world has witnessed an exponential increase in population growth; this growth has been accompanied by a sharp increase in food demand. To

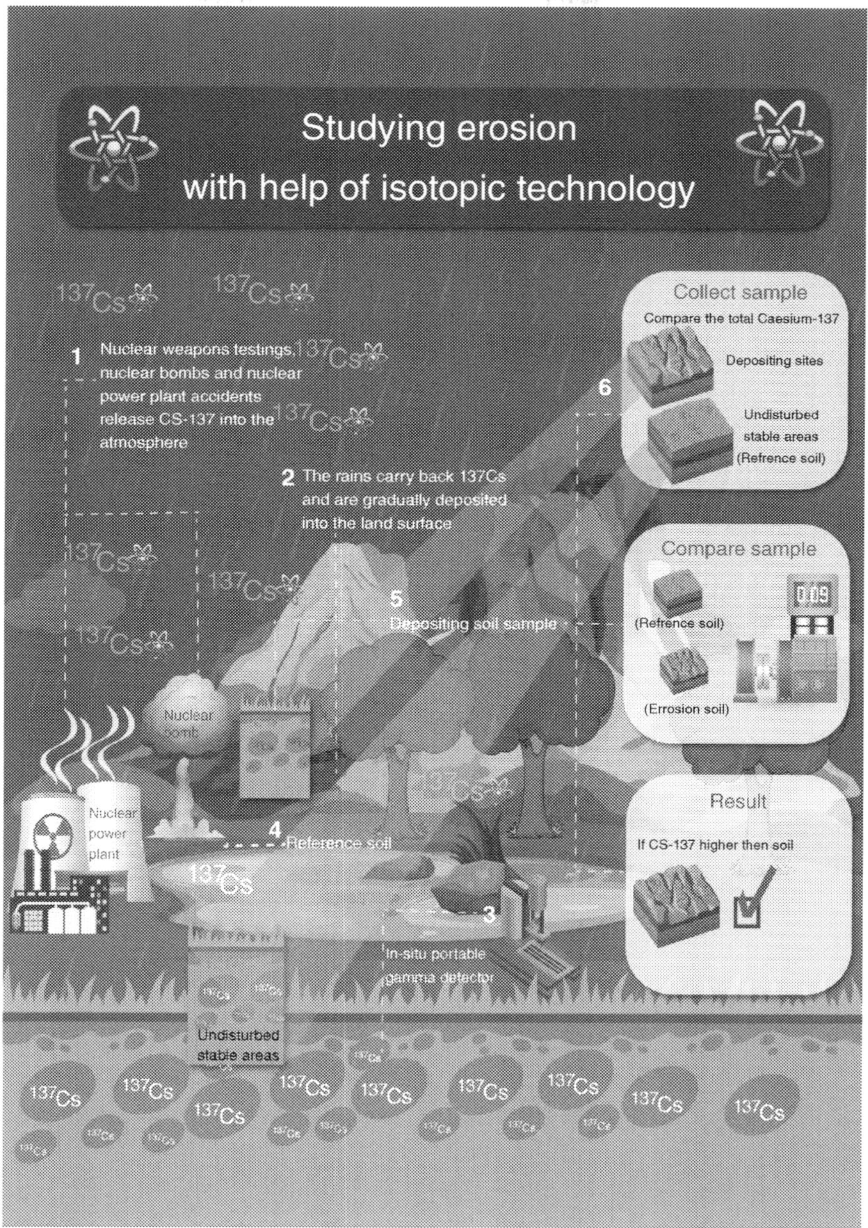

FIGURE 5.2 Soil erosion evaluation using nuclear isotopic Caesium-137 technology.

supply enough food for the entire growing world population, the demand for fertilizers continues to rise to achieve maximum crop production, which is becoming more expensive. Plant production is directly dependent on the proportion of water and nutrients absorbed by the plant because the plants need water, sunlight, and air, and soil nutrients of nitrogen, potassium, phosphorus, and others to grow and provide high yield production. In this sense, the application of radioisotope technology can assist the agricultural sector in the labeling of various fertilizers through the attachment of radioactive tracers in known varieties and quantities. This will enable farmers to directly explore the associated nutrients in the labeled products to determine critical absorption rates within the plants. Thus, the technology can be essential in the reduction of the amount of fertilizer needed to achieve a robust production hence minimizing costs and environmental damage which might affect the production. Not only that, but isotopic technology also helps farmers to add the required amount of fertilizers and water through drip irrigation (fertigation) that saves water, fertilizer quantities.

Moreover, Nitrogen-15 isotopes are also used to determine the amount of nitrogen from the atmosphere through biological nitrogen fixation in some crops such as legume crops. While the carbon-13 isotope helps to determine the amount of crop residue to stabilize soil and improve fertility. Furthermore, Carbon-13 isotope can also be used to trace and determine the exact food consumption and absorption for the animal through the evaluation of Carbon-13 isotope with their food waste. This technique can also assess the effects of conservation measures, such as the integration of crop residues on soil moisture and soil quality. This information enables the identification of the origin of the different crop species and their relative contribution to soil organic matter.

Plant during its early growth extends its roots to absorb other nutrients. Nitrogen and phosphorus are very essential elements for plant growth and production. According to FAO (2017), the total world demand for nonfertilizer nutrients including nitrogen is estimated at 45 million tonnes annually. Nitrogen annual demand itself is about 35 million tonnes of the total world demand. Although nitrogen is an essential element in all agricultural systems, it should be used appropriately since the excessive application of the product can result in land, air, and water pollution, consequently, causing global warming and climate change.

Similarly, maintaining the ability to keep its availability in the soil at recommended levels aids in keeping the environment safe and soil more agricultural fertile. Nitrogen-15 and phosphorus-32 isotopes are commonly used to track the movements of nitrogen and phosphorus from fertilizers or from soils, or from water and in the crops themselves. Isotopic technology provides quantitative data on the efficiency of use, movement and residual effects, and conversion of these fertilizers, water, and soil quality. Such information is valuable for smart agriculture in designing improved fertilizer application strategies. Thus, improving production, saving money, effort, and time. Scientifically, scientists consider Nitrogen-15 that is used widely as the most convenient nitrogen tracer since it is the most stable isotope that can easily be measured in plants, soil, air, and water in both natural and self-regulated environments. Notably, Nitrogen-15 accounts for 0.37% of the total nitrogen in the atmosphere compared to Nitrogen -14 that its abundance in the atmosphere is 99.63% with a high composition (IAEA, 1983).

Isotopes technology provides the only direct method for measuring the amount or proportion of a given nutrient in a fertilizer that is taken up by a plant. In this case, they can be used to determine, for instance, the efficiency with which plants take different forms of nutrient from different fertilizers, such as nitrogen in urea CH_4N_2O versus that in ammonium sulfate $(NH_4)_2SO_4$, or how different fertilizers and behavioral practices affect the uptake of a given nutrient. The fertilizer must be labeled with an isotope of the element of interest, but the labeling should not affect the physical and chemical characteristics of this fertilizer. Henceforth, both the isotopically labeled product and the nonlabeled fertilizer would behave identically concerning their availability for plant uptake. The extent to which the isotope is detected in the plant sample gives a measure of the proportion of the nutrient taken up from the fertilizer. The plant and soil samples labeled with ^{15}N are examined for total N and the ^{15}N:^{14}N ratio or ^{15}N abundance where samples are collected and codified by treatment and replication (Fig. 5.3). The ^{15}N abundance or the stable N isotope ratio, ^{15}N:^{14}N is determined in N_2 gas generated from the samples, by either optical emission spectrometry or either mass spectrometry. The procedures for the chemical and isotopic measurement techniques such as mass spectrometry or optical emission spectrometry were described in detail by Pruden et al. (1985), Buresh et al., (1982), and IAEA manuals (2001).

According to IAEA (1983), the effective performance of a nitrogen fertilizer depends on the fraction or proportion of a particular nutrient derived from an applied isotopically labeled fertilizer, which is computed according to the equations below (5.1 – 5.5):

$$\text{Specific Activity of Standard} = \left[\left(\text{Net counts per unit time in the aliquot of fertilizer standard}\right)/\left(\text{Total nutrient in the aliquot } (\mu g)\right]\right. \tag{5.1}$$

$$\%\text{ of nutrient derived from fertilizer } (\%\text{ Ndff}) = \left(\text{Specific activity of plant / specific activity of fertilizer}\right) \times 100 \tag{5.2}$$

$$\%\text{ of Nutrient derived from soil } (\%\text{ Ndfs}) = 100 - \%\text{ Ndff} \tag{5.3}$$

$$\text{Fertilizer nutrient uptake} = (\%\text{ Ndff}/\,100) \times \text{Total nutrient uptake} \tag{5.4}$$

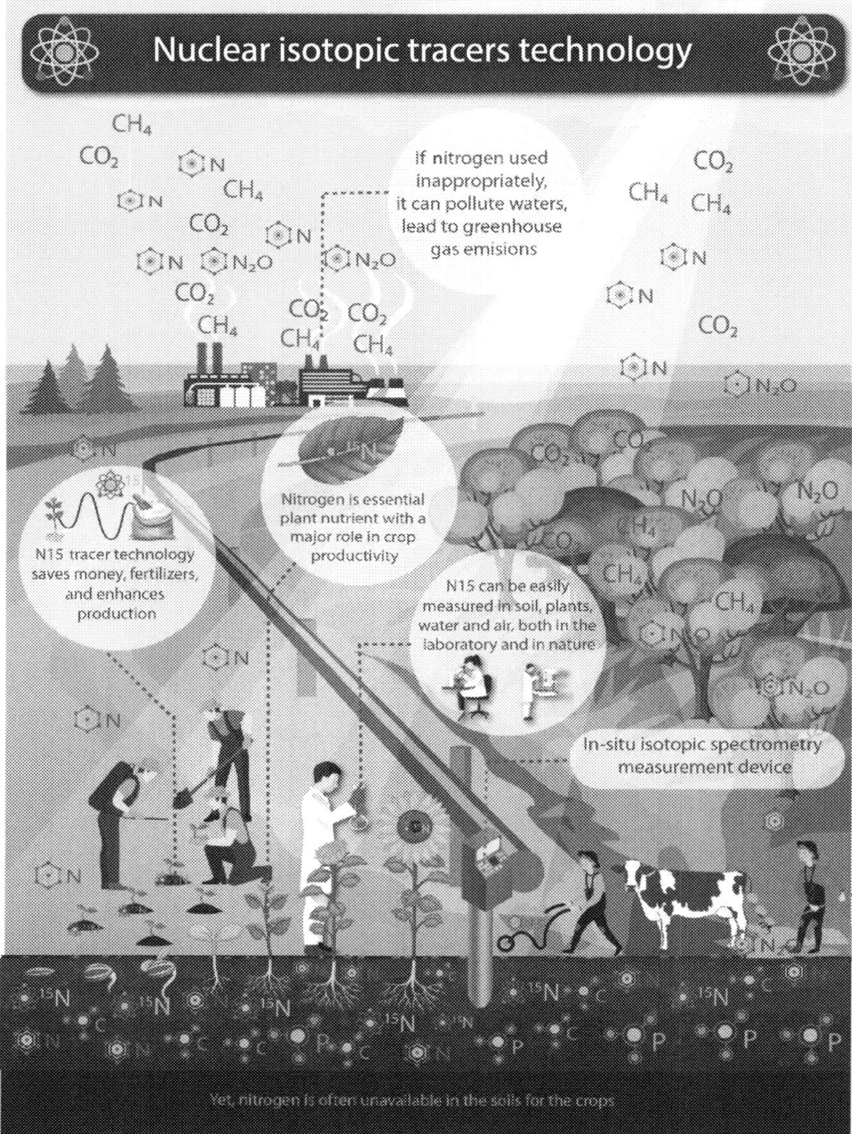

FIGURE 5.3 Labeled isotopes are used to track nutrient uptake from fertilizer, soil, or water in crops.

$$\begin{aligned}&\%\,\text{Utilization of Fertilizer nutrient (NUE)}\\&\quad=\left[(\%\,\text{Ndff}\times\text{Total nutrient uptake (kg/ha))}/\right.\\&\quad\left.(\text{rate of fertilizer application (kg/ha))}\right]\times 100\end{aligned}\tag{5.5}$$

For example, if total Phosphor (P) uptake = 65 kg/ha and % PdfF = 30%, then the proportion of P in the plant obtained from fertilizer based on the fourth equation is:

$$0.3\times 65\,\text{kg/ha}=19.5\,\text{kg P/ha}$$

However, if the plant-derived P from the soil too then, P is derived from soil = 65 − 19.5 = 45.5 kg/ha. The same formula and steps can be followed in the calculation of the amount for N or any other nutrient in the place of the specific activity used with P or other significant radioisotopes excluding the circumstance where ^{15}N is at excess shown by equation 5.6 below:

$$\begin{aligned}&\%\,\text{of nutrient derived from fertilizer (\%Ndff)}\\&\quad=\left(\%\,^{15}\text{N at. excess in plant}/\%\,^{15}\text{N at. excess in fertilizer}\right)\times 100\end{aligned}\tag{5.6}$$

Various methods have been introduced to aid in tracing the rate of nitrogen fertilizers in the environment and soil, and necessary steps have been taken to promote environmentally sustainable agricultural practices. The quantity of nitrogen can be adjusted and minimized to meet the desired levels for farmers. Toward that end, soil fertility can now be increased and reduced based on the application of nitrogen tracer to regulate how crops capture this nutrient from the atmosphere and the soil. Therefore, the application of Nitrogen-15 as a tracer element continues to play a critical role in minimizing the negative impact of nitrogen and maintaining food security across the world.

5.5 Breeding genome mapping (DNA marker) technology based on nuclear mutation

Plant and animal breeding technology that is based on genome mapping mutation—DNA Marker is a critical advancement in the agricultural sector as it is challenging to amount to sustainable yields while depending on conventional methods alone to meet the huge food demands of the accelerated global population. Mutation induction using nuclear radiation is defined as the procedure of exposing plants or animals to artificial radiation in an attempt to come up with desirable new genes to breed with other plants or animals. This mechanism assists in the production of enhanced crops and animals' productivity as well as new varieties that can withstand adverse weather conditions, mature at a faster rate, in addition to improving client acceptance, countries economy. According to the available literature, the history of the mutation technology dated back to 1928 whereby scientists realized mutagenesis which involved the reposition of chromosome traits and was first applied in the treatment of cereals by Sweden scientists Nilsson-Ehle and Gustafsson (Van Harten, 1998). Shortly after that, geneticists familiarized themselves with the radiation technology where they acknowledged as the most efficient and effective way to change plants' or animals' genes. Gamma rays and X-rays, and fast and thermal neutrons are the most applied technologies for altering crops' or animals' cell characteristics. Since then, many countries around the world have invested in developing these techniques to enhance species based on genome mapping by combining required DNA markers. For instance, in relation to plant mutation, nearly 27% of Chinese crops were developed using radiation techniques. About 11.5% of the Indian crops were also developed using radiation technology. On the other hand, about 5%–10% of Russian, Dutch, American, and Japanese crops have been grown using radiation technology (Waltar, 2003).

The genetic mutation by nuclear induction technology is an essential tool for any economy as part of sustainable development. For example, many countries have a lack of rainfall or drought or humidity or cold or heat or flooding, which all hinder the production of sufficient food to meet the needs of their growing populations. Nevertheless, other countries have high populations with special feeding programs that require high amounts of certain foods, which ultimately lead to shortages due to more consumption. The implications of such a situation are starving and a decrease in productivity thus negatively affecting the economy. Accordingly, the application of genetic mutation by nuclear induction technology helps to secure food demands. According to Waltar (2003), the production of more than 50% of the wheat used for pizza consumed in Italy and rice that is consumed by more than 50% of the global population are direct results of this technology. According to the 2015 data of the Food and Agriculture Organization and International Atomic Energy Agency (FAO & IAEA), over 3000 mutant varieties including wheat, rice, sunflower, soybean, tomato, and tobacco were subjected to mutation breeding programs in over 70 countries (Çelik and Atak, 2017).

The scientific fundamentals of the genetic mutation by nuclear induction technology are inclusive of the fact that breeders prefer physical mutagens due to their convenience, fast reproduction, and environment friendliness. The most prevalent physical mutagen is IR that is characterized depending on the nature of the particles or electromagnetic waves including gamma rays, X-rays, alpha-particles, beta-particles, and neutrons that lead to different ionization and biological applicability effects. IR can be either be direct or indirect that induce changes or aberrations or damage in DNA base molecules and eventually result in mutation. Direct radiation involves the interplay of atoms leading to free radical fabrication that makes changes in DNA molecules. On the other hand, in indirect radiation, the process of radiolysis forms free radicals that react with the target to produce brand radicles. The various types of IR display different capabilities making their application and programs diverse. IR leads to the change in structure and purpose of the DNA particle which translates to macroscopic phenotyping changes. As a result, gamma radiations are more preferred than chemical mutagens.

Gamma Radiation is the most preferred type of IR that is prevalent in biological, medical applications, and induced mutations in breeding studies. For radiation mutation to work, the source of the mutated cell is a critical factor that must be established, and usually the source either ^{137}Cs or ^{60}Co. The reason for this that some cells such as the somatic bays are hard to trace which makes it difficult to pass its traits to their offspring. On the contrary, embryonic cells are the best as they transfer their characteristics to the future generations. Furthermore, spontaneous mutations are random and free from human man interference with relatively moderate frequencies. However, mutagens provide an alternative to random mutations as they lead to the formation of preferred traits that are currently absent or were lost during evolution.

For plant mutation technology, the ratio of mutation per locus of mutant plant per M_2 generation that is known by mutation rate (mutation frequency) is a very important factor in achieving more success. It changes due to per dose and mutagen. Thus, the determination of the best dose for inducing mutants is more important rather than its type. According to Çelik and Atak (2017), it has been concluded that efficient doses in mutation breeding programs range between LD_{50} and LD_{30} (doses lead to 50% and 30% lethality). In this context, according to joint studies conducted between IAEA and FAO (1977), it has been reported in their "Manual on Mutation Breeding" as well as by Çelik and Atak (2017) that the average doses for crop species for rice, for instance, range from 120 to 250 Gy, while for common wheat (*Triticum aestivum*) and Barley (*Hordeum vulgare*) range from 30 to 70 Gy; while for the soybean (Glycine max) range from 100 to 200 Gy; while for, beans (*Phaseolus vulgaris*) range from 80 to 150 Gy; for tobacco (*Nicotiana tabacum*) range from 200 to 350 Gy, and for Alfalfa (*Medicago sativa*) range from 400 to 600 Gy. The selection of the desired mutants' characteristics should be in the second and third generations (M_2 and M_3) that will comprise of homozygotes mutant alleles. However, it must not be from M_1 because are not heterozygous plants and only one allele is affected by the mutation. Therefore, it impossible that mutation can be distinguished at M_1 (Çelik and Atak, 2017).

For plant mutation technology, for instance, various steps and procedures ought to be followed, as described in Fig. 5.4. Firstly, breeders must

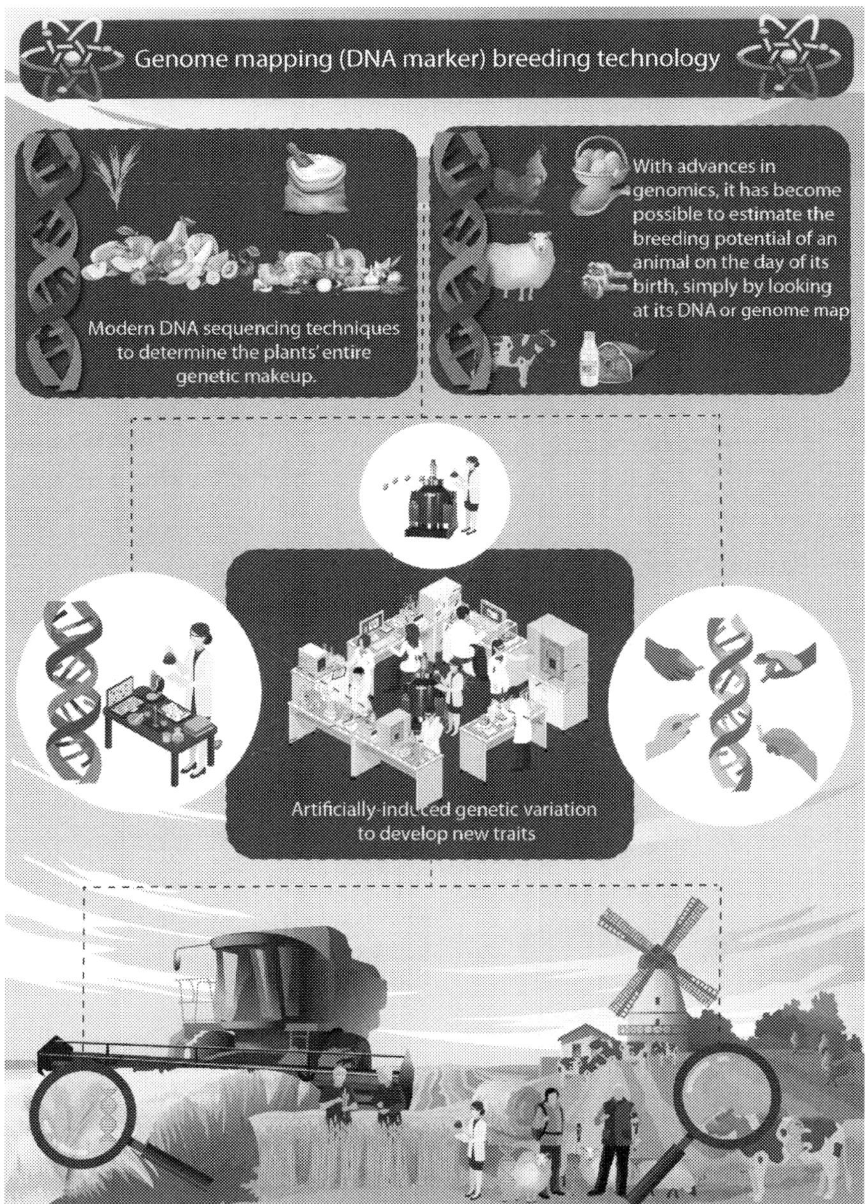

FIGURE 5.4 Breeding technology based on genome mapping (DNA marker) and nuclear mutation.

point out the species they intend to mutate. Secondly, plant breeders must determine the mutation frequency which is the ratio of mutant products per the second generation. Since plants have different degrees of sensitivity depending on species or environments, a preparatory experiment is necessary to outline the best dose according to the required environmental conditions (biotic/abiotic stress tolerance). The significance of this process is to identify the most appropriate dosage to meet the required environmental conditions to achieve excellent results with desired features. The third step involves the selection of suitable mutants from the second as well as the third generation. In this stage, geneticists will choose the mutants that are treatable with the most mutation rate from the heterozygous plant. The methods of choice vary from physical to phenotypic in addition to mechanical ways. Then mutants are assessed for the required features and environment (biotic/abiotic stress tolerance) to establish a mutant line by adding certain agents that are responsible for certain tolerance—for instance, adding a polyethylene glycol agent for drought tolerance.

After establishing a mutant line, plant breeders multiply the seeds to form a new product possessing enhanced properties in comparison with the original plant. The next stage is referred to as the mutational genomic evaluation step whereby breeders can tell the difference between plants produced through the technology. Furthermore, geneticists are able to determine the molecular source of crop stress response which assists in the realization of the similarities as well as the characterization of the differences between mutants using their DNAs. The methods used in this process consist of Random Amplified Polymorphism DNA (RAPD), intersimple sequence repeat, and Amplified Fragment Length Polymorphism (AFLP) (Çelik and Atak, 2017). RAPD is the most applied mode in the identification of genetic variations in breeding programs by the use of arbitrary primers. On the other hand, intersimple sequence repeat mode is more reliable, inexpensive, and informative than the RAPD due to its ability to amplify various fragments and provide supplementary data.

On the other hand, for animal breeding technology, the genetic mutation using nuclear induction technology plays a significant role in improving animals' productivity. Traditionally, farmers used primitive methods to increase local livestock productivity by hybridizing them with highly productive foreign breeds, mostly from temperate to tropical countries. This method was not successful or effective because these breeds were more susceptible to diseases than local animals because of the difficulty of adapting to local environments as well as it is a very slow process that takes many years to find the best potential for breeding. However, with the advances in nuclear science and genomics, this process becomes faster and can be decided from the first day of the animal's birth through the genome map, which helps to understand and identify the location of the desired characteristics for breeding such as milk, meat eggs production simply by looking at its DNA markers (chromosomes). Genome maps can also be used in many different applications in addition to selecting the optimum required characteristics such as early detection of human, animal, and plant diseases and early drug making.

Radioimmunoassay of hormones uses radiolabeled molecules such as iodine-125 of immune complexes that are used to measure the level of substances, usually measuring antigen levels such as hormone levels in blood or milk by the use of antibodies for improving outcomes of artificial insemination services. The (^{60}Co) source is used to generate Irradiation that is used to irradiate some cells from the studied animal. The purpose of this process is to randomly break up the DNA into short strands that will be later transplanted into rodent cells to reproduce more strands and generate more radiation hybrid maps panels. Radiation hybrid maps panels are used later for mapping the genomes of several DNA markers on to each of the broken DNA pieces that will form together with the genome map, which will contain thousands of DNA markers of the livestock and correctly organize genes and genetic markers along chromosomes (Fig. 5.4). Eventually, a genome map is used to select the best animals with a specific criterion, such as milk production or potential breeding. Moreover, during the early pregnancy diagnosis in animals, a radioisotope-based detection labeled nucleotides of biomarkers such as phosphorus-33, or sulfur-35 are used to validate the developed genome map with the selected animal for the potential breeding.

A number of IAEA member states started to adopt nuclear-induced genetic mutation technology due to its capability of enhancing the quality of crops and animals' productivity and admitted its success. Additionally, breeders currently prefer this method compared to conventional breeding methods. The primary reason is the production of new generations with better yields and characteristics. However, more research is still needed to be carried out by geneticists and nuclear scientists to gain more information to improve the production of a wide range of crops. Thus, more explanatory and comparative experiments should be conducted around the world to achieve this aim and to transfer the know-how of this technology to the developing countries.

5.6 Food irradiation technique

In light of the excessive use of chemicals in agricultural and animal production processes and poor storage conditions, food safety has become a growing public health concern for all countries of the world due to the high percentage of human-borne pathogens that are usually transferred to human through the food poisoning, the foodborne illness caused by bacteria and viruses including *E. coli*, Listeria, and Salmonella. It should be noted here that the volume of losses in agricultural and food products does not correspond to the steady increase in food demand due to the rise in the population, which constitutes an obsession that threatens food security. Losses in agricultural and food products usually caused by microbes, insects, and pests, which have become a substantial economic burden and a significant source of losses.

Food irradiation has succeeded in responding to this dilemma and, in particular, to the growing concerns about food-related diseases and food losses. The food irradiation technique is not a new technology. The history of food irradiation technology dates back to the early twentieth century. The first American and British patent to use IR to kill bacteria in foods was issued in 1905. Food irradiation gained significant momentum in 1947 when scholars discovered that meat and other foods could be sterilized with high energy. Food irradiation technique has been recommended to be a good choice to keep food healthy and fresh for military forces in the field. Accordingly, in the early 1950s, the US military laboratory began a series of experiments on fruits, vegetables, dairy products, fish, and meat (Bruyn,1997).

Furthermore, in 1958, the US Food and Drug Administration (FDA) approved food irradiation including wheat, potatoes, ham, spices, poultry, fruits, vegetables, and red meat. In 1962 the United Nations recognized food irradiation and established a Joint Committee of Experts on Food Radiation. Bruyn (1997) argued that in 1980, the committee concluded that "irradiating foods up to a ten kGy dose does not cause any special food or microbiological problems." In the year 1986 the FDA has permitted for expanded use of radiation in the US food supply. According to Bruyn (1997), the food irradiation technique was a part of the space program's success. In 1995 the FDA approved to irradiate frozen meals that are used by NASA astronauts with a dose of 44 kGy. Accordingly, Astronauts in AstroAmerica, in Apollo on the moon, and on the joint US-Soviet space flight, Apollo-Soyuz started drinking irradiated water and eating irradiated food such as roast beef, pork, smoked turkey, and other foods that are preserved fresh for long periods and free from microbes. The food irradiation technique enhanced the health of astronauts to perform their missions to the fullest.

Today food irradiation technology has been used in many countries of the world, where food is exposed to a specific dose from gamma rays or X-rays, or electron beam for a particular period of time to eliminate insects, bacteria, microbial contamination, and parasites in grains, dates,

FIGURE 5.5 "Radura" is the internationally agreed slogan that indicates food product has been irradiated (Food irradiation, https://en.wikipedia.org/wiki/Food_irradiation#/media/File:Radura_international.svg).

and legumes, spices, red meat, poultry, and fish, and thus prolong their shelf life. This technique is also used to stop the germination process in potatoes, onions, and garlic, to ensure their quality and safety.

It should be noted here that irradiated food has become available in wide markets along with nonirradiated counterparts. As a result of nuclear awareness, a lot of consumers prefer to prefer the purchase of irradiated food due to its quality, safety, and easy to distinguish through the "Radura" logo shown in Fig. 5.5. Radura is an internationally agreed slogan that indicates that the food product has been treated with irradiation.

Food irradiation technology has been approved by the US FDA that considers food irradiation technology to be safe, and that has been adopted by many international organizations, associations, and institutes, such as the World Health Organization (WHO) and the Food and Agriculture Organization because irradiation is better than traditional methods used to preserve food such as heating or chemical treatment or freezing. Radiation does not involve any raise in the food temperature, nor does it affect the texture and taste of the food. Irradiation does not leave harmful residues such as chemical treatment, for example, the use of ethyl and methyl bromide to sterilize grains and nuts from Insects, which are harmful to the public health and the environment. Accordingly, food irradiation technology supports the national economy in terms of import and export. In sum, the food irradiation technique using nuclear techniques contributes successfully to achieving food security. In this context, the WHO and the

Food and Agriculture Organization of the United Nations stated that about 33 countries have declared the use of irradiated food technology for human consumption, which are listed in Appendix 1 (World Health Organization, 1988).

The basic working principle of the irradiation food facility is that during the irradiation process, food is exposed to a particular form of IR. Studies and experiments proved that IR can kill spoilage organisms, including bacteria, molds, and yeasts responsible for food spoilage as well as destroys disease-causing organisms, including parasitic worms and insects pests, that damage food in storage. The radiations source used for the food irradiation process are either:

1. Gamma rays are produced from radioactive material such as Cobalt-60 or Cesium-137, which continuously produce high-energy gamma rays. They have the ability to penetrate foodstuff to a deep extent. In this context, Akinloye et al. (2015) argue that food irradiation using Cobalt-60 sources has some advantages over Cesium-137, such as its ability to penetrate deeply, uniformity of the dose distribution over the food product, and has low risk to the environment.
2. X-rays are produced from an electron accelerator. They have the ability to focus on small areas of food.
3. Electron beams are produced from an electron accelerator. They and can penetrate to a thickness of 1–2 inches.

During the food irradiation process, food is exposed to a precise and specific absorbing dose by taking into account the energy output of the source per unit of time and the spatial relationship between the source and the target. The radiation doses ranging from 50 Gy to 10 KGy depending on the food type and the desired effect. Table 5.1 shows the recommended doses by the World Health Organization (1988) for food irradiation.

Food irradiation plants usually classified into two types: either batch or continuous facilities. Fig. 5.6 depicts the modern type of multipurpose irradiation plant that can work as either batch or continuous plants and able to treat different types of products, including food, medical equipment, industrial products. In a batch facility, certain quantities of food are irradiated in batches for a precise dose for a certain period. When the first batch of food is irradiated is then unloaded, and another batch is loaded and irradiated. The main advantages of batch plant facilities are low cost, simple design, easy to operate, and flexibility. While in the continuous facilities, large quantities of a single food can be irradiated at a given dose in one go. The main advantages of continuous facilities are more economical, faster, able to treat large quantities of food in less time. Once the irradiation process, which is done by using a gamma-ray source,

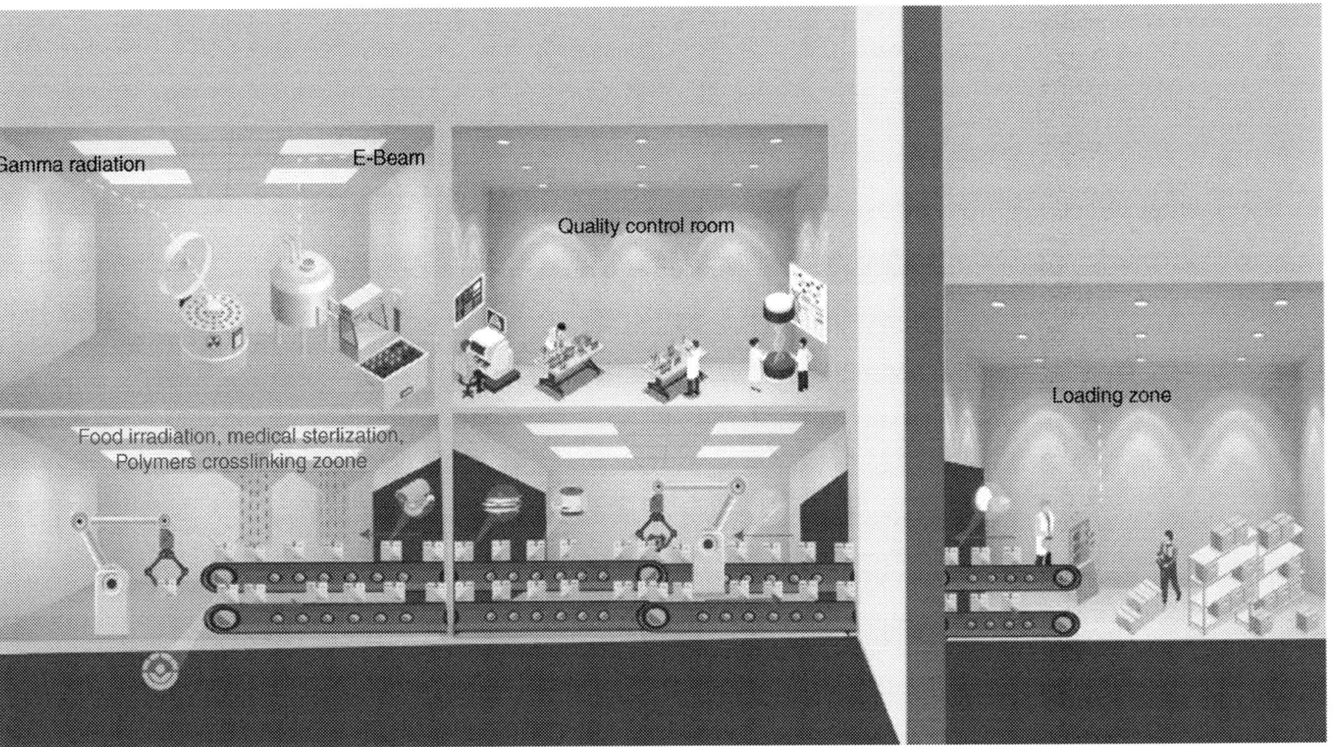

FIGURE 5.6 Multipurpose irradiation plants.

TABLE 5.1 Recommended doses by the World Health Organization and the Food and Agriculture for food irradiation.

Dose level (kGy)	Food type	Purpose
0.05–0.15	Potatoes, onions, garlic, gingerroot, etc.	Inhibition of sprouting
0.15–0.50	Cereals and pulses, fresh and dried fruits, dried fish and meat, fresh pork, etc.	Insect disinfestation and parasite disinfection Delay of physiological process (e.g., ripening)
0.50–1.0	Fresh fruits and vegetables	Delay of physiological process (e.g., ripening)
1.0–3.0	Fresh fish, strawberries, etc.	Extension of shelf-life
1.0–7.0	Fresh and frozen seafood, raw or frozen poultry and meat, etc.	Elimination of spoilage and pathogenic microorganisms
2.0–7.0	Grapes (increasing juice yield), dehydrated vegetables (reduced cooking time), etc.	Improving technological properties of food
30–50	Meat, poultry, seafood, prepared foods, sterilized hospital diets	Industrial sterilization (in combination with mild heat)
10–50	Spices, enzyme preparations, natural gum, etc.	Decontamination of certain food additives and ingredients

is complete, the radioactive bars should be submerged in an underground water pool. But if the irradiation process was done by using the source of the electron beams or X-rays, then the electron source is locked from the electricity.

5.7 Conclusion

The birth of the peaceful nuclear age came more than half a century ago as a result of The Treaty on the Non-Proliferation of Nuclear Weapons (NPT), which established a framework for increasing access to the peaceful uses of nuclear energy. Under Articles III and IV of the NPT, the IAEA assists States by their NPT obligations to adopt nuclear and theoretical technology in a wide range of peaceful nuclear applications and harness them to serve some sectors of sustainable 3D development (economic, social, and environmental). Accordingly, a lot of studies, efforts, and experiments have been made to harness the atom for peace and development. Since then, the world has witnessed a shift in peaceful nuclear energy applications in different sustainable development, which has grown exponentially. It includes but not limited to nuclear power, health, efficient plant and animal production, pest control, climate change reduction, improved water and soil quality, desertification control, wildlife improvement, sustainable environmental protection, nuclear forensic science in crime investigations, as well as many industrial applications such as food irradiation to keep it fresh, oil spill tracking, characterization of the oil reservoir in the geological formation and increasing the efficiency of oil production, sterilization of medical instruments, food imitation control, water desalination, and many other applications.

The dilemma of food security and fighting hunger is among the global challenges due to global population growth. This problem arises especially in the developing world, where the population growth rate is high in light of the severe shortage of food and water resources due to the impact of climate change and global warming. Reports indicate that the most important challenges in the agricultural sector lie in the efficiency of fertilization, loss of irrigation water, optimal absorption of nutrients, soil quality, and associated erosion problems.

This chapter has cast light on the advancement of nuclear and isotopic technology in the agriculture sector that contributed to securing food in sufficient quantities and distinct types to meet this shortfall. Where nuclear scientists have been able to use the nuclear properties of labeled radioisotopes in measuring and monitoring soil quality, soil erosion, and improving the quality and abundance of agricultural and animal production.

Nuclear technology and radioisotopes have managed to tackle all the issues mentioned above related to the agriculture sector. Accordingly, it has played an essential role in measuring absorption rates, measuring soil quality and its moisture, soil erosion, and rationalizing the use of fertilizers. The technique of using the labeled nitrogen probe helps in measuring the soil moisture and therefore determining the amount of water required, rationalizing irrigation, and saving water. This process is often combined with the technique of measuring absorption rates that helps to know exactly the required quantities of water and fertilizers for the plants. Fertilizers that farmers use mostly contain nitrogen or phosphorous, which are very expensive and often used in large and random quantities, amounting to millions of tons and dollars annually. Excessive use of these fertilizers has direct and indirect effects on human health and the environment that also play a significant role in increasing the emission of greenhouse gases. On the opposite, nuclear scientists have been able to harness atomic energy and radioisotopic technology by using labeled phosphorous or labeled nitrogen radioisotopes, to determine the amount of fertilizer needed for the plants, which contributed in reducing the impact of the global warming issue as well as contributed in saving a lot of money and fertilizers.

Moreover, nuclear technology and radioisotopic technology helped in the early detection of the factors that cause disease. This helped a lot in changing the genetic structure of plants and animals through the use of nuclear mutation technology in such a way that enhances their immunity system, increases the speed of plant and animal growth, and abundance of production. This approach relies on DNA markers in establishing a genomic map, by causing genetic changes in plant or animal genes as a result of exposing them to certain radioactive doses of gamma rays or other IR.

In addition to its prominent role in protecting the environment by reducing the level of chemical pesticides to control pests. Scientists have tended to think of an environmentally clean alternative, which is the use of nuclear sterile insects technique. The basic principle of this technique is that exposing the males of that pest in the late cocoon phase to specific doses of radiation to make them sterile. This is usually achieved by using gamma rays Issued by Cesium-137 and Cobalt -60. This method has been successful in a number of countries to eliminate the Mediterranean fly and many other pests that cause significant losses in crops according to many international reports

This chapter has shed light on some important roles of peaceful nuclear applications and radioisotopic technology in providing competitive solutions to the sustainable development of agriculture and food. Often unique to help fight hunger, reduce malnutrition, improve environmental sustainability, ensure food safety and authenticity, address soil erosion and irrigation water scarcity problems, and pest control. Finally, we can conclude that nuclear applications can be an effective and low-cost solution to many of the dilemmas facing the agriculture sector in the world to improve the agricultural level, supporting national economies, and supporting global food security. Read more about other essential applications of the peaceful nuclear and isotopic technology that plays a vital role in some other development sectors in the coming chapters.

References

Akinloye, M., Isola, G., Solasunkanmi, S., Okunade, D., 2015. Irradiation as a Food Preservation Method in Nigeria: Prospects and Problems, XI. IJRASET, pp. 85–96.

Bakri, A., Mehta, K., Lance, D., 2005a. Sterilizing Insects with Ionizing Radiation, In Dyck, V., Hendrichs, J., and Robinson, A. (Eds). Sterile Insect Technique: Principles and Practice in Area-Wide Integrated Pest Management. The Netherland, Springer, pp. 233–268.

Bakri, A., Heather, N., Hendrichs, J., Ferris, I., 2005b. Fifty Years of Radiation Biology in Entomology: Lessons Learned from IDIDAS. Entomological Society of America, 98

(1), 1–12. https://nucleus.iaea.org/sites/naipc/ididas/Relevant%20Library/Bakri%20 Fifty%20Years.pdf, accessed February 12, 2019.
Buresh, RJ, Austin, ER, Craswell, ET, 1982. Analytical methods in15N research. Fert. Res. (3) 37–62.
Bruyn, I., 1997. The application of high dose food irradiation: atomic energy corporation of South Africa. In: National Food Irradiation Seminar, pp. 32–40.
Çelik, Ö., Atak, Ç., 2017. Applications of ionizing radiation in mutation breeding. New Insights on Gamma Rays. InTech http://dx.doi.org/10.5772/66925.
Clark, M.C., 2018. Coexisting on Earth Homo sapiens Quagmire. Michael C. Clark, USA, pp. 1–500.
FAO and IAEA, 1977. Manual on Mutation Breeding. IAEA., Vienna, pp. 1–290.
FAO and IAEA, 2017. Use of 137Cs for Soil Erosion Assessment. Food and Agriculture Organization of the United Nations, Rome, pp. 1–64.
FAO, 2017. World Fertilizer Trends and Outlook to 2020. Food and Agriculture Organization of the United Nations, Rome, pp. 1–28.
FAO. 2019. Restoring the land. http://www.fao.org/3/u8480e/U8480E0d.htm. accessed February 17, 2019.
Gillman, H., 1992. The New World Screwworm Eradication Programme: North Africa, 1988–1992. Food and Agriculture Organization of the United Nations. Rome, Itly, pp. 1–19.
Hatzakis, E., 2019. Nuclear magnetic resonance (NMR) spectroscopy in food science: a comprehensive review. Compr. Rev. Food Sci. Food Saf. 18 (1), 189–220.
IAEA. 2001.Use of isotope and radiation methods in soil and water management and crop nutrition, IAEA Training Course Series No. 14, Vienna.
IAEA. 1983. A guide to the use of Nitrogen-15 and radioisotopes in studies of plant nutrition: calculations and interpretation of data. IAEA-TECDOC-288. International Atomic Energy Agency, Vienna. 1–65.
Jennings, C., 2014. The Deadly Air: Genetically Modified Mosquitoes and the Fight Against Malaria. Guardian Books, pp. 1–70.
Khanal, D., Munankarmy, R., 2009. Nuclear and related techniques for enhancing livestock and agriculture productivity. J. Agric. Environ. 10, 150–155.
Pruden, G., Powlson, D.S., Jenkinwson, D.S., 1985. The measurement of 15N in soil and plant material. Fert. Res. (6) 205–218.
Oatsvall, N., 2014. Atomic agriculture: policymaking, food production, and nuclear technologies in the United States, 1945–1960. Agric. Hist. 88 (3), 368–387.
Sahoo, S., Sahoo, S., 2006. Production and applications of radioisotopes. Phys. Edu. 5–11.
Van Harten, A., 1998. Mutation Breeding: Theory and Practical Applications. Cambridge University Press, Cmbridge, UK, pp. 1–353.
Waltar, A., 2003. The Medical, Agricultural, and Industrial Applications of Nuclear Technology, pp. 22–33 New Orleans, LA.
World Health Organization, 1988. Food Irradiation. A technique for preserving and improving the safety of food. World Health Organization in collaboration with the Food and Agriculture Organization of the United Nations, Geneva, pp. 1–88.

CHAPTER

6

Applications of nuclear science and radioisotope technology in advanced sciences and scientific research (space, nuclear forensics, nuclear medicine, archaeology, hydrology, etc.)

OUTLINE

Applications of Nuclear and Radioisotope Technology: The Atom for Peace and Sustainable Development
DOI: https://doi.org/10.1016/B978-0-12-821319-3.00001-4

6.1 Introduction

The birth of peaceful nuclear applications began when President Eisenhower announced on December 8, 1953, through a speech he delivered in the United Nations Council under the title of "Atoms for peace," and he stated that the atom, which divided the world would reunite the world. The US President Eisenhower aimed to end the nuclear arms race and harnessing the atom for peace. Since then, atomic scientists have intensified their research in developing nuclear and isotopic technology for peaceful applications to contribute to advancing sustainable development in several sectors including but not limited to health, agriculture, the sustainability of water resources, generation of clean energy, hydrogen economy, industry, archaeology, climatology, hydrology and geology and meteorology, and nuclear forensics.

In the healthcare sector, nuclear and isotopic technology has contributed to saving the lives of millions of patients through the provision of advanced health care that traditional medicine has been unable to provide. Modern nuclear medicine has contributed to accurate and early diagnosis of intractable and complex diseases such as cancer, heart, arteries, and brain, neurological disorders such as dementia, Alzheimer's disease, acquired immunodeficiency syndrome, and many other diseases. This is achieved through analyzing genomic mapping techniques and 3D nuclear imaging techniques (Positron emission tomography (PET), single-photon emission computed tomography (SPECT), molecular imaging) that measure the efficiency rate of Organ physiology in contrast to X-ray images that present only anatomical images. Diagnostic radiopharmaceuticals also played an important role in diagnosis, radiotherapy, improving nutrition, measuring absorption rates, predicting disease at an early stage, responding to it, and treating it. Nuclear technology is also used in the medical sector to sterilize a wide number of medical equipment and surgical instruments such as surgical scissors, gloves, bandages, kidney filters, medical ointments, blood solutions devices, catheters, and pharmaceutical raw materials using irradiation techniques, such as gamma rays or electron beam, which eliminate pathogens such as harmful bacteria, viruses, and parasites. Not only that but nuclear technology serves the medical sector in analyzing the pharmaceutical components by using mass spectrometry and thermal analyzer to identify their physical and chemical properties.

Nuclear forensics medicine plays a significant role in uncovering the circumstances of crimes and solving the mystery of many complex criminal cases such as radiological murdering, murdering using guns, and drug crimes. Nuclear irradiation technology is used in the analysis of microscopic and macroscopic evidence that is not seen with the naked eye that reveals the cause, time of death, the tool of the crime, and determining at high accuracy the existence of any criminal suspicions in the crime through irradiation of the collected evidence and tracing its sources until reaching the perpetrator. Nuclear forensics methods used in criminal investigations are nondestructive methods because the investigated sample will not be destroyed by radiation exposure compared to the conventional methods. This is very important for forensics' investigations when it is necessary to use the only available sample as a means of proof, or when the process requires additional tests on the only available evidence. Therefore, this cannot be achieved by using chemical methods that may consume the available sample during the evidence examination process, which may consume it all. Accordingly, the use of nuclear methods in crime laboratories will still be useful because it provides conclusive evidence that in many complicated cases cannot be provided using other methods due to its high sensitivity and accuracy, in addition to being a nondestructive analysis tool. Nuclear forensics medicine revealed the secret of the death of the French emperor Napoleon Bonaparte, who was killed by arsenic poisoning, as well as the cause of the death of Alexander Litvinenko, a former officer in the Russian Federal Security Service who was killed by radiological poisoning, and identification of the sniper who assassinated President John F. Kennedy in 1977 when the investigation into the assassination was reinvestigated.

Nuclear and isotopic technology provides an important achievement in preserving the history and understanding its chronology to prove or disprove many unknown theories in geology, hydrology, geophysics, climatology, oceanography, and palaeontology, all the way to biomedicine. Nuclear and isotopic technology has provided many radiometric dating methods such as carbon dating method, rubidium–strontium dating method, potassium-argon dating method (K–Ar), and many other methods to estimate the life span of organisms from tens to thousands of years based on the amount left of radioisotope materials in these organisms. Such lifetime is estimated by measuring the amount of decay of the radioisotope present in the sample and comparing it with a reference scale. The great results achieved by the radiometric dating technique enabled it to be ranked among the best discoveries of the twentieth century. No method so far has revolutionized human understanding of events that occurred thousands of years ago better than radiometric dating technology.

On the other hand, nuclear science and radioisotope technology play a significant role in the management and sustainability of water resources. Stable and radioactive isotope tracers in hydrological studies is a new technique used by scientists to evaluate the amount, the origins, the sustainability of water, and making an isotopic mapping that provides comprehensive information concerning the management of water resources, such as the aquifer size, recharge sources, interaction, and connectivity between geological formations, flow characteristics, water quality, water age, leakages through dams, storage, and potential water extraction rates. This information helps to avoid exhaustion of water sources by knowing the quantification of the running water balance, runoff, and interactions with groundwater and sources of recharge.

The new generation of nuclear technology is a giant step in humanity's future space exploration. Nuclear fission energy is the optimal option for high propulsion's missions in terms of technical maturity and energy density required for space missions. Three types of space propulsion are powered by nuclear fission energy namely nuclear electric propulsion (NEP), nuclear pulse propulsion (NPP), and nuclear thermal propulsion (NTP). Nuclear thermal-based rockets (NTRs) utilize nuclear fission reactions and works on a principle similar to nuclear power plants and propulsion ships. Nuclear scientific progress has proven that spacecraft works on nuclear fission propulsion system able to reach Mars at a great speed, ranging from 3 to 4 months, which is half the time needed by the fastest spacecraft available with chemical propulsion technology. Rapid travel will reduce the amount of radiation that astronauts may be exposed to during their travel between Earth and the red planet. The application of nuclear technology is not limited to providing NTP to the spacecraft but also supply the required electricity to power satellites orbiting around the Earth for many years and others, space equipment, and international space stations (ISSs) through converting the energy produced by the radioactive decay of plutonium or americium-241 into electrical energy using radioisotope thermoelectric generator (RTG).

6.2 Nuclear science and radioisotope technology applications in health

6.2.1 History of nuclear medicine

Society of Nuclear Medicine and Molecular Imaging—SNMMI (2020) argues that nuclear medicine has a very long history that is dated back to the end of the 19th century when Henri Becquerel first discovered mysterious rays of uranium in 1896, which Marie Curie called "radioactivity" in 1897. In 1903, Alexander Graham Bell proposed placing sources containing radium in tumors or closer to it, which help in the curing of these tumors. After nearly ten years of intensive nuclear studies and medical research, Frederick Proescher published the first study on the intravenous injection of radium for the treatment of various diseases. In 1924, a group of scientists including Georg de Hevesy, Christiansen, and Lomholt conducted experiments for using lead-210 and bismuth-210 nuclides in different animals. In 1936, Ernst's brother Lawrence made the first clinical therapeutic application of synthetic radionuclides when he used phosphorous 32 to treat leukaemia. In the same year, Joseph Gilbert and Robert Spencer used sodium 24 for the treatment of a leukaemia patient. In 1937 Hertz, Roberts, and Evans were able to study the physiology of the thyroid gland using iodine-128. Livingood, Seaborg, and Segre discovered iodine-131 and cobalt 60 and technetium 99M in 1938. According to Society of Nuclear Medicine and Molecular Imaging—SNMMI (2020), 1940 is a significant year in the history of nuclear medicine that brought a significant scientific revolution in nuclear science, where the Rockefeller Foundation funded the first cyclotron dedicated for the production of biomedical isotopes at Washington University. The first therapeutic dose of iodine-130 was given to a patient in 1941, and in 1947 Benedict Cassin used radioactive iodine to determine if thyroid nodules were accumulating iodine, which helps to distinguish between benign and malignant nodules. George Moore also used iodine-131 labeled diiodofluorescein to examine the brain's tumors. In the 1950s, nuclear imaging techniques were discovered, and in 1962 David Kuhl introduced different types of emission tomography including SPECT, PET, and CT. Nuclear medicine was officially recognized as a medical speciality by the American Medical Association in 1971. In 1978, David Goldenberg, used radiotracer antibodies to imaging tumors in humans. William Eckelman and Richard Reba performed the first successful imaging of SPECT for neurotransmitters in humans in 1983. 2001 was a historic year for nuclear medicine as a result of tremendous success made, 16.9 million nuclear medicine treatments were performed in the United States alone.

6.2.2 Nuclear medicine

There is an unrivaled use of nuclear and isotopic techniques in medicine that brought a scientific medical revolution in the early detection and treatment of incurable diseases that traditional medicine has been unable to provide. Nuclear medicine statistics in developed countries indicate that there is an increasing demand for patients for nuclear and radiological treatment, as about one person in 50 uses diagnostic nuclear medicine each year. Nuclear and radiological medicine can provide very accurate and three-dimensional information about the organ's physiological function that conventional medicine cannot provide, and is limited to anatomical imaging of the organs.

Radioisotopes have vast applications in the field of nuclear medicine. Dar et al. (2015) argue that radioisotopes have been used in radiotherapy, nuclear imaging, and diagnostic procedures. Diagnostic radiopharmaceuticals are playing a crucial role in the examination of blood flow to the brain, assessing the growth of bone, the functioning of the heart, liver, kidneys, and lungs (Dar et al., 2015). One of the most crucial uses of diagnostic radiopharmaceuticals is the prediction of the impacts of surgery and assessment of changes made by the treatments. The dose of radiopharmaceuticals that are given to a patient is sufficient to get the required info to answer all questions. Medically this radiation dose is minimal and safe. Moreover, radiation has an important advantage in nuclear medicine. It can destroy a cancerous tumor. This can be achieved by exposing the tumor area to either an external radiation source that is known medically by teletherapy or by exposing it to internal irradiation or what is known by brachytherapy, which is short-range radiotherapy.

The external radiation therapy, or what is known as teletherapy or gamma knife radiosurgery, is an effective treatment in eradicating tumors rather than removing them, as annually more than 30,000 patients worldwide are treated in this way. External radiation therapy is done by shedding a gamma beam from a radioactive cobalt 60 sources or from a high-energy source to the target area where the tumor is cancerous.

As for internal treatment with radionuclides, it is done by implanting a radioactive source or radioactive nanoparticle that is widely used in nuclear nanomedicine. Usually, a gamma-ray source or a beta emitter is used in the targeted cancerous area. Dar et al. (2015) argue that iodine-131 is widely used for treating thyroid cancer, nonmalignant thyroid disorders, and other abnormal conditions such as hyperthyroidism (an over-active thyroid), Polycythaemia vera (excess of red blood cells is produced in the bone marrow). Phosphorus-32 also provides excellent control to this excess. Lutetium 177 octreotate is also used for neuroendocrine tumors. Erbium-169, Yttrium-90, Holmium 66 are used for the treatment of non-Hodgkin's lymphoma, liver cancer, and for arthritis treatment, which all have proven their success in treating these cancers.

Moreover, Catheter-based Iridium 192 is used for the treatment of carcinoid tumors of the head, breast, and neck (Othman, 2019). While, Iodine-125 or palladium-103 is useful for the treatment of prostate cancer (Othman, 2019). Another example is lead 212 that is widely used in the treatment of pancreatic, ovarian, and skin cancers using the Neutron Capture Enhanced Particle Therapy Approach or Targeted Radionuclide Therapy, which involves injecting the patient with a neutron capture agent before it is irradiated with protons or heavy ions. Farhood et al. (2018) argue that Neutron Capture Enhanced Particle Therapy Approach is not only limited to lead 212, but it applies to boron 10 or gadolinium 157 that concentrate in malignant brain tumors. These agents are then irradiated by thermal neutrons or protons, which are highly absorbed by these agents and result in the emission of high-energy alpha particles that kill targeted cancer. In parallel, it is worth mentioning that according to the research made by Dadachova (2008), radiation immunotherapy is also a promising approach in treating HIV by using nuclides Bismuth-213.

On the other hand, Parr and Fjeld (1994) argue that nuclear isotopic techniques in medicine also help in measuring the level of fats in the body and nutrient absorption rates with high accuracy. For instance, the fat level in a body can be measured by using labeled water with deuterium (^{2}H) isotope. Drinking this labeled water help in evaluating the total water content in the body by taking samples of a human's saliva or urine before and after drinking the labeled water. Fats are naturally free of water, therefore, when drinking this labeled water, it is distributed evenly in the lean tissues of the body. This helps to determine the amount of fat-free weight by calculating the difference and therefore finding out how much fat the body stores.

6.2.3 Diagnostic and therapeutic radiopharmaceuticals

Radioisotopes have vast applications in the field of medicine. They are used for medical diagnostic or therapeutic purposes. Dar et al. (2015) argue that biologically and chemically each organ in the human body absorbs a specific type of chemical element. For instance, calcium is consumed by the bones, the thyroid gland consumes iodine, and the brain consumes amounts of glucose. This feature helps in attaching a specific and labeled radioisotope to a specific organ based on its biological activity, which will interact with the targeted organ's biological and metabolic processes. Based on this scientific fact, radiopharmaceuticals have been developed for the examination of many complex diseases such as blood flow to the brain, assessing the growth of bone, and the functionality of the heart, liver, kidneys, or lungs. Moreover, they are used to evaluate if the surgery and prescribed treatment are effective and successful. From the safety point of view, radiopharmaceuticals doses are safe because they have very low radioactivity and a very short half-life. Technetium-99 (Tc-99), for instance, is widely used in nuclear medicine. Over 80% of nuclear medicine applications rely on Tc-99 because of its ideal features to be used in medical nuclear imaging. All metabolic processes can be easily scanned by Tc-99 due to its short half-life of six hours. It decays by an isomeric process. Many radiotracers can be easily produced and easily incorporated into a wide range of biologically active substances. For instance, Tc-99 and thallium-201 chloride are used in Myocardial Profusion Imaging for the detection of the coronary artery.

On the other hand, Alauddin (2012) argues that Fluoro-deoxy glucose (FDG) that incorporates F-18 is another vital radiopharmaceutical that is widely used for PET imaging because of its low radioactivity and has a short half-life of fewer than two hours. In this process, cell metabolism is examined by incorporating FDG into the cell and identify the targeted affected cell from non-effected. One of the most complex applications of fluorodeoxyglucose is the examination of brain perfusion to measure the amount of blood taken in specific areas of the brain, which helps provide information about how the brain functions. The most important types of brain perfusion tests are single-photon emission computerized tomography (SPECT) or positron emission (PET) scanning combined with radiotracers. These techniques are crucial for clinicians to also diagnose various brain disorders such as Alzheimer's disease and other forms of dementia.

In view of the spectacular results achieved by nuclear medicine and radiopharmaceuticals in the treatment of cancer and HIV, which have puzzled scientists for nearly half a century are the strongest scientific and medical evidence of the superiority of nuclear and radioisotopes sciences in overcoming intractable diseases. In this regard, radiation immunotherapy has deservedly succeeded in killing different types of viruses and infected malignant cells permanently in such a way that it is unable to automatically rebuild its immune system compared to the antiretroviral therapy. While the radionuclide-antibodies induced by nuclear radioisotopes technology are designed to trace and identify particular antigens of certain cells at a particular position, which are then recruited by the immune system to identify and blockade required antigen whether it is a virus or bacteria.

In this context, Bismuth-213 is a good example of the radioimmunotherapy that has been designed at an extremely precise level to target exclusively HIV-feeding protein without harming any other healthy cells even if the virus has arrived blood–brain barrier within the central nervous system

(Dadachova, 2008). There are many other radionuclides used in treating a large number of diseases that ordinary medicine was unable to treat according to what was confirmed by many scientific studies.

6.2.4 Positron emission tomography & single-photon emission computed tomography—nuclear imaging

PET is a nuclear medicine tomographic imaging technique that provides three-dimensional images with the main focus on how tissues and organs physiologically work (Mieny, 2003). This is achieved by injecting or ingesting or inhaling a certain labeled radioactive tracer depending on the targeted organ or tissue to be investigated. Griffeth (2005) argues that ^{18}F-2-fluoro-2-deoxyglucose (FDG) is commonly used in PET (Fig. 6.1).

The injected radiotracer accumulates at the affected organ or tissue (disease area) that will react at higher levels of chemical activity. It is detected by the gamma camera detectors that are mounted on a rotating gantry. These detectors detect the positrons emitted by the injected radiotracer. The positron is a particle that has a mass similar to the electron but oppositely charged. The positrons emitted by the radiotracer react with electrons in the body and when they meet, they annihilate each other and release photons. These photons accumulating in the cancer cells and are shown in the PET images as luminous patches because they have a higher metabolic rate than normal cells (Griffeth, 2005). Furthermore, Griffeth (2005) argues that PET has numerous advantages over other available classical tests that are unable to reveal; the most important of which are: the ability to detect disease at a very early stage; the ability to detect whether cancer has spread or not and if so, up to what extent it has spread; the ability to check whether the cancer treatment was successful or not, the ability to detect the recurrence of cancer; the ability to detect and evaluate many complex pathological cases related to cancer, heart disease, brain disorders, Alzheimer, and seizures; and the ability to accurately detect many types of tumors, such as brain, cervical, colon and rectal tumors, breast tumor, oesophagus, lung, lymphoma, pancreas, prostate, thyroid, and heart disease. Furthermore, PET can detect areas of low blood flow in the heart as a result of clogged arteries in the heart and has the ability to examine brain perfusion. PET can be combined with other medical imaging techniques such as computerized tomography or magnetic resonance imaging for better diagnosis and evaluation of the disease (Fig. 6.2).

SPECT imaging (Fig. 6.3) is also a nuclear medicine tomographic imaging technique that offers three-dimensional (tomographic) images (Campeau and Fleitz, 2016). According to the National Institute of Biomedical Imaging and Bioengineering—NIBIB (2020), the working principle of SPECT is similar to PET the only difference between SPECT and PET scans is the type of radiotracers and the released particles. In the case of SPECT

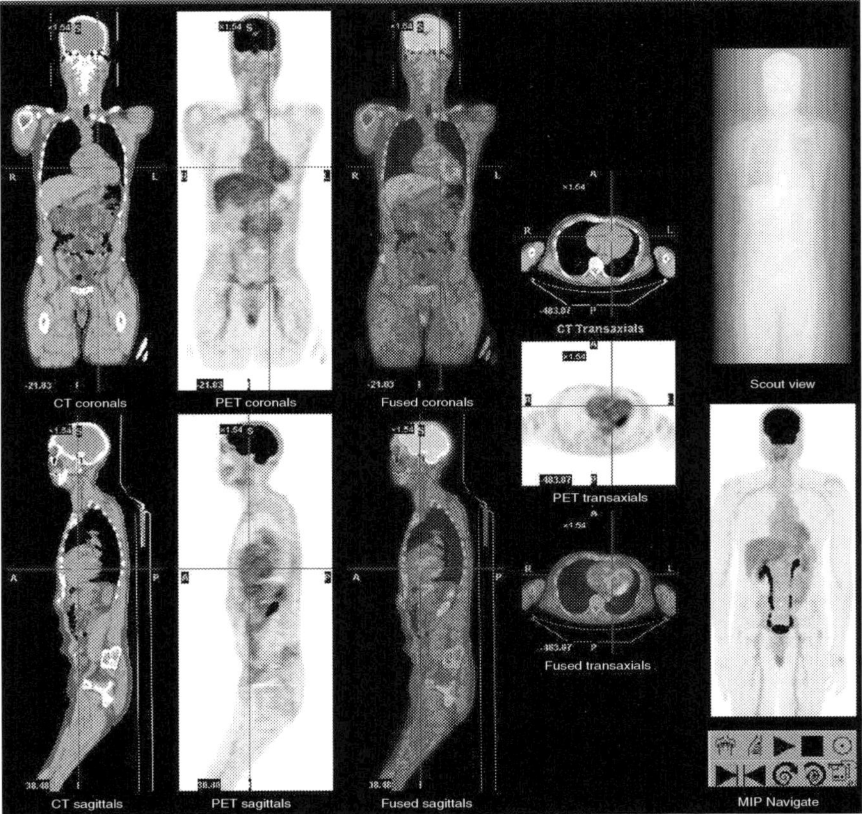

FIGURE 6.1 PET/CT scan using FDG-18 to examine cancerous tumor and organ's functionality (Wikipedia, 2020, Nuclear medicine, available from https://en.wikipedia.org/wiki/Nuclear_medicine#/media/File:Nl_petct.jpg). *PET*, positron emission.

scans, detectors measure gamma rays that decay from the radiotracers used. But in the PET scan, detectors detect positrons that release photons' energy.

6.2.5 Antibiotic irradiation

According to the available literature, biological studies, and the scientific research related to the manufacturing of antibiotics, they all agree that the manufacture of antibiotics is based on a common scientific principle, which is to determine the type of disease, whether viral, bacterial, or any other type and returning it to its origin through the separation of microorganisms and genes (DNA /RNA) that make up the disease using

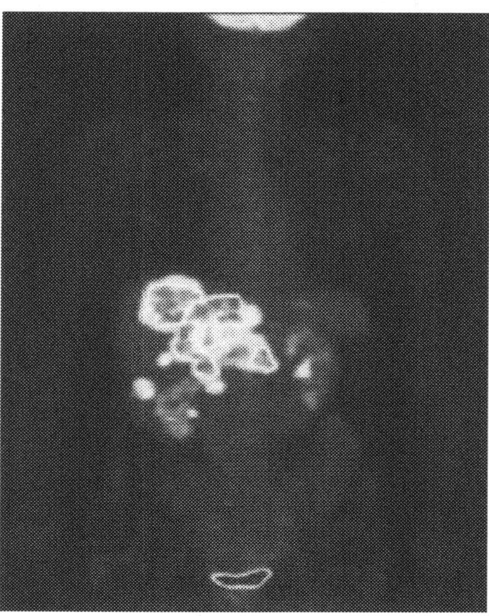

FIGURE 6.2 PET scan coupled with labeled radioactive (18F-FDG) shows a large metastatic tumor mass of colon cancer in the liver (Positron emission tomography, available from https://en.wikipedia.org/wiki/Positron_emission_tomography#/media/File:PET-MIPS-anim.gif).

several techniques such as Reverse or Quantitative Polymerase Chain Reaction (RT-QPCR) to know the genetic origin and DNA of the virus or bacteria (Baron, 1996, O'Connell, 2002). This also helps the scientists to determine whether the virus is natural or induced, studying its behavior, understanding its development and characteristics. Finally, separating the genes as an introductory step in culturing the aetiology and derivatives of the disease. According to the College of Physicians of Philadelphia (2020), this is often what biological scientists follow at the most antibiotic laboratories to produce the required antibiotic or vaccines. Unfortunately, most of the available antibiotics or vaccines are treated by a chemical or thermal inhibition process to inactivate the reproduction of pathogens, which may only work for a limited spectrum of pathogens. Accordingly, there is a high probability that antibiotics or vaccines may become ineffective due to the high level of uncertainty associated with pathogens' reproduction as a result of the inefficiency of the inhibition process (FAO/IAEA, 2019). Therefore, it is expected that the reproduction of the antigen inside the body may occur and consequently the failure of the antibiotics or vaccines. Nuclear technology can tackle this dilemma and provide an effective solution to inhibit all kinds of pathogens in the antibiotics using the irradiation approach. The basic principle of this approach is to expose the

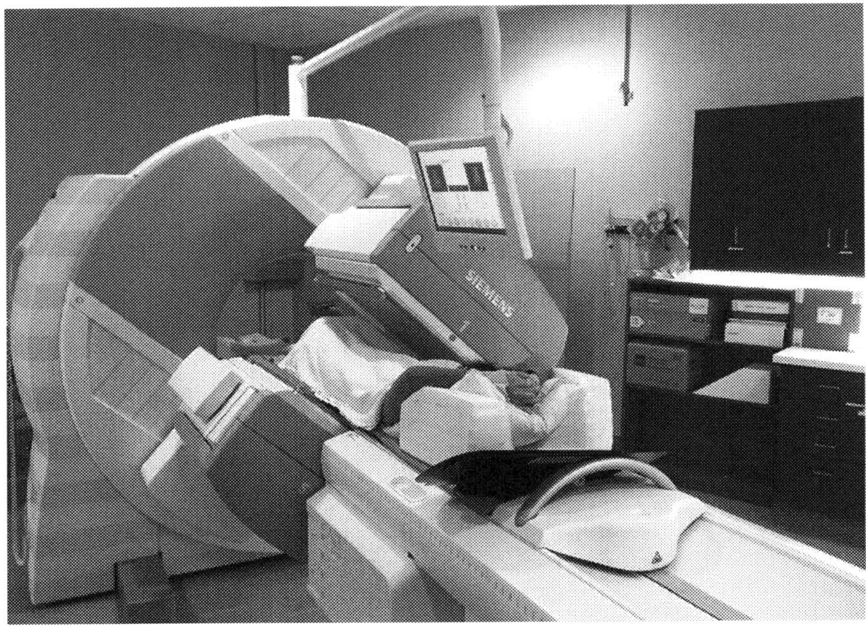

FIGURE 6.3 Single-photon emission computed tomography (*SPECT*) (Wikipedia, 2020, Single-photon emission computed tomography, available from: https://en.wikipedia.org/wiki/Single-photon_emission_computed_tomography#/media/File:SPECT_CT.JPG).

pathogens to a certain radiation dose of gamma rays or X-rays that cause sterility. Therefore, irradiated pathogens remain alive and unable to reproduce, thus allowing the body to build the required immunity safely. This does not only apply to human vaccines, but rather to animal vaccines. The vaccine irradiation technique has proven to be efficient and its ability to expand the vaccine portfolio. Not only that, but nuclear and isotopic technology also can trace the source and the country of origin of the diseases and viruses, the countries where the virus is likely to spread, the ways of transmission either via human infection or migratory birds such as avian influenza. The path of the virus can be traced using the labeled radiotracer approach. Scientifically, every organism even if it is a microorganism is carrying a special biological code in its genes and has a special isotope that can be tracked and monitored through it.

6.2.6 Sterilizing medical products

Radiation sterilization for medical applications emerged in the second half of the twentieth century and has grown in popularity dramatically since then (Heldman, 2003). The basic principle of radiation sterilization relies on ionizing radiation, mainly gamma rays from Co-60 or Cs-137 sources, X-rays, or electronic radiation. Radiation sterilization is used to sterilize many medical products. This technique is cheap and effective compared to thermal or chemical sterilization techniques. Sterilization with ionizing radiation, as shown in Fig. 6.4, inactivates microorganisms, kills germs, and keeps medical products sterilized for a long period. It also allows already-packaged products to be sterilized. Radiation sterilization is not limited to medical tools such as cotton wool, burn dressings, surgical gloves, heart valves, bandages, plastic and rubber sheets, and surgical tools. But it can be used to sterilize a wide range of heat-sensitive items since it is a "cold" process, such as powders, ointments, solutions, and biological preparations such as bones, nerves, and skin for use in tissue grafting. Not only that but it is used for blood sterilization to ensure that blood transfusion processes are clean. IAEA (2020) argues that "More than 160 gamma irradiation plants around the world are operating to sterilize medical devices. Around 12 million m3 of medical devices are sterilized by radiation annually and more than 40% of all single-use medical devices produced worldwide are sterilized with gamma irradiation."

6.3 Nuclear forensics and crime management

Nuclear science has gained much importance since the mid of 20th century. The intensive research on the peaceful uses of nuclear technologies apart from military applications started after the Second World War. Since then, peaceful nuclear science and technologies have been widely used in power generation, health fields, agriculture fields, industrial applications, and many more. Interestingly, nuclear technology has helped to uncover the threads of crimes and trace their sources using both quantitative and qualitative nondestructive methods in analyzing microscopic and macroscopic samples that are obtained from the crime scene after its occurrence. Nuclear technologies help to solve the mystery of many complicated cases in the courts and provide important and absolute information that can be used as conclusive evidence during the trial. For example, in 1977 when the investigation into the assassination of President John Kennedy was reinvestigated, nuclear technology was used to investigate the available evidence and it provided conclusive evidence and conclusions in this accident. Results from nuclear activation analysis proved that Oswald's rifle assassinated president John Kennedy (University of Rhode Island, 2004). The use of atomic science in criminal investigations is witnessing a remarkable development in light of the development of crime tools. Criminals try their best not to leave any evidence on the crime scene, but some pieces of microscopic or macroscopic evidence are always left behind. Nuclear science helps forensics to reach accurately these evidences.

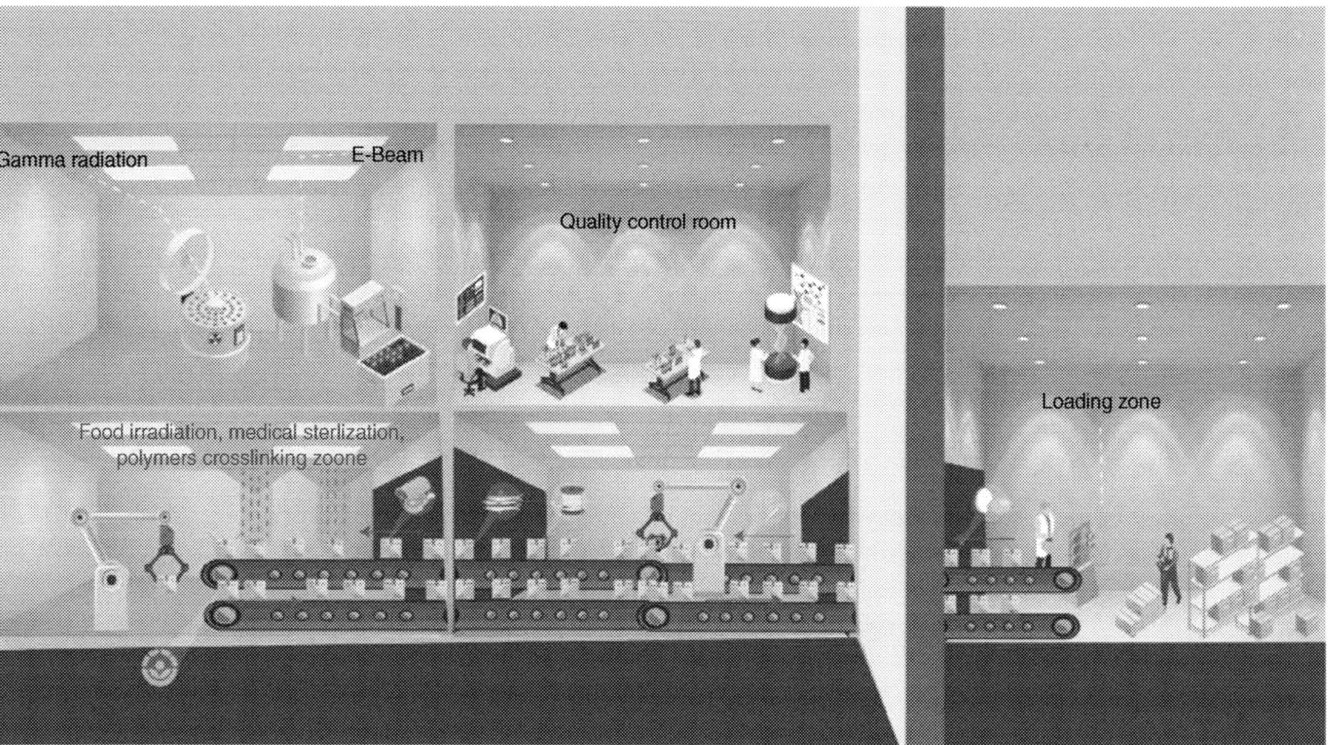

FIGURE 6.4 Sterilizing medical products using ionizing radiation.

Nuclear technology including Ions, X-rays, and neutrons activation has shown many benefits over other conventional forensics techniques. For example, X-ray scanning technologies that are used overall airports around the world that are well known to many travelers are trusted technology that can detect and recognize metallic objects, sharp machines, weapons hidden in luggage, drugs, and others illegal stuff. This simple technology has helped to stop many planned crimes before they happened, and it has the credit for protecting the lives of millions of travelers around the world. Therefore, nuclear technology has contributed to many decades in reducing crime and detecting it in different ways. Not only that, but nuclear technology assists nuclear forensics in analyzing very complicated criminal cases by finding a single particle source in several million with remarkable accuracy as crucial evidence in a similar way to the accuracy provided by transmission electron microscopy (Kristo, 2018). Ionic beams technique is another useful tool used in nuclear forensics to trace the source of materials used in the crime. The composition of the material at the crime scene may interact with ionic beams. Therefore, its source can be traced by this technique. Vij (2008) argues that nuclear forensics techniques including neutron activation, flameless atomic absorption spectrometry, SEM-DEX, and many other all help in revealing the murders' crimes using firearms. The information provided using these techniques including but is not limited to the manufacturer of a bullet, its type, distance of shootout, bullet' powder type used, and many other of precise details. These tools help in solving crimes, spotting forgeries, and detecting fraud.

Usually, the chemical analysis used in conventional forensics analyses provides evidence that is primitive and often qualitative such as the color and the size. In contrast, the scientific advancement in nuclear forensics provides a large scale of quantitative and qualitative information such as correlating the materials' physical and chemical characteristics with its origin and its production history (IAEA, 2006). It has been found that many of the evidence collected from the crime scene that is analyzed by nuclear forensics having similar physical and chemical compositions of the original tools used in executing the crime. This homogeneity of substances has led to the use of isotopic composition techniques to discriminate against many forms of materials that have been taken at different times and places to provide detailed information about the investigated material including its physical properties, chemical, elemental composition, isotopic ratios, reference information, and other signatures of the material (IAEA, 2018).

6.3.1 Why nuclear forensics?

The gunshot residues, hair, single-rooted teeth, a shred of fiber, and any paint clip are one of the most crucial objects for nuclear forensics investigators (Fig. 6.5). Many different types of forensics techniques are

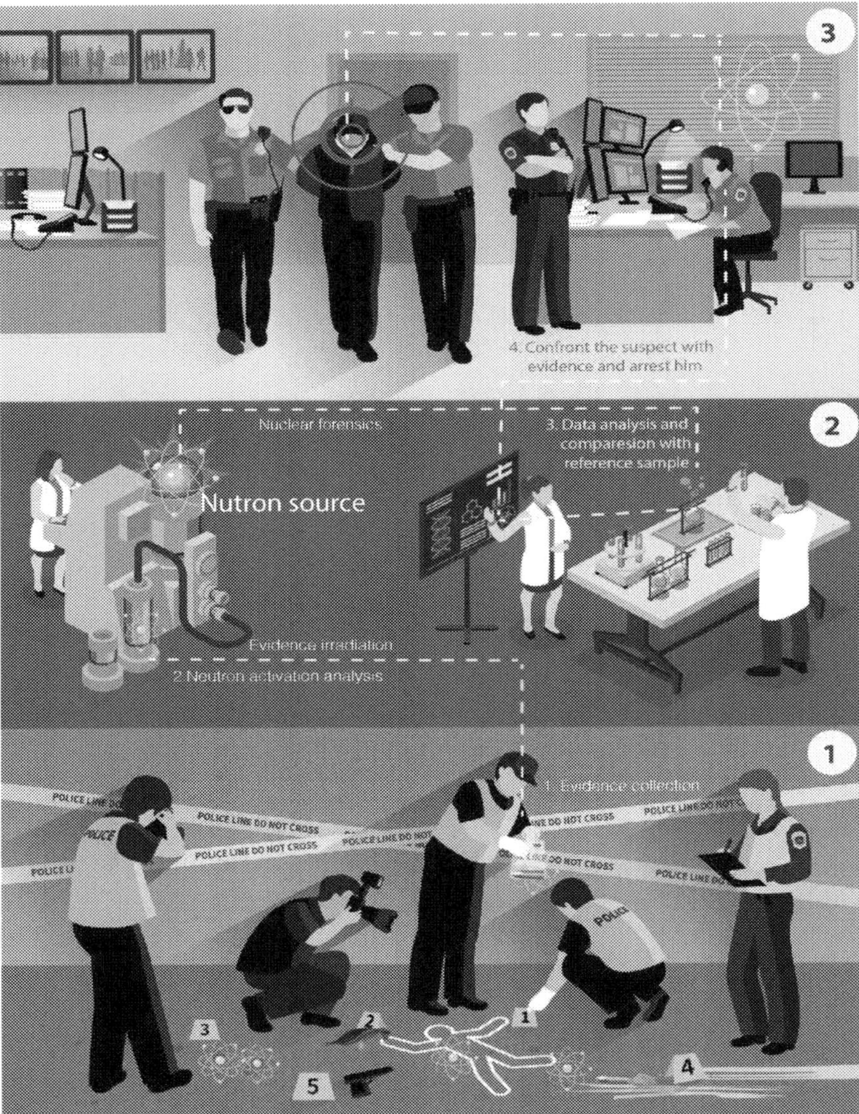

FIGURE 6.5 Examination of crime evidence using nuclear forensic techniques.

used by investigators to solve the mystery behind many complicated cases presented in front of the courts and to provide conclusive evidence that condemning criminals to achieve justice. However, in many cases, investigators are unable to do so due to the lack of appropriate tools to examine the evidence without destroying them. Therefore, the main problem with traditional forensics techniques is that they are all destructive analytical techniques.

Extensive scientific studies have been conducted to find new, nondestructive techniques to assist prosecutors in pursuing the most complex cases against criminals in the courts. Therefore, nuclear scientists have developed nuclear techniques that are nondestructive for evidence's examination. Hence, it plays a vital role in helping judges and juries in giving correct verdicts. Nuclear forensics methods used in criminal investigations are nondestructive methods because the investigated samples are not destroyed by radiation exposure. This is very important for forensics investigations when it is necessary to use the only sample as a means of proof, or when the process requires additional tests on the only available evidence. Therefore, this cannot be achieved by using traditional methods such as chemical analysis that may consume or destroy the only sample available during the evidence examination process. Therefore, criminal investigators, crime laboratories, and judges owe a lot to atomic science in solving very complicated crimes.

6.3.2 Nuclear forensics technologies in support of crime investigation

6.3.2.1 Isotope ratio mass spectrometry

Isotope ratio mass spectrometry (IRMS) is used in the comparison of DNA evidence or fingerprints for many nonbiological materials of multidimensional isotopes such as drugs, flammable liquids, explosives, fibers, textiles, paints, varnishes, inks, plastics, and many other materials. Isotopes' ratio determination approach helps to reveal the material's identity with a high level of accuracy equivalent to one in 1.47 billion (Yinon, 2003). For example, when heroin or cocaine are seized from different locations, it is crucial to determine their geographical origin. This makes it easy to identify drug producers or traffickers. EA-IRMS helps to identified precisely the ratio difference in nitrogen and carbon isotopes contained in the seized heroin or cocaine. These details help in identifying precisely their geographical origin based on the soil in which these were grown correlated with the international isotopes' signature map (Finnigan, 1995). The isotopic signature can be identified with high accuracy with the support of the availability of moisture and nutrients. Not only that but synthetic controlled materials like amphetamines may also be detected using this technique. The conditions employed in synthetics help to create its isotopic signature. The packaging materials of different substances, especially drugs, are of great interest to all law enforcement agencies. These materials do not have distinct physical markings. Traditional techniques are unable to provide details about their isotopic signatures compared to the IRMS technique that can precisely provide details of their isotopic signatures.

Moreover, many ways have been developed by which the IRMS technique is employed in the analysis of many volatile liquid samples. The GC-combustion-IRMS technique, for instance, is used to identify any combination of isotopes in any volatile liquid samples (Gentile et al., 2015). Not only that, but the use of isotope ratio technique plays a significant role in studying the fragments that are found from the bombs' blast scene. Usually, it is hard to find the type of material used in the bomb after detonation. However, by using nuclear forensics analysis of explosives using IRMS, the isotopic ratios of fragments can be identified even after a complete explosion. Usually, the results obtained from IRMS analysis will show the same isotopic ratios as used in the primary explosive (Benson et al., 2009).

Hobson (1999) argues that the use of isotope ratio analyses such as δ13C, δ15N, δ34S, δD, δ87Sr are used to trace the trophic origin and animal migration. This method is considered to be very useful and advanced compared to traditional tracing methods in controlling the smuggling of endangered or banned animals as well as predicting infectious diseases that animals transmit across borders. This approach relies on the fact that each type of food has a unique isotope signature that is reflected in the tissues of living organisms. These signatures differ according to the geographical location and biochemical processes. Therefore, the smuggling of endangered animals can be monitored through the examination of isotope signatures, which provide information on the location of previous feeding and the motherland.

6.3.2.2 *Particle induced X-ray emission or proton-induced X-ray emission*

Particle induced X-ray emission or proton-induced X-ray emission (PIXE) is a nondestructive technique was discovered by Sven Johansson in 1970 (Ali, 2017). This method is widely used in determining the elementary composition of a substance or sample to understand its source, history, and originality. PIXE able to identify about 72 inorganic elements from Sodium through Uranium (Ali, 2017). The scientific principle of this technique is that when exposing the sample to an ionic beam, or protons, an atomic reaction occurs. The active protons excite the inner shell electrons. As these inner shell electrons are expelled, X-rays of the characteristic energy of the element with some other specific electromagnetic radiations are emitted with different wavelengths in the electromagnetic spectrum. The produced X-rays and other electromagnetic wavelengths are then recorded and measured using energy dispersive detector that identifies the elemental composition of the substance or the sample.

Sen et al. (1982) argue that PIXE is a useful tool being used in nuclear Forensics to analyze the bullet theory through the analysis of the firearms residues. This helps in finding the distance between the gun and the victim.

This distance plays a significant part in understanding the reason for murders. PIXE gives specific information about the type of gunpowder and the type of bullet used in the attack. During the investigation, this information is correlated to the wound found on the body of the victim. The literature reveals that most gun powders analyzed with PIXE technology contain mainly copper, barium, and lead. These results were obtained by identifying the peaks of these different elements not only that, the reproducibility and consistency of the tests done by PIXE precisely can identify the type of firearm used in the crime. According to the amazing and accurate results provided by PIXE technology, experts recommend using this technology to detect many criminal cases such as analysis of narcotic drugs, paint clips, forgery of documents, and glass fragments. Accordingly, PIXE technique plays also a significant role in revealing the document's forgery by analysing the components of the two different kinds of inks that were produced by various companies. Similarly, the peaks of elements found in drugs assist in finding the origin of drugs.

6.3.2.3 Neutron activation analysis or atomic absorption spectroscopy method

Neutron activation analysis is used to determine elements concentration in a vast amount of material regardless of the chemical shape of the sample but its main focus only on the nucleus. This method relies on activating the nuclei of the sample by bombarding it with a neutron from a reactor or any other neutron source or by electromagnetic radiation of high energies. The nuclei of this atom become excited and unstable that starts to release gamma radiations of certain energies. These energies are marked for each element in the same sample. For instance, microscopic samples of gunpowder can be taken directly from the crime scene or from the parts extracted from the victims' bodies or from the hand of the suspect, where usually gun powders concentrate between the index finger and thumb at the back of the hand. This can be done using a mean of a cotton swab moistened with a solution of 5% nitric acid or by pouring molten paraffin wax on the hands. The sample is then taken to the lab for activation analysis using either the neutron activation analysis method or the atomic absorption spectroscopy method to identify the basic components of the gunpowder in that sample. Nuclear activation analysis method helps to provide conclusive evidence in many homicides. The results obtained from nuclear activation analysis often indicate that the barium isotope 139 (half-life 83 minutes) and the antimony 122 (half-life 2.7 days) are the main isotopes of gunpowder (Charles and Midkiff, 1977). Nuclear activation analysis has a wide range of applications in uncovering the secrets behind many different crimes, including but not limited to analyzing paint samples founds in the suspects' bodies, clothes, cars, hair, papers, and metals. This helps to provide definitive evidence at high accuracy of their sources and matching it with the collected evidence from the crime scene (IAEA, 1990).

Neutron activation analysis is a nondestructive method that is a reliable method used to identify different elements in the sample by identifying the concentration levels of the constituent elements of taken samples. These samples include foods, cosmetics, soil, paper, drugs, and vital tissues, especially in poisoning cases in which a sample is taken from poisoned drinking water or food or hair or teeth or nails to be analyzed to determine the elements of arsenic and mercury. Arsenic and mercury are the two common elements used in poisoning murders. According to Mari et al. (2004), it is likely that French Emperor Napoleon Bonaparte was killed by an arsenic poisoning before arriving at Saint Helena, where Neutron activation analysis shows that arsenic was found in Napoleon Bonaparte's hair samples. Results in the autopsy were consistent with the diagnosis of gastric ulcer cancer. Many other studies and investigations found high concentrations levels of arsenic, and antimony in the samples taken from Napoleon Bonaparte's hair, and all agreed that arsenic poisoning was the cause of Napoleon's death (Mari et al. 2004). Usually, the victim who is assassinated by arsenic or mercury poison does not feel it because this element has no color and or taste and the human body will try to expel as much venom as it could before it is spreading. The central excretion spot for accurate analysis is the hair. when a small hair fragment as 1 mm found in the crime spot, it makes a big difference in the investigation path. That tiny piece of hair can be important and conclusive evidence that reveals the reason for death, the time of poisoning if analyzed using the neutron activation method or other nuclear approaches. Not only that but also when activating the food that contains mercury or arsenic extracted from the victim's stomach or a sample from the suspect's nails with neutron isotope, the arsenic-76 that has a half-life of 26.3 hours is formed. Therefore, poison can be easily detected by this type of analysis. In this context, this method was also used to investigate the death of Alexander Litvinenko the former officer of the Russian Federal Security Service, who was been murdered by poisoning. On the other hand, Musazzi and Perini (2014) argue that neutron activation analysis besides many other nuclear methods such as PIXE can also be used to detect counterfeiting of paper money. This can be achieved by nuclear activating the metal chips or alloys contained in the money that will give distinctive spectrums through which the safety of the paper money can be verified or not.

6.3.2.4 Gamma-ray spectrometer analysis

Gamma-ray Spectrometer analysis is a useful method in nuclear forensics that provides spectral analysis for each element in the sample based on their wavelengths that are different from each other because each element has a chemical imprint (optical spectrum) that makes it distinguished from another. Gamma-ray spectroscopy is a rapid analytical technique that can be used to identify various radioisotopes in the collected sample. This method has many applications, especially in cases that require nondestructive analysis such as forensics laboratories in drug cases and the analysis of hair, blood, and urine. Besides the gamma-Ray Spectrometer Analysis technique, Harper et al. (2017) argue that mass spectrometry also can be used and considered as the most discriminatory of the drug testing techniques.

6.3.2.5 The role of radiometric dating in detecting homicides

The radioisotope dating method is one of the most important discoveries of the 20th century. It is a popular method and has wide applications in archaeology, climatology, hydrology, geology, meteorology, and nuclear forensic sciences. This method helps to determine the age and date dates of the death of organic matters. Therefore, this technology is a useful tool in criminal investigations. This technique helps forensic scientists to know some important information such as the age of the victim, the date of death. This can be achieved by measuring the levels of isotopes available in the tissues of people who were killed in accidents and crimes, especially those that occurred long ago. For example, the carbon radioisotope ratio can be analyzed by examining the stool or bones (Alkas et al., 2010). The next section will discuss in more detail the scientific principles of the isotope historiography method and its principle of operation.

6.4 The role of radiometric dating in archaeology, geology, paleoclimatology, and hydrology

Radioisotopes dating methods are a vital branch of nuclear science that is widely used in many applications, such as archaeology, climatology, hydrology, geology, meteorology, and nuclear forensics science. Nuclear dating methods are very useful and precise tools for the determination of the age of rocks, the ancient artworks, cave paintings, engravings, and the dates of death of the organic matters. Accordingly, geochemists, geologists, archaeologists, and archeologists have joined hands with nuclear scientists to widen the research on archaeology and other related fields to maximizing the use of nuclear peaceful applications.

The history of radioisotope dating is linked with the concept of using nuclear science for peaceful purposes. Professor Willard Libby is the first one who introduced the idea of ^{14}C during the first United Nations Conference on the Peaceful Uses of Atomic Energy in 1955 at Geneva (IAEA, 1963). Along with his team of researchers, working on developing the concept of theory for radiocarbon dating. They found radiocarbon extracted from sewage sludge in Baltimore. The necessary counting procedure and equipment were developed by conducting a worldwide assay of woods. Then they proceeded to do actual dating of those samples having a known age. These procedures validated that their technique worked. Accordingly, their findings were published in 1952. Nuclear physicists interested in archaeology developed the findings of Libby in the 1950s for using ^{14}C dating as the chronometer. Later this concept was used in the determination of the age of different objects related to earth sciences such as origin and nature of rock formation and the history of meteorites. Willard Libby was awarded the Nobel Prize in 1960 for his scientific contribution to the use of carbon 14 for dating (Taylor and Bar-Yosef, 2016).

Scientifically, it has been concluded that the carbon that Libby used as a source for dating constitutes the basic building block for all organic compounds. It is found in nature in the form of two stable isotopes, which are carbon ^{12}C and ^{13}C, while ^{14}C is an unstable radioactive isotope with a half-life of 5730 years (Cockell, 2020). Wicander and Monroe (2012) argue carbon is constantly formed in the atmosphere when cosmic rays that contain high-energy protons collide with gas atoms in the atmosphere. At this stage, their nuclei are divided into protons and neutrons. When a neutron collides with the nucleus of a nitrogen atom (atomic number 7 and atomic mass number 14), it is then absorbed by the nucleus and followed by a proton release. Accordingly, the atomic number of the atom decreases by one, but the atomic mass number remains the same. As a result, the atomic number changes, and a new element is formed, which is carbon 14 (atomic number 6, atomic mass number 14). The newly formed ^{14}C is replenished continuously in the carbon cycle and is absorbed along with carbon isotopes ^{12}C and ^{13}C by almost all organisms at the same proportion as a result of eating other organisms or eating plants or by inhaling the air that contains carbon (Fig. 6.6). Therefore, carbon is renewed in their tissues, but when the organism dies, it stops renewing carbon in its tissues, and ^{14}C:^{12}C ratio decreases due to the decay of ^{14}C into nitrogen by releasing a single beta-decay. Accordingly, the time of the organism's death can be found by measuring how much ^{14}C is left undecayed inside the body at that time.

The archaeologists measure the remaining ^{14}C using different techniques including optical spectroscopy or mass spectrum by taking samples from the bones or fabric or wood or plant fibers. Then, the sample of matter is burned to emit carbon dioxide, and then the ions extracted from this gas are arranged according to their mass in the accelerated mass spectrum. Vuong et al. (2016) argue that the mass spectrum can measure the ratio of ^{14}C to ^{12}C up to one part in each quadrillion (10^{-15}), which is correlated to a sample's lifespan of 50,000 years. While, the optical spectroscopy, which is a highly sensitive technique can detect very small amounts of matter by measuring how much light it absorbs. The wavelengths range between 2.5

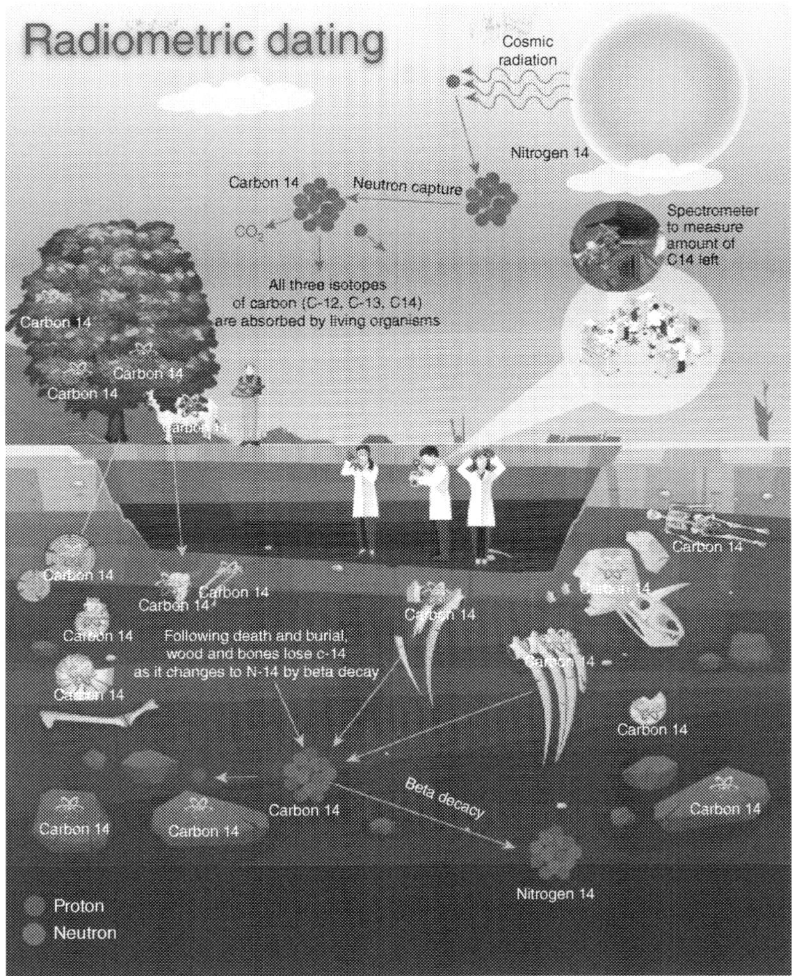

FIGURE 6.6 Radiometric dating method.

and 10 microns are the range in which the gas molecules have the strongest absorption.

Radioactive dating techniques are based on determining the decay of specific radioisotopes. The half-life of that radioactive isotope must be known for accuracy purpose; therefore, the measurement of the radioactivity and isotopes ratio is very crucial for the success of this process. Thus, unstable radioisotopes formed by cosmic bombardment and naturally available stable isotopes of long half-lives are used for this purpose. According to Britannica Educational Publishing (2010), the most common nuclear dating methods are the rubidium–strontium method, the uranium–thorium method, and the potassium–argon method, the samarium–neodymium method, the rhenium–osmium method.

6.4.1 Common radiometric dating methods

6.4.1.1 Rubidium–strontium dating method

The isotope of Rubidium 87 decays to an isotope of Strontium 87. The half-life of this isotope is approximately 48.8 billion years. This method is useful for estimating the age of rocks, minerals, and meteorites by measuring the amount of Strontium 87 present in these rocks (Britannica Educational Publishing, 2010). This method dates materials aged from 10 million years to 10 billion years. Historically, the strontium isotopes' ratio collected from the teeth and bones of the king Yax K'uk' Mo', who ruled the most important city of Mayan dynasty Copan helped to determine his origin (Bentley, 2006).

6.4.1.2 Uranium–thorium dating method

Scholars such as Ayliffe et al. (1997) and Edgell (2006) argue that U234/Th230 is another excellent method for dating beyond the ˜50,000–500,000 years. This method can be used to date the age of lake sediments, cave deposits, corals. Corals, speleothem, and porous materials like bones have a characteristic to absorb uranium (Greene, 2002). In the case of the corals, for instance, uranium is converted to thorium, which gathers in their skeleton, therefore the ratio of U/Th indicates the age of these corals that may range up to 500,000 years.

6.4.1.3 Potassium–argon dating method

Kelley (2002) argues that this method is one of the earliest isotopes dating techniques and the most popular isotopic dating techniques used in geochronology. This method allows to date the materials that are beyond the limit of radiocarbon dating method and goes up to billions of years ago (Peppe and Deino, 2013). When ^{40}K decays to ^{40}Ar in a rock, ^{40}Ar remains inside a rock in gaseous form unless it is heated. The amount of argon in rock gives information about the time since it last cooled. This method is useful for estimating the age of volcanic activity, galaxies, and the Earth. The Potassium–Argon Dating Method revealed that our galaxy's lifespan ranges between 10 and 20 billion years and that the Earth's age is 4.6 billion years. The universe must be older than our galaxy. Within experimental error, this estimate agrees that the age of the Universe is about 15-billion-year. The age equation for the K–Ar isotope is described as follows:

$$t = \frac{1}{\lambda} \ln \left[1 + \frac{\lambda}{\lambda_e + \lambda'_e} \frac{Ar^{40}}{K^{40}} \right] \tag{6.1}$$

where

t is the time.

λ is the total decay of ^{40}K.

($\lambda e + \lambda' e$) is the partial decay constant for ^{40}Ar.

$^{40}Ar/^{40}K$ is the ratio of radiogenic daughter product to the parent ^{40}K.

6.4.1.4 Radiocarbon dating method

The carbon dating methods are the most popular radiometric methods that are used to determine the age of any object contains organic materials (Hamblin, and Burns, 2005). These organic materials include fossils, plants, rocks, scripts, and many other items that contain stable and radioactive carbon isotopes. In the determination of trees' age, radiocarbon measurements are calibrated using different techniques such as tree rings of known age or a spectrophotometer or calibration curve. The rings' variation of the trees provides the calendar age of the tree. This is usually done by taking a horizontal section of a tree's main trunk. Researchers analyze the visible rings that are caused by the natural growth of a tree. Often these rings are affected by the seasons, which reflects the change in the growth of these trees. This method is could be suitable for the temperate climate but still not 100% reliable due to the weather changes or the missing or damaged rings. While scientifically, the radiocarbon dating method is still considered a more reliable and accurate process than the tree rings method. Moreover, the radiocarbon dating method is also used to date prehistoric cave paintings and engravings that may date back as long as 30,000 years ago. There are many artworks in the form of paintings, engravings, and sculptures that may date to the Stone Age. The history of these art paintings and cave engravings has always been a controversial topic until the emergence of radiocarbon dating method that helped in extrapolating the history of the old nations such as the Mayan civilization, which are symbolically important for historians. This can be done by taking a sample of graphite or charcoal extracted from the pigments. graphite or charcoal are both carbon-based used as art materials. A study presented by Mithen (2005), shows that the earliest date found at the Chauvet cave in the Ardeche in France was for a paint of rhino that is dated back to about 32,417years. Smith (2009) argues that the radiocarbon dating method able to date to approximately 50,000 years ago while the K–Ar method able to date from 100,000 years old up to billions of years ago.

The basic principle of nuclear dating process depends on the half-life of radioisotopes, which describes the amount of the time it takes for half of that isotope found in the sample to degrade. In the case of radiocarbon dating, the carbon half-life is 14 5,730 years. After 5730 years, the amount of ^{14}C remaining in the body is half the original amount. So, to determine the lifespan of an organism we need to consider the half-life of ^{14}C as well as the decay rate, which is -0.693 as per ^{14}C dating Eq. 6.2:

$$t = [\ln (N_f/N_o)/(-0.693)] \times t_{1/2} \quad (6.2)$$

where ln is the natural logarithm, N_f/N_o is the percent of ^{14}C in the sample compared to the amount in living tissue, -0.693 is the decay rate, and $t_{1/2}$ is the half-life of ^{14}C (5700 years).

For instance, to determine the age of a fossil that contains 12% of ^{14}C compared to a live sample, Eq. 6.1 is used as following:

$$t = [\ln (0.12)/(-0.693)] \times 5700\,\text{years}$$

$$t = [(-2.120)/(-0.693)] \times 5700\,\text{years}$$

$$t = [3.059] \times 5700\,\text{years}$$

$t = 17{,}436.3$ years old, this means that the organism died 17,436.3 years ago.

Historically, the radiocarbon dating method has shown great results to discover the real age of the Iceman (Ötzi), who was living in the Alps and died a long time ago. His body was found in 1991 when the ice moved and melted. The man's body was recovered, and pieces of tissue were studied to examine carbon-14's content using accelerator mass spectroscopy. The best estimate from this dating technique says that the man lived between 3350 and 3300 BC (Verma, 2007). Not only that, but the radiocarbon dating method was used to date the boat of a pharaoh that was discovered in a sealed crypt and reassembled in a museum near the pyramids. Its wood was dated using this method and found to be about 4500 years old. The basic principle of radiocarbon dating applies to other isotopes as well. Potassium 40, for instance, is another radioactive element that is found naturally in the body of organisms, specifically the human body, and has a half-life of 1.3 billion years.

6.5 Water resources management using isotopic technology

Carroll (2012) argues that almost 70% of the earth is covered by water and the remaining 30% is dry land. In total, 97.41% of the earth's water is salty water found in the oceans, 2% is in ice caps and glaciers, and only 0.59% of the remaining water is the water available to humankind and found in form of underground water or surface freshwater (Pipkin et al., 2012). Population inflation has increased the global demand for water, food, and water-dependent industries in their operational processes. Consequently, this has led to an increase in water consumption in agricultural and industrial activities, which has led to a rapid depletion of the underground water reservoirs and threatens their sustainability (FAO, 2017). This valuable resource is often poorly managed due to lack of related knowledge and information. So far, there are no comprehensive assessments of groundwater in many countries of the world. This makes these countries unable to manage their water resources. Nuclear hydrology using the isotope approach is one of the most advanced and the cheapest methods that has proven its success in providing a reliable system for sustainable water management and long-term water supply compared to other conventional hydrological monitoring approaches.

6.5.1 Isotopic composition of water

Scientifically, the water molecule is composed of two hydrogen atoms and one atom of oxygen. This oxygen and hydrogen have different isotopes such as ^{16}O, ^{17}O, ^{18}O,^{1}H, ^{2}H, ^{3}H, respectively (Hölting and Coldewey, 2018). Isotopes have the same number of electrons and protons but different numbers of neutrons and therefore different atomic mass. Therefore, water is not the same and differs from one place to another (IAEA, 2019). Some water carries heavier atoms while some carry lighter atoms depending on the isotopic composition of the water (Galewsky, 2016). For instance, some water molecules comprise one oxygen atom of mass number 17 or 18 with two hydrogen atoms of mass number 1 ($H_2{}^{17}O$ & $H_2{}^{18}O$). While some water molecules may comprise one oxygen atom of mass number 16 with two hydrogen atoms of mass number 2 ($HD^{16}O$). This difference in isotopic composition is used as the biological signature of water for other organisms that live on the water (IAEA, 2019). In any hydrology cycle, water molecules containing both the heavy and the light isotope rise as water evaporates from the sea (Fig. 6.7). The heavier isotopes molecules fall first, and they usually fall near the coast due to their heavy atomic mass. While the molecules with lighter isotopes are easily moved by clouds far distances (IAEA, 2019). Therefore, the origin of different waters can be found by measuring the difference in the proportions between the light and heavy isotopes. Accordingly, different recharge components can be identified through the stable isotope compositions of groundwater because evaporation of water from surface water bodies, in particular under semiarid and arid conditions, leads to an increase in the proportion of the heavy isotopes such as deuterium and oxygen-18 (Silveira and Usunoff, 2009). Not only that, but the recharge mechanism, its type, and the age of that groundwater that may range from a few days

FIGURE 6.7 Water resources management using isotopic technology.

to thousands of years can also be identified by measuring the abundance of naturally occurring radioactive isotopes in water like carbon 14, tritium, and isotopes of noble gas that usually dissolve in the water (Cartwright et al., 2017). The flow of the groundwater gives an estimate of the age of the groundwater. For example, if the groundwater flow is very slow; then it means the groundwater is tens of thousands of years old. Thus, if this water is not properly extracted and managed then it can take tens of thousands of years to replenish again. Accordingly, nuclear sciences play a major role in managing and sustaining water resources by defining the sources of water, recharge mechanism, age, quantity, quality, and interaction.

6.5.2 How old is water?

Determining the age of water is a crucial factor to address the issue of shortage of water resources that result from the excessive extraction of water from the underground. It also helps to know water quantity, quality, and recharge time. IAEA (2013) argues that the presence of certain naturally occurring radioactive isotopes tracers assist in knowing the age of water, which may range from a few months to millions of years (Fig. 6.7). For instance, the presence of ^{3}H indicates that this water is less than 100 years old (Holland and Turekian, 2010). While the presence of ^{14}C indicates that the water age is around 40,000 years old. On the other hand, the water that is containing ^{81}Kr, ^{4}He, ^{129}I, and ^{36}Cl indicates that age goes from 40,000 years up to millions of years. Not only that but, the isotopes composition of the groundwater in hydrological environments indicate significant information about the regional climatic conditions at the time of recharging. Moreover, stable and radioactive isotopes, including ^{18}O, ^{14}C, ^{2}H, and ^{3}H, provides vital information about the physical and chemical processes acting on groundwater and the geological formations, and hydrological features of aquifers including the rate of the recharge, spatial and temporal movement, aquifer interconnections, mixing time, and the origin (IAEA, 2004). These parameters are difficult to be measured using any other technique other than isotopic technique. This information is vital for optimal management and assessment of the groundwater resources, especially in the mountainous regions and in developing countries. In these areas, hydrological information is scarce, and there is no long-term observed data.

6.5.3 What else do nuclear science and radioisotope technologies tell us about water?

On the other hand, the sustainable management projects of groundwater conservation resources are crucial to many countries in the world. Quantitative and qualitative characterization of groundwater recharge of groundwater also plays a key role in their sustainability. Water level monitoring of aquifers only shows small fluctuations of groundwater, so these data are not reliable. While isotopic composition of recharge (expressed as abundance of oxygen-18 and deuterium) gives more reliable information the groundwater. For instance, if the source of recharge is precipitation isotopic composition found in the underground water then this will provide the data of the precipitation such as rain type and their frequency. But, if the source of the recharge is the surface water, then this will provide information about the recharging source whether it is a river or a lake. This helps to understand the seasonal variation in isotopic composition and therefore finding of recharge composition for optimal water resources management. Radioactive tracers such as ^{14}C, ^{3}H, ^{36}Cl, and noble gases are used in the determination of the residence time and recharge mechanism (Cartwright et al., 2017). For instance, ^{14}C can be used to determine the residence time of the water using Eq. 6.3:

$$T = -8267 \ln\left({}^{14}C_{t}/{}^{14}C_{i}.q\right) \tag{6.3}$$

where:

T is the residence times.

q is the proportion of Dissolved inorganic carbon (DIC) introduced via recharge.

$^{14}C_i$ is the initial ^{14}C of DIC in the recharging water.

IAEA (2011) argues that some radioisotopes such as anthropogenic krypton-85, bomb chlorine-36, and bomb tritium are used to determine residence time and flow velocity due to their radioactive decay and their transient nature. Flow velocity is estimated by measuring the decrease in the concentration of radioisotope along the flow path. It estimates the groundwater flow rate and helps to understand the connectivity and interaction between geological formations. According to IAEA (2011), the movement of the groundwater is considered to be very slow and usually not exceeding a few meters per year. However, if the water is found moved a few kilometres away, it means that it is thousands and hundreds of years old, which is the estimated time for the water to move that far. ^{14}C is widely used in a hydrology radiometric approach due to its half-life of 5730 years old and therefore has the ability to date groundwater between 1000 and 40,000 years old (Aggarwal et al., 2013).

Nuclear techniques offer significant information to maintaining water sources' quality that can be polluted by heavy metals or petroleum by-products or drainage water from mining activities or fertilizers or other pollutants (IAEA,1993). This technique detects the movement and solubility of the pollutants in underground aquatic systems. Groundwater is mainly a mixture of local precipitation and recharges from surface water. The proportion of recharge is crucial because it provides information about different recharge components that may be polluted, especially under arid and semiarid conditions or in areas close to industries. Many scholars including Torres-Martínez et al. (2020) argue that isotopes are useful tracers to collect this information and tracing the source of pollutants such as sulphate or nitrate or uranium or arsenic. Furthermore, the precipitation in recharge and the relative proportion of surface water can be estimated by using the isotopic balance equation. Deuterium and oxygen-18 can be calculated in rivers, and a fraction of river water in groundwater can be estimated by finding the difference in isotopic composition. Radiotracers are important to trace any contamination in the water flow into lakes, rivers, the ocean, and the underground. These radiotracers can detect and determine if the water is contaminated as a result of nuclear plants explosions such as Chernobyl and Fukushima or as a result of nuclear weapons tests. For instance, Uranium-238, and Uranium-235 radiotracers are used to assess whether uranium in the water is of natural origin or from nuclear activities. Accordingly, radiotracers play a significant role in predicting and identifying the pathways of contaminants by studying the radionuclides, the outbreaks of harmful algal in the freshwater, levels of arsenic in the groundwater, which all cause major health and environmental risks.

Isotope tracers in hydrological studies is a new and cheap technique used by scientists to have a better understanding and comprehensive water resources management, which provides significant information such as the aquifer capacity, interaction, and connectivity between geological formations, flow characteristics, water quality, water age, leakages through dams, storage, and potential water extraction rates. This information helps to avoid exhaustion of water resources through quantification of the running water balance, runoff, and interactions with groundwater (Shamsuddin et al., 2018). For instance, isotopes of water (hydrogen and oxygen), radioisotopes tritium (^{3}H), dissolved carbon (^{14}C), and noble gas such as ^{3}He, ^{4}He, and ^{81}Kr are used to estimate the ages of groundwaters, their sources, interaction, and determine the sources of groundwater pollution. The use of ^{15}N, ^{13}C, ^{3}H, ^{238}U, and ^{235}U help to detect pollutant sources in groundwater reservoir systems, and their biodegradation levels. Furthermore, the use of oxygen, hydrogen, nitrogen, carbon, and sulphur isotopes with other geochemical parameters helps to model recharge and sediment dynamics in river basins.

6.6 The role of nuclear science in space exploration

6.6.1 The history of nuclear applications in space

According to the US Department of Energy (2015), space-age started in 1958 when the Soviet Union launched the first unmanned spacecraft Sputnik 1. that was carrying a Soviet satellite into orbit. On January 31, 1958, The United States Launched successfully its Explorer 1. On October 1, 1958, NASA was officially formed. On April 12, 196, the Soviet Union launched the first manned space mission in space. On May 1, 1961, Alan Shepard becomes the first American arrived into space through the Freedom 7 mission. President Kennedy announced on May 25, 1961, the goal of sending astronauts to the moon before the end of the 1960s. On June 29, 1961, the first use of a nuclear power system in space started when American Navy launched its first nuclear-powered spacecraft "The US Navy's Transit 4A Navigation Satellite," which was powered by an RTG. On February 20, 1962, astronaut John Glenn became the first American orbited the Earth within the 5-hour flight of Friendship-7. The first nuclear-powered spacecraft "Nimbus III" was successfully launched by NASA on April 14, 1969, as the first US satellite to measure sea ice and the ozone layer. On July 16, 1969, the Apollo 11 scientific experiments began on the moon. Two 15-watt radioisotope heating units were used to keep their tools warm enough to operate. The American astronauts Edwin "Buzz" Aldrin and Neil Armstrong were the first humans to walk on the moon on July 20, 1969. On December 7, 1972, Apollo 12-17 Lunar Surface experiments were powered by 70-W RTGs. On March 2, 1972, Pioneer 10, which is a nuclear-powered spacecraft, was the first spacecraft to fly past Mars, visit Jupiter, cross the asteroid belt. On April 5, 1973, Pioneer 11, which is a nuclear-powered spaceship took the first up-close pictures of Saturn, discovered two new moons, and an additional ring around the planet. Viking 1 and Viking 2 spacecraft that were powered by RTGs were the first spacecraft to successfully land on Mars on August 20, 1975, & September 9, 1975, respectively. Voyager 2 was the only spacecraft used to study Jupiter, Saturn, Uranus, and Neptune—at close range. Voyager 2 continued to work for more than 30 years after was launched on August 20, 1977, and powered by the nuclear system. Similarly, Voyager 1 was powered by three multi-hundred-watt RTGs. On November 20, 1988, the construction of the ISS Began. The Galileo orbiter was sent on a mission to Jupiter on October 18, 1989, was powered by two RTGs in addition to 120 radioisotope heater units to keep its scientific instruments warm. The mission to study the heliosphere by Ulysses started on October 6, 1990, which continued to work for more than two decades and was powered by a general-purpose heat source RTG. Moreover, On July 7, 2003, NASA scientists announced that the Opportunity rover that was powered by eight radioisotope heater units had found direct evidence of water on Mars. On January 14, 2005, Huygens successfully landed on Titan's surface and its system was kept warm by radioisotope heater units. On November 26, 2011, NASA started its first mission "Curiosity" to tweet from space, relying on a single multi-mission RTG. Finally, the Mars 2020 rover mission launched on July 30, 2020, to take coring samples from Martian rocks and soil.

During the last 70 years, the exploration of space is one of the most complex jobs for humans due to the harsh conditions and complex challenges in space. However, this is seen as a significant advancement in human history. The cost of space exploration is very expensive. However, nuclear science has been playing significant roles in making space exploration more successful, reliable, and cheaper since the launch of the first nuclear-powered spacecraft in the 1960s. The capability to provide energy of solar power, fuel cells, and many other traditional systems is not reliable as compared to nuclear-powered systems. Thus, nuclear technology able to provide sustainable and long-term operational electricity required for space missions alongside the necessary propulsion power to penetrate the gravity forces.

6.6.2 The role of nuclear energy in space

The only two practical power sources in the space are rays of sun and heat produced by natural radioactive decay for many-year space missions.

The radioisotope power system is a system that converts heat generated by the decay of radioisotopes such as plutonium-238 or Americium-241 into electric power (Maidana, 2014). This system is one of the most vital systems for long missions in a distant solar system where solar power cannot provide the required power to the spacecraft. Maidana (2014) argues that the reasons for using Plutonium-238 (^{238}Pu) to power spacecraft are its stability at high temperature, can produce a huge amount of energy in small amount, its half-life is 88 years, which means a sustainable source of heat and electricity production for 88 years, and emitting of low levels of radiation that do not damage instruments or cause radiological health risks to the astronauts. The space plutonium battery contains plutonium in ceramic form for safety reasons, so in the event of an explosion during the launch accident, it breaks into large pieces like a torn coffee cup, instead of evaporating and spreading, and this not only prevents environmental damage but also protects people.

Americium-241 (^{241}Am) is another radioisotope that is also used to generate power in the space via RTGs. ^{241}Am has one-fourth of the energy as compared to ^{238}Pu; however, it is less expensive and widely available. It has a longer half-life than ^{238}Pu that goes up to 432 years (Maidana, 2014). Americium-241 has a gamma activity of 8.48 mSv/hr/MBq at one meter so it cannot be used unless a proper shielding system is provided (Maidana, 2014). Accordingly, more research has been developed to improve the safe use of ^{241}Am via RTG systems in an attempt to use ^{241}Am as a cheap and sustainable alternative to ^{238}Pu due to its long half-life.

6.6.3 Nuclear propulsion in space

Nuclear energy proved its efficiency to provides the required propulsion energy for space missions. Nuclear fission energy is the best option for high propulsion mission's in terms of technical maturity, energy density, and reliability. According to Nam et al. (2015), there are three types of space propulsions are powered by nuclear fission energy, namely NEP, NPP, and NTP.

NTRs utilize nuclear fission reaction and works on a principle similar to nuclear power plants and propulsion ships. In NTRs, the liquid hydrogen is heated to a high temperature as a result of the high thermal energy released by the fission reactor and then hydrogen propellant expands through the rocket nozzle that creates thrust at a superior speed. This allows carrying three times the load capacity compared to the chemical propellants. The advantages of NTR are high propellant efficiency, bi-model functions, reliability for more than 50 years, and high level of safety. Furthermore, NTR plays significant role in the exploration of the solar system, transportation, and supplying power to the ISS. The new design of the electricity generation system (EGS) such as the 100 kW-EGS converts

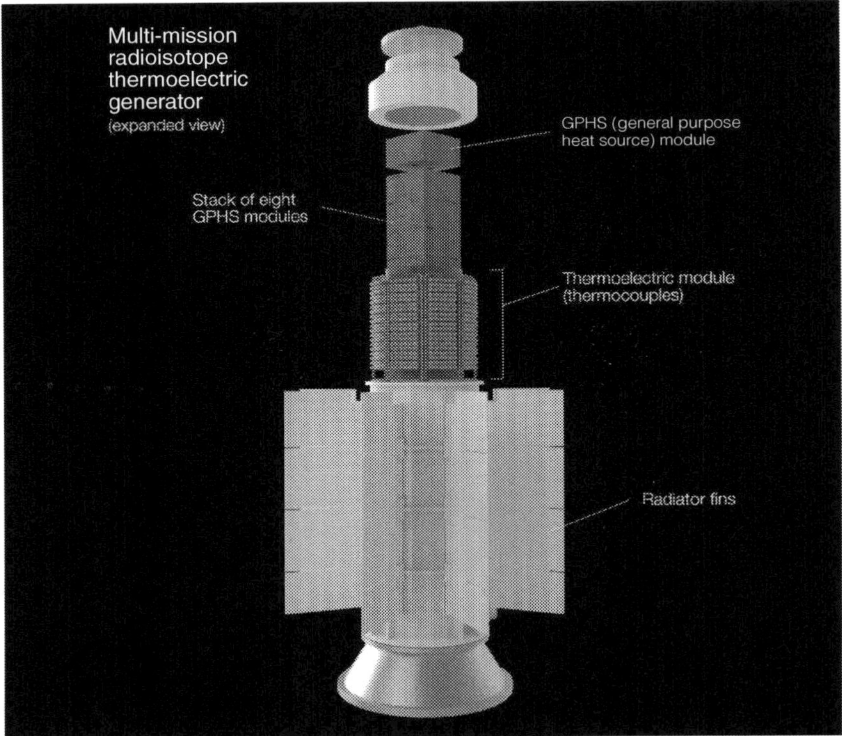

FIGURE 6.8 Multi-Mission Radioisotope Thermoelectric Generator (NASA, 2020, Power and Thermal Systems, available from https://rps.nasa.gov/power-and-thermal-systems/power-systems/current/).

the thermal energy generated from the extremely high-temperature gas-cooled reactor (EHTGR) into electric power that is required for spacecraft operations. This system eliminates the need for heavier electric power sources such as RTGs or solar system.

As mentioned earlier, the first US spacecraft "the US Navy's Transit 4A navigation satellite" that was powered by nuclear energy was based on RTG. RTG was developed by the Atomic Energy Commission, which was the predecessor to the US Energy Department. The US Air Force, the US Navy, NASA, and the Energy Department have now developed eight generations of RTG. RTG converts heat from the decays of Plutonium-238 decay into electricity through nonmoving thermocouples integrated into RTG (Fig. 6.8). The electrical power generated by the RTG installed on the Navy's Transit 4A managed to produce about 2.7 watts of electrical power. The success of RTGs led Transit 4A to make a record for the oldest broadcasting spacecraft for its first decade.

FIGURE 6.9 A conceptional design from NASA of Orion spacecraft, powered by nuclear pulse propulsion (Wikipedia, 2020, Nuclear pulse propulsion, available from https://en.wikipedia.org/wiki/Nuclear_pulse_propulsion#/media/File:NASA-project-orion-artist.jpg).

The power conversion options in electricity power generation for space missions include thermodynamic heat cycles (Rankine, Brayton, and Stirling cycles) that ensure both small radiator size and high efficiency in addition to another two technologies that are Thermoelectric and Thermionic systems (Mason, 2002). Brayton's and Rankine's cycles yield high efficiency and power according to proven technologies. The Stirling cycle can offer great potential at low power under a few kW, but the main disadvantage of using this cycle is that this technology is immature yet, and it may result in very high risks for its long-term space applications.

Scientists argued that the model developed to design NPP rockets is not recommended because of its lower propellent efficiency, the radiation that is emitted is enormous, the high chance of uncontrolled explosion. Hafizbegovic (2015) argues that the NPP model (Fig. 6.9) is a hypothetical method proposed by Stanislaw Ulam in 1947 and developed as Project Orion, which aims to design a nuclear pulse rocket. The basic principle of NPP is that the spacecraft will be propelled by several reactions resulting from a series of atomic explosions in the propulsion engine. Between 1958 to 1962, the United States and the Soviet Union detonated about twenty nuclear bombs, including thermonuclear bombs, in space at altitudes ranging between 50–500 km, but the project was closed in 1965, which was considered illegal according to the Partial Nuclear Test Ban Treaty (PTBT).

FIGURE 6.10 Nuclear thermal propulsion (NTP) (Nuclear Thermal Propulsion: Game-Changing Technology for Deep Space Exploration, available from https://www.nasa.gov/directorates/spacetech/game_changing_development/Nuclear_Thermal_Propulsion_Deep_Space_Exploration).

On the other hand, NEP uses nuclear fission reaction where the thermal energy produced from a nuclear reactor is converted to electrical energy by expelling plasma out of the nozzle that propels spacecraft (IBP, 2007). NEP is not recommended due to its lower acceleration and low thrust to weight ratio. While, NTP technology has a huge potential in space propulsion, especially for distant space explorations (Fig. 6.10). The thermal energy generated from the fission reaction of uranium-235 atoms is utilized for heating the propellent containing low molecular-weight such as hydrogen to generate the required thrust via the thermodynamic nozzle (Nam et al., 2015). The impulses provided by an NTP engine are as twice much as that of the best chemical rocket engine having half lift mass capacity. Nam et al. (2015) argue that NTRs are considered to be one of the most vital rockets in the modern space propulsion due to its high propellent efficiency resulting from the high efficiency of nuclear reactor used and known by "EHTGR."

In NTR, hydrogen propellent is heated to the maximum temperature by fission energy without fuel melting and then cooled by hydrogen propulsion. Enthalpy is absorbed by propellent from EHTGR and is then expelled out via a converging-diverging nozzle at a very high temperature. The engine of an NTR has a thermodynamic nozzle, a propellent feeding system that has a Turbo-Pump Assembly (TPA), and an EHTGR (Fig. 6.11).

Insulant propellent tanks contain the liquid hydrogen and come out via a pump of the TPA. The secondary reactor parts include moderator, nozzle, and reflector that receives the hydrogen flow to obtain heat required for feeding the rocket's thrust. For the driving turbine, the heated hydrogen runs up to the TPA when secondary parts are cooled. According to Burns and Johnson (2020), the fuel development under the Space Nuclear Thermal Propulsion program shows that the primary fuel zone of a core receives the hydrogen and is then expelled through the thermodynamic

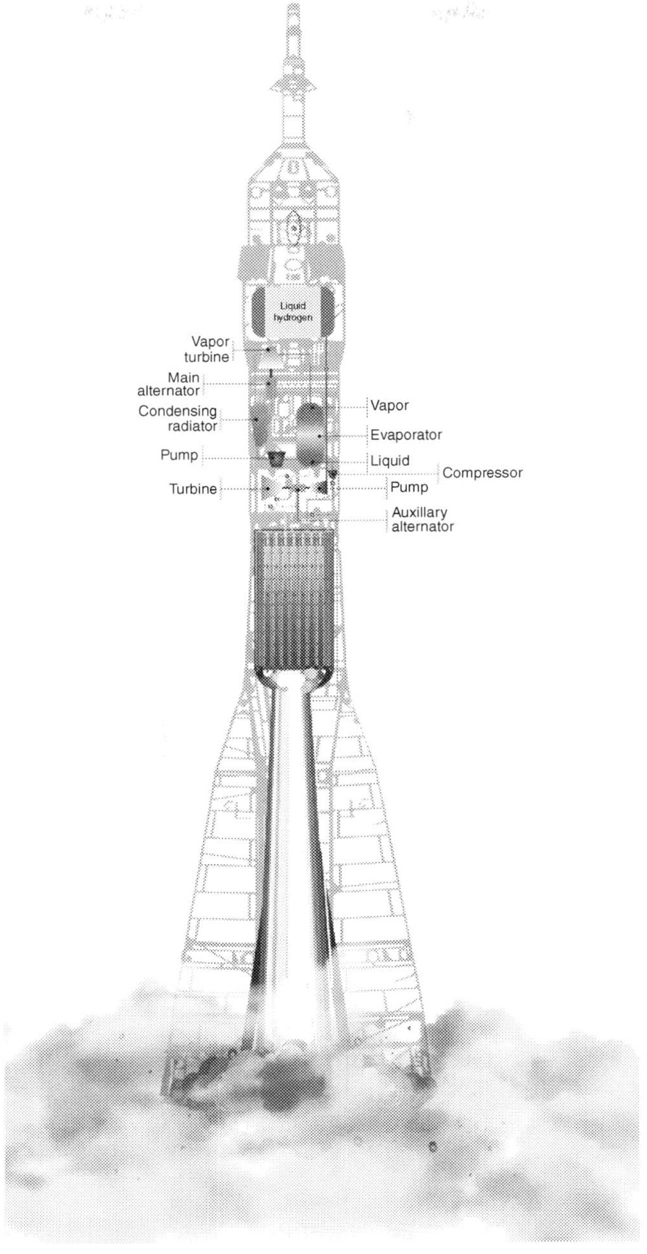

FIGURE 6.11 Nuclear thermal rocket operating using extremely high temperature gas cooled reactor (*EHTGR*).

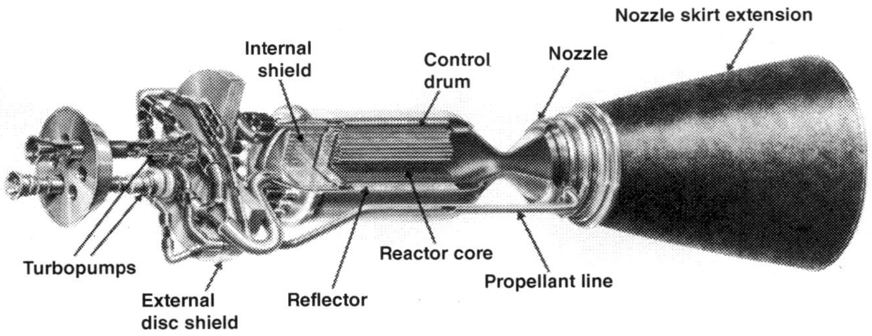

FIGURE 6.12 An explanatory drawing of the NERVA thermodynamic nuclear rocket engine (NASA, 2020, Nuclear Thermal Propulsion: Game Changing Technology for Deep Space Exploration, available from https://www.nasa.gov/directorates/spacetech/game_changing_development/Nuclear_Thermal_Propulsion_Deep_Space_Exploration).

nozzle at approximately 2400–3000 K, which generates the necessary and efficient rocket thrust. Thus, the transit time from Earth to Mars can be cut from 6 to 3 or 4 months by using an NTR engine according to NASA'S Marshall Space Flight Center (Whitehouse, 2020). This helps to reduce the crew's exposure to cosmic radiations and the microgravity that cause serious health effects. The energy density makes nuclear propulsion a fundamental technology in the exploration of space. Borowski et al. (2012) argue that the main parameter of the propellent efficiency of a rocket engine is the specific impulse (I_{sp}) and NTR's able to achieve (I_{sp}) of ~900 seconds (s) or more and are far better than what can be achieved so far by the chemical rockets. (I_{sp}) is related to the molecular weight of the propellant and the temperature of the core. The launch cost of an NTR is cheap because NTR accomplishes longer space missions at much lower requirements since the ratio of the thrust is high over the rate of propellant consumption (Borowski et al., 2012).

The NTR system is characterized as a dual-function system that is capable of generating electrical power and propulsion energy at the same time (Nam et al., 2015). It can generate electric power with a dynamic power system that uses the existing reactor's heat. The electrical power of a spacecraft is primary for communication provides the required warmth to the instruments. One of the main advantages of the bimodal NTRs is that they eliminate the need for additional sources of electric power like a solar electric system or RTG. A lot of research was developed in this regard as a part of two major projects for nuclear propulsion in space that are NERVA and ROVER (Fig. 6.12). These projects were conducted from 1955 to 1973 in the United States (Nam et al., 2015). Twenty different reactors were built in these projects and brilliant technical successes demonstrating the practical feasibility of NTR. Studies show that the probability of system failure and the risks of radiation exposure from NTR technology are very low and therefore, make it safer and reliable as compared to chemical rockets (Nam et al., 2015). The radiation is not released from the rocket at the time of its launching into the atmosphere of the Earth because these are not used when the rocket is taking off from the Earth. The transit time of NTR is short in deep space explorations. Moreover (Nam et al., 2015), revealed a number of the advantages of using NTR, which includes its low cost of launching, for example, the cost of launching two NTR missions are equivalent to the price of launching one chemical rocket, its bimodal function that provides both the required propulsion force and the required electric power, and finally NTRs are two times faster than chemical rockets with the same amount of propellant.

6.7 Conclusion

Nuclear science and its peaceful applications witnessed great importance after World War II, specifically on December 8, 1953, when the US President Eisenhower announced in a speech delivered to the United Nations Council under the title of "Atoms for Peace." Since then, atomic energy scientists have focused their research on developing the use of nuclear science and its peaceful applications to contribute to advancing sustainable development and achieving world well-being. The applications of nuclear science and radioisotope technology have played a prominent role in advanced science and scientific research such as space exploration, nuclear medicine, nuclear forensics, archaeology, geology, palaeontology, hydrology, and many more.

Nuclear and isotopic technology has played a prominent role in predicting and treating millions of patients around the world who suffer from serious and incurable diseases, such as cancer, heart, and brain diseases that traditional medicine has been unable to treat. Diagnostic nuclear imaging techniques such as PET and SPECT, the use of radiotracers, and internal or external irradiation techniques examine precisely the efficiency of the work of human body organs and identify the presence of any malignant tumors of cancer cells and eliminating them without harming healthy cells. Nuclear and isotopic technology boosts the progress of medicine in predicting neurological diseases such as dementia and Alzheimer's. Ionizing radiation is another crucial application of nuclear science and radioisotope technology that is used widely in the medical sector to sterilize medical equipment and products.

It is noteworthy that nuclear technology helped in solving the mystery of many cases in the courts and provided important and absolute information that could be used as conclusive evidence during the trial. The use of atomic sciences in criminal investigations has witnessed a remarkable development in light of the development of crime's tools, as criminals try their best to leave no evidence at the crime scene, but some microscopic and macroscopic evidence is always left behind. Nuclear sciences assist forensics medicine in gaining access to this evidence and uncovering clues of crimes by tracking this evidence using nondestructive quantitative and qualitative radiological analysis methods of samples obtained from the crime scene. Analysis of nuclear activation showed that Oswald's rifle was used to assassinate President John F. Kennedy in 1977 when the assassination accident of President John F. Kennedy was reinvestigated

Radioisotopes dating methods are a vital branch of nuclear science, which are widely used in many applications such as archaeology, climatology, hydrology, geology, meteorology, and nuclear forensics science. Nuclear dating methods are very useful and precise tools for the determination of the age of rocks, the ancient artworks, cave paintings, and engravings, and the death date of the organic matters. Accordingly, geochemists, geologists, geophysicists, and archeologists have joined hands with nuclear scientists to widen the research and maximizing the use of nuclear peaceful applications in these fields. In archaeology, for instance, the history of art paintings, cave engravings, and skeletons has always been a controversial topic until the emergence of radiometric dating method that helped in extrapolating the history of the old nations such as the Mayan civilization.

Not only that but nuclear science and radioisotope technology play a significant role in water resources management and its sustainability. Isotope hydrology is one of the modern scientific research methods used by experts in many countries around the world to trace the effect of freshwater movement according to the water cycle. This helps to provide a better understanding in assessing and identifying ground and surface water sources, the age, recharging mechanism, pollutants, movement, and interactions of this water within different geological formations. Isotope hydrology relies on the analysis of various labeled isotopes in the natural water, to generate a detailed hydrological map. This helps scientists to make evidence-based decisions on sustainable resource management for sustainable development.

Chemically, the water molecule is composed of two hydrogen atoms and one atom of oxygen. Oxygen, as well as hydrogen, have different isotopes such as ^{16}O, ^{17}O, ^{18}O,^{1}H, ^{2}H, ^{3}H, respectively. Scientifically, this proves that water is not same and is different from one place to another based on the scientific fact that isotopes have the same number of electrons and protons but different numbers of neutrons and therefore different atomic mass. Accordingly, some water may carry heavier atoms while some others carry lighter atoms. Scientific studies reveal that the presence of certain naturally occurring radioactive isotopes tracers assists in knowing the age of water that may range from a few months to a million years. For instance, the presence of ^{3}H indicates that water is less than 100 years old. While the presence of ^{14}C indicates that the water age is around 40,000 years old. On the other hand, the water containing ^{81}Kr, ^{4}He, ^{129}I, and ^{36}Cl indicates that age range from 40,000 years up to million years. Moreover, the isotopes composition of groundwater in hydrological environments provides significant information about the regional climatic conditions at the time of recharging. Stable and radioactive isotopes, including ^{18}O, ^{14}C, ^{2}H, and ^{3}H, can give vital information about physical and chemical processes acting on groundwater and hydrological features of aquifers that include the rate and mechanism of recharge, aquifer interconnections time, and the origin. These parameters are difficult to be measured using any technique other than isotopic technique. This information is vital for the sustainable management and assessment of water resources, especially in the dry regions and in developing countries.

The peaceful application of nuclear science and radioisotopic technology was not limited to the Earth, but also reached the moon and neighboring planets. During the last 70 years, the exploration of space was considered to be very expensive and one of the most complex jobs for humans due to the harsh conditions and complex challenges in space. However, this is seen as a significant advancement in human history. Nuclear science has been playing significant roles in making space exploration more successful, reliable, and cheaper since the launch of the first nuclear-powered spacecraft using RTG that has been launched on June 29, 1961, by the American Navy and named "Transit 4A navigation satellite." Nuclear fission energy is the best option for high thrust propulsion mission's in terms of technical maturity and energy density required for long space missions. There are three types of space propulsions are powered by nuclear fission energy, namely NEP, NPP, and NTP. NTRs utilize nuclear fission reaction and works on the principle similar to nuclear power plants and propulsion ships. Fission reaction provides the necessary propulsion power to penetrate the gravity forces that also help in reducing the exposure time of space radiation that astronauts may be exposed to during their travel between Earth and the red planet. The capability to provide reliable and sustainable long-term operational electricity required for space missions that conventional systems such as solar power, fuel cells are unable to provide as compared to nuclear-based systems. This is achieved by converting the energy produced by the radioactive decay of plutonium or americium-241into electrical energy via RTG. This energy is used to power the satellites orbiting around the Earth for many years, space equipment, and ISSs.

References

Ali, A., 2017. Failure Analysis and Prevention. BoD—Books on Demand. In Tech, Croatia, pp. 1–216.

Aggarwal, P., Araguas, L., Choudhry, M., Duren, M., and Klaus Froehlich, K., 2013. Lower groundwater ^{14}C age by atmospheric CO_2 uptake during sampling and analysis. National Ground Water Association, Ohio, USA, pp. 20–24.

Alauddin, M., 2012. Positron emission tomography (PET) imaging with 18F-based radiotracers. Am. J. Nucl. Med. Mol. Imag. 2 (1), 55–76.

Alkass, K., Buchholz, B., Ohtani, S., Yamamoto, T., Druid, H., Spalding, K., 2010. Age estimation in forensic sciences. Application of combined aspartic acid racemization and radiocarbon. Mol. Cell Proteom. 9 (5), 1022–1030.

Ayliffe, L., Marianelli, P., Moriarty, J., Moriarty, K., Wells, R., McCulloch, M., Mortimer, G., Hellstrom, J., 1997. Rainfall variations in southeastern Australia over the last 500,000 years. In: Proceedings of the Seventh Annual V. M. Goldschmidt Conference, Tucson, p. 2265.

Baron, S., 1996. Medical microbiology. University of Texas Medical Branch, Galveston, Tex, USA, pp. 1–1273.

Benson, S.J., Lennard, C.J., Maynard, P., Hill, D.M., Andrew, A.S., Roux, C., 2009. Forensic analysis of explosives using isotope ratio mass spectrometry (IRMS)—discrimination of ammonium nitrate sources. Sci. Just. 49 (2), 73–80.

Bentley, R., 2006. strontium isotopes from the Earth to the Archaeological Skeleton: a review. J. Arch. Method Theory (13) 135–187.

Borowski, S.K., McCurdy, D.R., Packard, T.W., 2012. Nuclear thermal propulsion (NTP): a proven growth technology for human NEO/mars exploration missions. In: Paper presented at the 2012 IEEE Aerospace Conference. https://ieeexplore.ieee.org/document/6187301. (Accessed 26 August 2020).

Britannica Educational Publishing, 2010. Geochronology, Dating, and Precambrian Time: The Beginning of the World as We Know It. Britannica Educational Publishing, New York, pp. 1–248.

Burns, D., Johnson, S., 2020. Nuclear Thermal Propulsion Reactor Materials. Intech Open, London. Avilable from: https://www.intechopen.com/online-first/nuclear-thermal-propulsion-reactor-materials.

Campeau, F., Fleitz, J., 2016. Limited Radiography Delmar Cengage Learning. USA, pp. 1–640.

Carroll, D., 2012. Rock Weathering. Springer, pp. 1–204.

Cartwright, I., Cendón, D., Currell, M., Meredith, K., 2017. A review of radioactive isotopes and other residence time tracers in understanding groundwater recharge: possibilities, challenges, and limitations. J. Hydrol. 555, 797–811.

Charles, R., Midkiff, J., 1977. Detection of gunpowder/gunpowder residues—a state-of-the-art review and recommendations for further research. In: Prepared for the Committee on New Development and Research, American Society of Crime Laboratory Directors. Forensic Branch National Laboratory Center, U.S. Treasury Department, Washington, DC, pp. 1–33.

Carroll, D., 2012. Rock Weathering. Springer, New York- London, pp. 1–204.

Cockell, C., 2020. Astrobiology: Understanding Life in the Universe. John Wiley & Sons Ltd, Hoboken USA, pp. 1–632.

Dadachova, E., 2008. Radioimmunotherapy of infection with 213Bi-labeled antibodies. Curr. Radiopharm. 1 (3), 234–239.

Dar, M., Masoodi, M., Farooq, S., 2015. Medical Uses of Radiopharmaceuticals. PharmaTutor 3 (8), 24–29.

Edgell, H., 2006. Arabian Deserts Nature, Origin, and Evolution. Ch19: Dating methods as applied to Arabian deserts and deposits. Springer, The Netherlands, pp. 1–583.

FAO, 2017. Water for Sustainable Food and Agriculture. A report produced for the G20 Presidency of Germany. Food and Agriculture Organization of the United Nations, Rome, pp. 1–27.

Farhood, B., Samadian, H., Ghorbani, M., Zakariaee, S., Knaup, C., 2018. Physical, dosimetric and clinical aspects and delivery systems in neutron capture therapy. Rep. Prac. Oncol. Radiother. 23 (5), 462–473.

Finnigan, M., 1995. 15N/14N and 13C/12C by EA-IRMS. Forensic studies using the ConFlo II interface. Application Flash Report No. 15.12/1995 PL 0/1177.

Galewsky, J., Larsen, H., Robert, D., Field, R., Worden, J., Risi, C., Schneider, M., 2016. Stable isotopes in atmospheric water vapor and applications to the hydrologic cycle. Rev. Geophys. 54 (4), 809–865.

Gentile, N., Siegwolf, R.T., Esseiva, P., et al., 2015. Isotope ratio mass spectrometry as a tool for source inference in forensic science: a critical review. Forensic Sci. Int. (251) 139–158.

Greene, K., 2002. Archaeology: An Introduction. University of Pennsylvania Press, Philadelphia, pp. 1–334.

Griffeth, L., 2005. Use of PET/CT scanning in cancer patients: technical and practical considerations. Proc (Bayl Univ Med Cent) 18 (4), 321–330.

Hafizbegovic, E., 2015. Life Is a Challenge: Journey to Discover the Secret to Life. Balboa Press, Bloomington, IN, USA, pp. 1–180.

Hamblin, J., Burns, W., 2005. Science in the Early Twentieth Century: An Encyclopedia. ABC-CLIO. Santa Barbara, CA, USA, pp. 1–399.

Harper, L., Powell, J., Pijl, E.M., 2017. An overview of forensic drug testing methods and their suitability for harm reduction point-of-care services. Harm. Reduct. J. 14 (52), 1–13.

Heldman, D., 2003. Encyclopaedia of Agricultural, Food, and Biological Engineering, pp. 1–1208.

Hobson, K., 1999. Tracing origins and migration of wildlife using stable isotopes: a review. Oecologia 120 (3), 314–326.

Holland, H., Turekian, K., 2010. Isotope Geochemistry: A derivative of the Treatise on Geochemistry. Academic Press, London, UK, pp. 1–752.

Hölting, B., Coldewey, W., 2018. Hydrogeology. Springer, Berlin, Germany, pp. 1–357.

IAEA, 1963. Dating the Past with Radioactivity. IAEA. Vienna, Austria, (5–1), 18–21.

IAEA, 1990. Practical aspects of operating a neutron activation analysis laboratory. IAEA-TECDOC-564, IAEA, Vianna Austria, pp. 1–251.

IAEA, 1993. Nuclear techniques for sustainable development: water resources and monitoring environmental pollution. IAEA Bull. 1 7–13.

IAEA, 2004. Isotope Hydrology and Integrated Water Resources Management. Vienna, pp. 1–485.

IAEA, 2006. Nuclear forensics support. Technical Guidance Reference Manual. IAEA. Vienna, Austria, pp. 1-65.

IAEA, 2011. Using isotopes effectively to support comprehensive groundwater management—NTR 2011 Supplement. 55th IAEA General Conference Documents. IAEA. Vienna, Austria, pp. 1–10.

IAEA, 2013. Isotope Methods for Dating Old Groundwater. IAEA, Vienna, pp. 1–357.

IAEA, 2018. Development of a National Nuclear Forensics Library: A System for the Identification of Nuclear or Other Radioactive Material out of Regulatory Control. IAEA, Vienna, pp. 1–41.

IAEA, 2019. Isotope hydrology: an overview. IAEA Bulletin, Vienna, Austria, 4–5.

FAO/IAEA, 2019. Expanding the range of vaccines to fight human and animal diseases: how nuclear science helps. Using Nuclear Science to Expand the Vaccines Portfolio Video. YouTube https://www.iaea.org/newscenter/multimedia/videos/expanding-the-range-of-vaccines-to-fight-human-and-animal-diseases-how-nuclear-science-helps. (Accessed 11 August 2020).

IAEA. 2020. Medical sterilization. https://www.iaea.org/topics/medical-sterilization#:~:text=More%20than%20160%20gamma%20irradiation,are%20sterilized%20with%20gamma%20irradiation. (Accessed 13 August 2020).

IBP, 2007. Russia Nuclear Industry Business Opportunities Handbook Volume 1 Strategic Information, Developments, Contacts. IBP, Washington, USA, pp. 1–406.

Kelley, S., 2002. K-Ar and Ar-Ar dating. Rev. Mineral. Geochem. 47, 785–818.

Kristo, M., 2018. Chapter 21: Nuclear Forensics. Handbook of Radioactivity Analysis, Radioanalytical Applications. Livermore National Laboratory, Livermore, CA, USA, Vol. 2, pp. 1–59.

Maidana, C., 2014. Thermo-Magnetic Systems for Space Nuclear Reactors: An Introduction. Springer, New York, pp. 1–53.

Mari, F., Bertol, E., Fineschi, V., Karch, SB., 2004. Channelling the emperor: what really killed Napoleon? J. R. Soc. Med. 97 (8), 397–399.

Martínez, A., Mora, A., Knappett, P., Ornelas-Soto, N., Mahlknecht, J., 2020. Tracking nitrate and sulfate sources in groundwater of an urbanized valley using a multi-tracer approach combined with a Bayesian isotope mixing model. Water Res. 182, 115962.

Mason, L., 2002. Power Technology Options for Nuclear Electric Propulsion, pp. 114–121 NASA. IECEC 2002 Paper No. 20159.

Mieny, C., 2003. Principles of Surgical Patient New Africa Books. Souh Africa, pp. 1–1103.

Mithen, S., 2005. Creativity in Human Evolution and Prehistory. Routledge, London and New York, pp. 1–312.

Musazzi, S., Perini, U., 2014. Laser-Induced Breakdown Spectroscopy: Theory and Applications. Springer, New York, pp. 1–565.

Nam, S.H., Venneri, P., Kim, Y., Lee, J.I., Chang, S.H., Jeong, Y.H., 2015. Innovative concept for an ultra-small nuclear thermal rocket utilizing a new moderated reactor. Nuc. Eng. Technol. 47 (6), 678–699.

O'Connell, J., 2002. RT-PCR Protocols.Methods in molecular biology. Humana Press Inc, Totowa, NJ, USA, pp. 1–378.

Othman, S., 2019. Applications of Radiopharmaceutical in Nuclear Medicine Series 1. Universiti Tun Hussein Onn Malaysia. Penerbit UTHM, Malaysia, pp. 1–86.

Parr, R., Fjeld, C., 1994. Human health and nutrition: How isotopes are helping to overcome "hidden hunger". IAEA Bull. 4, 18–27.

Peppe, D.J., Deino, A.L., 2013. Dating rocks and fossils using geologic methods. Nature Edu. Know. 4 (10), 1.

Pipkin, B., Trent, D., Hazlett, R., Bierman, P., 2012. Geology and the Environment. Cengage Learning, USA. pp. 1–592.

Sen, P., Panigrahi, N., Rao, M., Varier, K., Sen, S., Mehta, G., 1982. Application of proton-induced X-ray emission technique to gunshot residue analyses. Foren. Sci. 27 (2) 330–9.

Shamsuddin, M., Sulaiman, W., Ramli, M., et al., 2018. Assessments of seasonal groundwater recharge and discharge using environmental stable isotopes at Lower Muda River Basin. Malaysia Appll. Water Sci. 8, 120.

Silveira, L., Usunoff, E., 2009. Groundwater—Volume II. EOLSS Publications, Oxford, UK, pp. 1–584.

Society of Nuclear Medicine and Molecular Imaging-SNMMI. (2020). Historical timeline. http://www.snmmi.org/AboutSNMMI/Content.aspx?ItemNumber=4175. (Accessed 6 August 2020).

Smith, C., 2009. Anthropology For Dummies. John Wiley & Sons, Hoboken, NJ, USA, pp. 1–384.

Taylor, R.E., Bar-Yosef, O., 2016. Radiocarbon Dating: An Archaeological Perspective. Routledge, New York, pp. 1–404.

The College of Physicians of Philadelphia. (2020). Vaccine development, testing, and regulation. https://www.historyofvaccines.org/content/articles/vaccine-development-testing-and-regulation. (Accessed 11 August 2020).

The National Institute of Biomedical Imaging and Bioengineering- NIBIB. (2020). Nuclear medicine. https://www.nibib.nih.gov/science-education/science-topics/nuclear-medicine. (Accessed 10 August 2020).

University of Rhode Island. (2004). Neutron Activation Analyses Proves Oswald Acted Alone In JFK Assassination. Science Daily. https://www.sciencedaily.com/releases/2004/10/041025131255.htm. (Accessed 15 August 2020).

US Department of Energy. (2015). The history of nuclear power in space. https://www.energy.gov/articles/history-nuclear-power-space. (Accessed 15 August 2020).

Verma, H., 2007. Atomic and Nuclear Analytical Methods: XRF, Mössbauer, XPS, NAA and Ion-Beam Spectroscopic Techniques. Springer, Berlin, pp. 1–376.

Vij, K., 2008. Textbook of Forensic Medicine and Toxicology: Principles And Practice. Elsevier, India, pp. 1–756.

Vuong, L., Song, Q., Lee, H., Roffel, A., Shin, S., Shin, Y., Dueker, S., 2016. Opportunities in low-level radiocarbon microtracing: applications and new technology. Future Sci. OA 2 (1) FSO74.

Whitehouse, D., 2020. Space 2069: After Apollo: Back to the Moon, to Mars, and Beyond. Icon Books, pp. 1–336.

Wicander, R., Monroe, J., 2012. Historical Geology. Brooks/Cole - Cengage Learnin, USA, pp. 1–448.

Yinon, J., 2003. Advances in Forensic Applications of Mass Spectrometry. CRC Press, USA, pp. 1–279.

CHAPTER

7

Nuclear safety & security

OUTLINE

Applications of Nuclear and Radioisotope Technology: The Atom for Peace and Sustainable Development
DOI: https://doi.org/10.1016/B978-0-12-821319-3.00009-9

7.1 Introduction

The peaceful uses of nuclear and isotopic technology have witnessed a great development in recent years, so the uses of nuclear technology have multiplied in pushing the wheel of sustainability development for many countries around the world. Many countries have adopted them in generating clean electricity in addition to their use in supporting various development sectors including but not limited to the industry, agriculture, medicine, pharmacy, space, archaeology, nuclear forensic, and many other vital sectors.

The spread of peaceful nuclear applications and the vertical proliferation of nuclear weapons by nuclear countries have led to increasing fears of the unlawful seizure of nuclear materials and equipment and their use in a manner that threatens human security. Accordingly, this called for necessary security precautions in light of the increasing risks of using these materials by criminal terrorist groups, especially after the events of September 11, 2001. Nuclear terrorists may adopt the method of direct attacks the nuclear facilities or recruiting insiders to support their terrorist activities from inside the nuclear facilities. ITDB (2020) reported that its illicit trafficking database contained a total of 3686 confirmed incidents reported by 139 participating States from 1993 until December 31, 2019. There are 290 confirmed incidents of these reported 3686 that involved malicious acts or illicit trafficking of nuclear and radioactive materials including highly enriched uranium. In total, 1023 incidents that are related to stolen or missing nuclear and radioactive materials, and 2373 incidents are related to uncontrolled sources, inadvertent unauthorized possession or shipment of nuclear or other radioactive material. These numbers cause concern and indicate that there is some defect in the nuclear security systems that should be reviewed and rectified immediately.

Accordingly, nuclear security is one of the axes of security that has received great attention in many countries' national defence policies. Nuclear security is an integral part of the economic security, environmental security, and human security system in light of the interference of peaceful nuclear applications to support the various sectors of development, which necessitated surrounding these uses with the necessary security precautions in light of the increasing threats of terrorism. Therefore, many countries of the world sought to strengthen their nuclear security systems to be within the defensive strategies at the national, regional, and global levels. The nuclear and radiological threat is no longer limited to the uses of nuclear weapons but has extended to many other fields, such as the unauthorized seizure and withdrawal of nuclear materials, illegal trafficking of nuclear materials. Nuclear terrorism is no longer one country's responsibility such as nuclear club countries, but rather concerns all countries by virtue of international transportation of nuclear materials, the spread of nuclear and radioisotopes technologies around the world. Therefore, it became necessary to create a nuclear safety and security system that guarantees the protection of nuclear installations from the inside, so that they can operate safely, efficiently, and has high human reliability free from risks that could result in the presence of insiders who may be recruited by terrorist groups to carry out terrorist plans. It is estimated that 80% of terrorist recruitment are originated based on social engineering and spear-phishing, which focuses on exploiting human weakness. For this purpose, this chapter will shed light on a number of essential topics such as nuclear safety that plays a vital role in the safety of the nuclear plant, its operators and therefore eliminating any potential of nuclear and radiological accidents that may pose a severe threat to nuclear security, national, and international security. This can be achieved by introducing a new methodology called "nuclear safety -based quantitative risk assessment and dynamic accident modeling using the SMART approach." Furthermore, any inherent safety system requires a reliable national emergency response plan to be in place and able to provide sufficient protection to the public in case of any nuclear or radiological accident.

Heinrich (1931) argues that human factors and human reliability are accountable for 88% of accident occurrence, therefore, this chapter will investigate the importance of human reliability assessment in the nuclear industry and related methods used for human reliability assessment. Human reliability failures may lead to having insiders who are a great threat to nuclear safety and security system. Such insiders can be easily recruited by terrorist groups and could be either a reliable nuclear operators who have significant authorities or could be a nuclear security guard. Therefore, it is crucial to distribute authorities and adopt a strong nuclear security system designed to avoid such a threat or risk. The statistics issued by the incident and trafficking database (ITDB) of illicit trafficking or malicious acts raise concerns about the stolen or smuggled quantities of nuclear and radioactive materials and the risk of reaching these stolen materials into the hands of terrorists or the nuclear black market that poses serious threats to both the national and international security. Accordingly, these statistics confirm the current nuclear security systems are still vulnerable and there is an urgent need to intensify international cooperation in the field of nuclear intelligence.

7.2 Nuclear safety-based quantitative risk assessment and dynamic accident modeling using the SMART approach

Numerous safety engineering and risk assessment management research have revealed the fact that operational and occupational misconducts pose significant risks to people involved in the nuclear industry, nuclear plants, and the environment. It is, however, feasible to mitigate accidents related to nuclear operations in its preliminary stages through preventive methodologies such as the improvement of human reliability and electromechanical machinery, adequate maintenance of appropriate safety measures, and the creation of safety barriers to reduce the dangers of risks and life-threatening situations. Operational and occupational risks from nuclear operations are easy to be identified in the early stages. The situation could be improved by developing adequate systems which aid the prediction of impending calamities, reliability levels, effective control measures, and mitigating risks at the source, as well as propagating the importance of early incident prevention awareness with the aim of making the nuclear operation inherently safer. It is important to carry out assessments and checks to ascertain the efficiencies and the reliability of safety barriers provided, confirming the adequacy of these provisions will further prevent the occurrence of major operational and occupational accidents thus promoting the safety and health of the workers. This section will introduce a new approach of quantitative risk assessment management and dynamic accident modeling in typical nuclear operations, this will be achieved by implementing the System Hazard Identification, Prediction, and Prevention (SHIPP) Methodology and Rational Theory (SMART approach) developed by Al Nabhani et al. (2017). The proposed approach is characterized by the following unique features: (1) dynamic modeling of operational and occupational accidents and risks considering the effectiveness and the reliability of the provided safety barriers, (2) uncertainty reduction by establishing a reliable system that facilitates the prediction of the likelihood that the safety barriers might fail, and (3) dynamic updating of any abnormal event probability occurrence as new information becomes available from the main system and its subsystems. This approach considers the occurrence of an accident as a result of the interaction of many random, physical, or latent components in the system or its subsystem in sequential or nonsequential order, which contribute to the degradation of the system reliability and therefore cause an accident. The proposed approach creates an integrated framework that aids the dynamic prediction and updating data related to abnormal events, reliability degradation using Bayesian updating theorem that update the risk assessment process when new evidence or any information that indicate of a deviation in the operation process is noticed. The essence of this approach is to develop a system that aids the monitoring of operational and occupational risks based on the reliability in a dynamic pattern. This will promote the development of effective and protective measures to reduce the imminent risks related to the nuclear industry.

7.2.1 The working principle of the quantitative risk assessment and dynamic accident modeling using the SMART approach

SMART's approach is a hybrid approach to the SHIPP methodology and rational theory. The SHIPP methodology is a generic framework with a purpose to identify the components of the main system and its subsystems, evaluate the effectiveness of the available safety barriers, and finally modeling accident occurrence process (Rathnayakaa et al., 2011; Rathnayakaa et al., 2013). According to the science of safety engineering and risk assessment management, accident modeling process can be performed using the following models: (1) sequential model where an accident is considered to occur in sequential order, and usually, this process can be represented by the "Swiss Cheese Model." In this model, cheese slices represent the safety barriers for the main system while the latent failures such as human errors and equipment failures that are hard to be quantified are represented by the random holes in cheese slices; or (2) epidemiological model where an accident is considered to occur because of a combination of physical and latent causes; or (3) systematic model, which considers the accident occurrence as a result of the interaction between different components of the system and its subsystems. Usually, this model is considered to be a complex model due to the complexities in the design and operational process of both the system and its subsystems that make it hard to track or predict the consequences.

The main objective of the rational theory is to systematically model the behavior of safety and prevention barriers set to prevent accident occurrence based on the logical, inductive, reliability, and probabilistic analysis. The inductive analysis reasoning approach, which examines the true reasons behind accident occurrence to reach a true conclusion (since we are not sure of the accident causes that contributed to its occurrence). That is why we are investigating the causes or hypothesizes behind accident's occurrence so that we can avoid or prevent accidents from occurring in the future. Such causes or hypothesizes are viewed as supplying strong evidence for the truth of the conclusion; therefore, the truth of the conclusion of an inductive argument is probable, based on the available evidence. For example, if you fall, you might be injured or might not be and this is in line with the probabilistic analysis approach that can be mathematically expressed as following: if an event A has occurred or is true then its consequence B, or C, or D probably be true to occur. On the opposite, the deductive approach means that we are sure and confident of the outcomes. Accordingly, the conclusion of a deductive argument is certain. Therefore, in the science of safety engineering and risk assessment management, the deductive approach can only be applied if and only if we are certain about the root cause behind the accident if and only if we are certain about the root cause behind the accident. For example, if you are shot by a gun directly in the heart, then you will die. Mathematically this can be expressed as following: if the occurrence of an event A is true and certain, then consequences occurrences of B, C, and D are true. Accordingly, the basic premise of the rational theory is that an accident occurs because of the occurrence of joint conditional behaviors of different parameters in the system and its associated subsystems. Therefore, the rational theory investigates logically all physical and latent causes and related reliabilities of the system and its subsystems that have contributed to the accident occurrence. Thus, it has the ability to predict any potential errors that may arise from the evaluation of the latent and physical causes, review, and update the assessment process.

By integrating the SHIPP methodology and rational theory, the SMART approach (Fig. 7.1) is, therefore, able to: (1) identify the reliability of the systems and their subsystems and their interactions; (2) identify and analyze all possible occupational scenarios; (3) model all possible accidents' scenarios based on the performance and the reliability of safety barriers using Monte Carlo simulation; (4) predict and update the failure probabilities of the identified safety barriers; (5) enable proactive management of risks using either adaptive risk management techniques or precautionary principle techniques.

Fig. 7.1 illustrates the flowchart of the procedure applied for the quantitative risk assessment and dynamic accident modeling using the SMART approach in the nuclear industry, which mainly comprises of five major parts, as follows:

1. SHIPP methodology part. In this process, the components of the system and its subsystems and their interactions are identified and analyzed. This helps to know how the system and its subsystems are functioning and interacting with each other. The hazards identification process can be started in connection with the system and its subsystem after identifying and understanding the relationship between the system and its subsystems. This helps to track and identify easily all possible hazards in connection with the components of the system and its subsystems. Finally, is to simulate all possible accident as a result of the system and its subsystems' interaction and reliability, and their behaviors for thousands of times so that all possible causes have been well-identified for ease reasoning and prevention purposes. Therefore, if the identified hazards caused major threats, then the rational reasoning part is executed and if it is not, then it is recommended to use any of classical qualitative risk evaluation for nonmajor hazards.
2. Rational reasoning part: the main objective of this step is to identify and evaluate the efficiency, reliability, and performance of safety and prevention barriers identified for the system and its subsystems. Usually, five

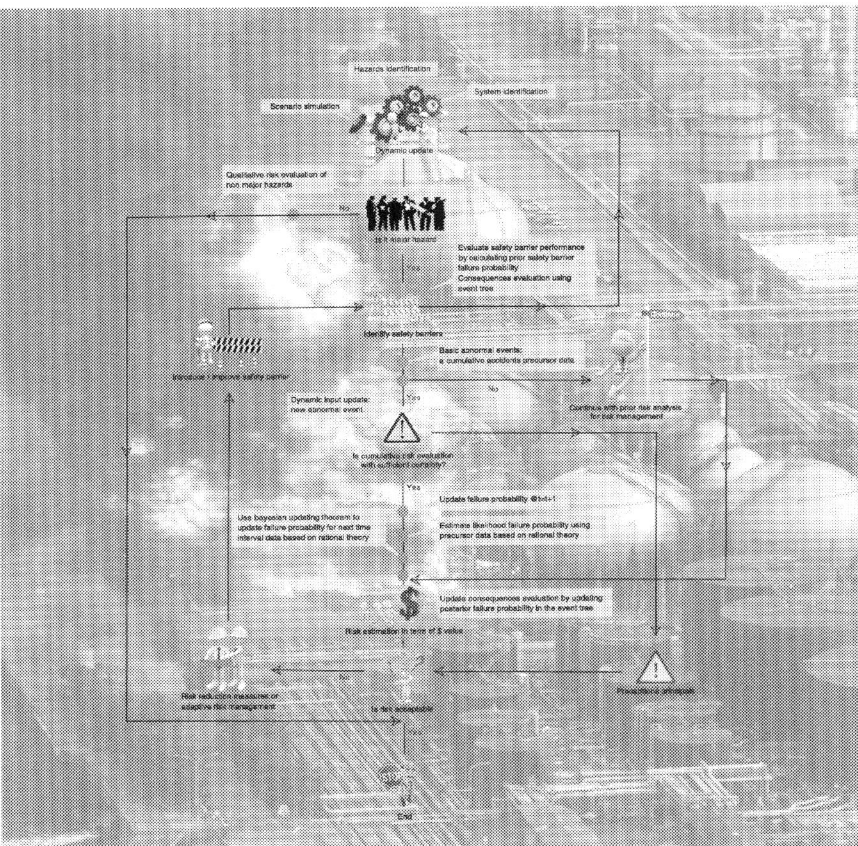

FIGURE 7.1 The flowchart of the procedure applied for the quantitative risk assessment and dynamic accident modeling using the SMART approach.

sequential and interconnected safety barriers for accident prevention are identified to prevent any nuclear accident, these are as follows:

I. Early Detection Safety & Prevention Barrier (EDSPB). This is the release prevention barrier that is responsible for preventing the initiating event of an accident release at the nuclear plant such as sensors, detectors, safety instrumented systems, and control process equipment. Safety instrumented systems and control process equipment are essential equipment in any high-risk process plants, such as petrochemical or nuclear plants. For instance, these plants are subjected to fire risk, explosion, tank overflow, gas release, chemical or radiation exposure. Usually, these plants are built with many safety and security layers including process design, process control system, and alarms. Despite this, these layers are still insufficient to provide enough protection. For instance, Ramsden (1996) reported that in 1974 a nylon plant in the United Kingdom exploded and 28 killed, 100 injured. After 10 years and exactly on December 3, 1984, a fertilizer plant in Bhopal in India also exploded and it has been reported that nearly 15000 people were killed during the first month after the explosion and about 200,000 were injured (Du Bois and Richeldi, 2009). The explosion of the refinery in Texas City on March 23, 2005, resulted in the death of 15 people and the injury of 170 people (Manuele, 2011). These plants were designed according to the highest safety standards and are operated by well-qualified people. Despite that still, something went wrong, which can be attributed to human errors and the lack of dynamic accident modeling and quantitative risk assessment. Accordingly, safety instrumented systems are important early detection safety and prevention barriers. They are an independent system designed to control abnormal events and composed of sensors, logic servers, and control equipment. They react immediately when there is a failure by the control process equipment that are designed to operate very complicated and risky operations such as nuclear plant as an alternative to manual operation to reduce human error. For instance, if a reactor in a nuclear plant is over-pressurized, then sensors will send signals to the logic server that will interpret received signals and will send the instruction to a process control equipment, which will activate accordingly, shut down the flow and release the pressure. However, still, the safety instrumented system is not reliable 100%. Thus, the safety instrumented system is subjected to failure probability and as failure probability goes down, the system becomes more reliable and its safety integrity level increases. So, this can be improved by adding a redundant safety system with high safety integrity level that makes the system safer and reliable.

II. Isolation Integrity Safety & Prevention Barrier. This is a dispersion prevention barrier at any process in the nuclear plant. It includes, but is not limited to, the following sub-barriers: equipment insulation; emergency shut down mechanisms, work permits, and many others.

III. Personal Protection Equipment and Exposure Duration Safety & Prevention Barrier (PPE&EDSPB). It includes, but is not limited to, the following sub-barriers: personal protection equipment (protective clothing, face mask, hand gloves, and safety boots), personal gas, and radiation detectors.

IV. Emergency Management Safety & Prevention Barrier. This safety barrier is considered as a mitigation barrier to control the hazards of any nuclear accident and its consequences. It includes, but is not limited to, the following sub-barriers: emergency preparedness, emergency intervention and response plan, emergency medical plan, emergency and safety drills, human reliability enhancement training.

V. Management and Organization Safety Prevention Barrier. This safety barrier intervenes positively or negatively in preventing accidents by interfering with the efficiency of all other barriers based on the behavior and awareness of the management and its responsibility toward strengthening these barriers. It includes, but is not limited to, the following sub-barriers: training, competency programs, safety policies, legislation, operating procedures, effective risk and safety management system, decision-making, management practices, and knowledge, leadership, and communication.

3. The efficiency and performance evaluation of the identified safety and prevention barriers can be done by gathering historical data related to their failures, abnormal events occurring during the operational process. Then consequences occurrence probabilities are evaluated using the event tree model. This process requires continuous update of any new information or arrival of new evidence or occurrence of any abnormal events or any deviation in the process. This helps to systematically model the behavior of all possible root, passive causes, human reliability within the system and its subsystem. These factors are found to be the main contributors to the accident occurrence and identify the reliability and the performance of the identified safety and prevention barriers. Once the cumulative risk evaluation based on rational reasoning provides sufficient certainty, then the third part of the SMART flowchart, which is the prediction part will take place. However, If the uncertainty level is still high, then precaution principles will be used. This will provide high-level safety protections in the absence of any details about the risk level.
4. Prediction part: The main purpose of this part is to predict the likelihood of failure probabilities of the identified safety barriers during the next time of interval (t+1) based on the precursor accumulative knowledge and historically gathered data. Monte Carlo simulation is used to simulate gathered data thousands of times to systematically model the behavior of all possible root and passive causes. These causes have been identified to be the main contributors to the accident occurred within the system and its subsystem based on the performance of the identified safety and prevention barriers. In this step, logical, inductive, and probabilistic analysis are used to improve the certainty level in order to have a better and accurate understanding of the accident process and its behavior.
5. Updating part: Once all likelihood of failure probabilities of the identified safety and prevention barriers have been calculated based on all expected scenarios during the next time of interval (t+1), Bayesian inference mechanism is applied to update the probability based on

the hypothesizes from all the scenarios created by the simulator using existing evidence or new information when become available. The Bayesian update is a key element in any dynamic accident modeling and quantitative risk that aids to take accurate decision-making related to safety improvement. Once safety failure probabilities have been updated based on the arrival of any new information or availability of new evidence or occurrence of any abnormal events/ these probabilities are posterior failure probabilities that are used to be fed in the event tree model to update the consequences occurrence probabilities. This provides a holistic idea about the system degradation in the next time of interval (t+1). Accordingly, it helps hugely decision-makers to take the right decision at a very early stage to prevent accidents from occurrence.

6. Risk management part: Obtained risk value is then converted into a dollar value to evaluate if the risk is acceptable or not. Accordingly, the decision is taken depending on which of the risk management techniques described below will be used to bring the risk to an acceptable low level:
   - Adaptive Risk Management Technique: This technique employs the available knowledge and data related to the risk certainty level to reduce the risk by introducing new changes in the system. It exclusively depends on the experience of the decision-maker to recommend the types of risk reduction to be applied.
   - ALARB Risk Management Technique: This technique aims to bring down the risk to an acceptable level through the adoption of a number of risk reduction measures as a function of the cost. The main hypothesis of this technique is to maintain the cost as low as possible compared to the benefit gained otherwise there is no point to implement it. For example, If the cost paid to introduce new safety barriers is higher than the benefit gained from the introduction of this risk reduction measure, then it is not recommended to implement based on ALARB's premise.
7. Finally, based on the decision that will be taken regarding which techniques to be adopted for risk reduction measures such as introducing new improvement or modifying the existing safety prevention barriers. Accordingly, SMART will automatically and dynamically update and evaluate the performance of the improved safety barriers and associated risks to make sure that the risk is reduced.

7.2.2 An overview of an emergency response plan to protect public in case of a nuclear or radiological emergency

An emergency response plan is primarily executed to assist organizations in tackling different emergency conditions that could take place on their premises. Thus, in the case of a nuclear or radiological accident occurrence, there is no time to set up or call for an emergency meeting to discuss the plan forward and how to deal with the emergency situation. Instead, the action plans should be ready to be executed instantly without wasting a single minute. Failure to act promptly may lead to adverse health issues and more deaths, which are otherwise possible to avoid. Accordingly, an emergency action plan called a "15 min—emergency plan" should be contrived and implemented to protect the public. In case of any nuclear or radiological accident, within the first 15 minutes, the concerned authority for initiating the emergency response plan should proclaim the general emergency. Then the decision-makers authorities should be approached in the next 15 min so that they can categorize the area nearby the accident into risk zones as per described classification below (IAEA, 2015). These classifications are determined according to the hazard index defined for the nuclear facility, which is based on the hazard index value, and the total stock of dispersible radioactive material in the facility, as shown in Eq. 7.1:

$$\text{Index of Danger} = \sum_i (A_i/D_{2i}) \tag{7.1}$$

where

A_i is the activity of the ith radionuclide.

D_{2i} reflecting the danger of ith radionuclide in the dispersible form defined by IAEA (2006).

1. A precautionary action zone: It is marked by an area of around 3–5 km radius from the actual locality of the radiological or nuclear accident, which arrangements shall be made for taking urgent protective actions and other response actions including but not limited to immediate safe evacuation, taking iodine thyroid blocking medicine. If the immediate evacuation is not possible for any reason, then shelter and take iodine thyroid blocking medicine until it is safe to evacuate.
2. An urgent protective action planning zone: It is margined by an area of 15–30 km radius from the actual locality of the radiological or nuclear accident. Urgent protective actions and other response actions include safe evacuation after precautionary action zone, if that is not possible, then shelter and take iodine thyroid blocking medicine until it becomes safe to evacuate.
3. An extended planning distance: This zoon marked by an area of 50–100 Km radius from the source of the radiological or nuclear accident beyond the urgent protective action planning zone. In such situation, arrangements shall be made to conduct monitoring and assessment of the radiological situation off the site in order to identify areas and this should continue from day one to a few weeks or months following the significant radioactive release. Under such circumstances, caution of contaminated food and drinks must be taken, evacuate/relocate where Operational Intervention Levels -OILs are exceeded within days to weeks (further details on OILs can be found in the IAEA safety

standard: Operational Intervention Levels for Reactor Emergencies and Methodology for their Derivation- EPR-NPP-OILs-IAEA, 2017).

4. An ingestion and commodities planning distance: This zone is marked by a 100–300 km radius from the actual locality of the nuclear accident, beyond the extended planning distance, which arrangements shall be made to take a response to protect/restrict ingestion of contaminated food and water and to continue sampling.

Within those 15 minutes, the off-site decision-maker must direct the public to take protective measures according to each classified zone as described earlier and the public must start taking their urgent protective actions accordingly. Following the public' protective actions, the emergency leader should hold a press briefing to address the public's concerns. Outside the "Urgent protective action planning zone," centers must be established to register and monitor those who exceed Operational Intervention levels OIL 8 and OIL 4, decontaminate and provide preliminary medical screening, and advise hospitals to immediately treat these people who are exposed to radiations in these zones. Areas, where OIL 1 has been exceeded must be evacuated within 24 hours. Zones where OIL 3 is exceeded, contaminated food, milk, and water, must be restricted and continuously sample them in accordance with OIL 4 for monitoring purposes. The public who are in the areas, where OIL 2 is exceeded, must be relocated to a safer place within a span of one week up to a month. In case of a nuclear or radiological emergency, the following challenges regarding safeguarding acts must be considered:

- Wind direction may change at any time.
- Large populations may be required to be evacuated faster but in a safe manner to avoid crowding.
- There are vulnerable people such as the elderly and the hospitalized who need to be evacuated first.
- Thyroid blocking agents are required to be administered in time and enough stock must be available.

For more information on the General Safety Requirements Nuclear Security Guidelines No. GSR Part 7 (Preparedness and Response for a Nuclear or Radiological Emergency) by IAEA-2015, is attached in Appendix 2.

7.2.3 Human reliability assessment in nuclear industry

Various scholars argue that human reliability is the main reason for the occurrence of accidents. For instance, McKinnon (2019) states that unsafe acts are responsible for most of the accidents. In the 1920s, Heinrich concluded that "of all the accidents, about 88% are due to unsafe act by human,

10% due to unsafe conditions, and only 2% happened because of the acts of providence" (Johnson, 2011). Accordingly, if we successfully identify and control unsafe acts and unsafe conditions, then the accidents can be reduced by a considerable deal of 98%, where human factors (anthropogenic agents) contribute 88% of accident prevention. Due to this reason, human errors and the monitoring of human reliability are of great importance. Human error has been defined in multiple ways in the literature. For instance, Rasmussen (1993) defines human error as the actions of humans that differ from reference action as they are personalized and keep on varying with time. While Moray and Senders (1991) defined human errors are as an outcome of detectible behavior, arising from different levels of various psychological processes. These kinds of behaviors can be estimated using production merits and various approaches of assessing human reliability, including quantitative and qualitative methods for measuring the contribution of humans' errors/factors to the risks known as "Human Reliability Assessment." Appendix 3 summarizes the common human reliability methods used by various industries including the nuclear industry. The terminologies 'human error' or 'human factor' are mostly considered interchangeable in the safety and risk assessment literature and can be distinguished as the underlying causes of accidents (human factors) and immediate causes (human errors) (Schonbeck et al., 2009).

Al Nabhani et al. (2017) argue that organizational and management safety barriers indulged in causing accidents because they have a strong influence in drawing the strategic plans related to the safety improvement of their organization. This is achieved by having proper training, motivation, leadership, safety culture, and safety management plans that play vital roles in fostering safety of the system and human reliability, whereas the absence of these key drivers in the agenda of Organizational and management, will lead to accidents occurrences.

7.3 Insiders threat to nuclear safety and security

To fulfill safety requirements, the outlines made for a safe situation require considering the aptitude for starting occurrences from humans' failure, the failure of the equipment, engineering design, components, the system, and its subsystem. Human failure might involve a violation of nuclear operating procedures, leading to incorrectly closing the main valve of the cooling flow line and therefore, temperature buildup. While an engineering failure might involve nonfunctionality of the safety instrumented system such as the temperature sensor that is responsible for sensing any high temperature, which is supposed to raise the alarm and act accordingly. Both failures might lead to the explosion and discharge of nuclear radioactive materials that pose a serious threat to nuclear safety and security, therefore, safety and security are two sides of the Same Coin.

Accordingly, there is an urgent need to couple risk analysis techniques that consider a series of failure events as the cause of the accident occurring in the system or subsystems with reliability theory that precedes the cause of electromechanical or human failure (Leveson, 2020).

In analyzing the sequence of failure or abnormal events in nuclear facilities, the inherent safety and security system is considered a robust system when it takes into account internal and external threats to nuclear safety and security. Safety and security system should eventually exhibit that the human-based and engineered protective barriers and their interconnections have reliable strength for defending and confronting against any potential abnormal event or failure that may result in a nuclear threat. The quantitative risk assessment and dynamic accident modeling procedure modeled by the reliability approach will address the threats of human error. Unfortunately, many of the available human reliability assessment methods do not consider the threat of insiders to nuclear safety and security because many perceive it to be purely a security issue. However, the insider's threat poses a serious concern to both safety and security functions because they are entwined and required careful risk management. Hunker and Probst (2011), identify three ways to address insiders' threat in any nuclear facility that consider sociological, socio-technical, and technical aspects. These aspects are incorporated in the taxonomy of human error in the assessment of human reliability, which is pivoted on three human taxonomy that are:

- Human error: Unintended slip either skilled-based error (such as execution error) or Rule-based or knowledge-based error such as (wrong decision making).
- Misuse: Can be either unintended violation due to unclear rules or not understanding rules or intended violation for shortcuts.
- Insider: intended and planned abuse under force to cause either safety or security harm.

Under these circumstances, it is important to conduct a joint safety and security risk assessment and management that comprises of five main steps, which are as follows:

1. Identify safety hazards and security threats including insider threats of the central system along with its subsystem.
2. Identify safety and security barriers for the system and its subsystems.
3. Evaluate dynamically and quantitatively the performance, efficiency, and the reliability of safety and security barriers by calculating prior safety and security failure probability and consequences evaluation using event tree technique, analyzing precursor data, dynamically update arrival of new evidence to have sufficient certainty, then update failure probability @t $= t + 1$ using Bayesian updating theorem using rational theory.

4. Estimate likelihood failure probability using precursor data, then update consequences by updating posterior failure probability in the event tree.
5. Estimate the risk by assessing the hazard or threat consequences and their likelihood using the event tree.
6. Check if risk/threat is acceptable or not, if not then introduce risk/threat reduction measure or adaptive risk/threat management.

The IAEA (2008) argues that the nuclear industry insiders pose direct threats to both nuclear safety and security, and intentional human errors and malicious attacks may be one of the most important of these threats. Bunn and Sagan (2017) report that insiders may tamper with safety equipment to cause massive damage to the plant and cause huge economical loss as part of their terrorist plot such as the case with unit 4 of Belgium's Doel nuclear power plan. Insiders may also forge records related to the quantities of nuclear or radioactive materials with the aim of stealing small quantities of them frequently so that they are not easily detected, and with time, the stolen quantities multiply, and thus the risk doubles (IAEA, 2008). Over the past two decades, the Insider Threat Team, which is part of the Computer Emergency Response Team, has monitored numerous data related to internal threat attacks from various sectors, as these attacks have targeted either the theft of confidential or sensitive information or nuclear materials for personal gain or for intelligence purposes; or it is sabotage attacks to destroy the nuclear system. Therefore, insiders pose a serious threat to the nuclear industry (Bunn and Sagan, 2017).

However, malicious threats may also occur as a result of the failure of different security barriers in the system and the subsystem including human failure that is of constant concern in the industry, or from failure in the design of systems and operating procedures. Human failure and malicious acts may merge, by chance or by a planned internal attack. Thus, a safety and security management system must be able to distinguish between them and to interact immediately and effectively based on available variables. Accordingly, human factors and errors constitute a threat to nuclear safety and security, especially when there is a threat by human factors to safety and security that falls under the classification of insiders. Further guidance on safeguards and protection against insider threats can be found in Appendix 4.

7.4 Insiders and nuclear cyberterrorism

The concept of cyberterrorism appeared for the first time in the early eighties of the last century. According, Cyber Operations and Cyber Terrorism (2005), Kevin Colman defined cyberterrorism as "the premeditated use of disruptive activities, or the threat thereof, against computers and/or networks, with the intention to cause harm or further social, ideological, religious, political or similar objectives, or to intimidate any person in furtherance of such objectives." Cyberterrorism may include the usage of the internet for recruiting terrorists by luring and extorting them in order to disrupt infrastructure, or gather sensitive information, or cause physical damage to achieve political or extreme ideological gains. Most basically, cyberterrorism, for instance, may infringe the control system of air or ship traffic to cause an accidental collision or banks' account hacking. The design could play a significant role in supporting cyber-terrorists to do a terrorist's attack, for example, the recreational system network on airplanes, which has a USB port, or any cyber entry point is considered as one of the more vulnerable systems that could result in full control on the plane system.

Cyberterrorists may attack any rigid nuclear security system by hacking computer network operations with the support of insiders or by using advanced electronic technologies. Cyberterrorists can counterfeit or actually control the nuclear arms of a state and attack another country by simply reaching and controlling the nuclear facility network system, and if this happens, then this might ignite a nuclear world war. This may be an easier alternative for terrorists to build or acquire a nuclear weapon or dirty bomb themselves. This would give them more power and give the terrorists the advantages of high speed, elimination of geographical distance, and relatively low cost in comparison to the classical suicide terrorist attacks. Persistent loopholes in the electronic systems and computer technologies in nuclear facilities, which are to some extent unable to identify any hacking attempts or trace the identity of hackers have led to an inherent vulnerability in the use of these networks to manage nuclear facilities and nuclear arsenals.

Every computing device that is attached to the international network is vulnerable to remote control and infiltration. Even though the computers operating under closed and secured air gap network might also be harmed by different hacker approaches with insiders' help, for example, wandering notebooks, wireless and wired access spots, maintenance entry points, hardware, and software, a USB flash drive, internal emails. All of these factors may help insiders to infringe on a protected network by installing a programmed virus in the closed and secured network to causing damage to the entire system or to stealing information and transmitting them through hardware combinations that use a number of different mediums to overcome the airgap secured system and this includes acoustic, light, seismic, magnetic, thermal, and radiofrequency waves using different techniques, such as AirHopper, BitWhisper, and GSMem (Abaimov and Ingram, 2017). Not only that but, data can be sent via impulse signals through pressurized water such as water in the fire sprinkler systems that are available in all critical buildings, where sent signals to be collected later by a receiver or radar outside the facility.

Despite, the great efforts made by militaries or nuclear facilities to modernize their safety and security systems and their operational devices by adopting the latest technologies of the information technology and artificial intelligence revolution available in the market, encompassing modernized security setups, automated systems, such as nuclear triad capability, still this provides multiple entry points for terrorists to breach these systems. It is easier for terrorist networks to have control over these electromechanical devices and programmed software systems. In this way, they can successfully take command of a nuclear-armed submarine or the control room of a nuclear power station, and therefore, taking control over critical equipment. In this regard, Carr (2010) argues that in January 2002, a firewall-protected network for defences at Ohio's Davis–Besse nuclear power plant was infiltrated by Slammer worm virus that managed to breach the network and disabled the safety monitoring system for about five hours. Abaimov and Ingram (2017) report that the Stuxnet virus targeted Iran's nuclear program in 2010. The Stuxnet virus-infects computer systems running on Windows and in this accident targeted the supervisory control and data acquisition systems for Iran's nuclear program, it controlled Programmable Logic Controllers, which allow the automation of electromechanical processes used to control machines and processes including centrifuges to separate nuclear materials and cause huge damage to the centrifuges. Not only that but, cyberterrorists, for instance, can control the cyber systems of China from another country and utilize them for launching a nuclear attack by the United States against Russia or attempt to change the launch coordinates of the missile. In this scenario, Russia would believe the United States is attacking them, and the United States would perceive it to be a nuclear attack conducted by China. Or give false orders to launch a nuclear strike to India or Pakistan and send misleading statements directly to military leaders, government officials, and media sites to declare war and claim responsibility to stimulate a swift military response. Moreover, hackers can take control of and disable emergency security systems and emergency response communication. During such a case, they can block traffic and give misleading information that might lead to a disaster and hindering the disaster relief effort. Hackers also can give false orders to an operator in a nuclear plant or to a security guard at a nuclear facility to open the gate for a truck that is containing smuggled nuclear materials. Therefore, a well-organized and properly managed collaboration of a cyberterrorist and an insider could be sufficient for launching automated nuclear weapons without being indulged in compromising control and command booths directly.

7.5 Design-based threat

As mentioned previously, the design vulnerability plays a vital role in supporting terrorism to carry out any terrorist attack, for example, the entertainment system network onboard aircraft, which contains a USB port, or any electronic entry point that is considered a serious gap and an easy way for launching a cyber-terrorist attack. It could lead to complete control over the aircraft's system and passengers. Before delving into the design components, it is important to know the difference between the concept of safety and security very clearly. According to the IAEA (2011), it defines nuclear security as "the process of prevention and detection of and response to the theft, sabotage, unauthorized access, illegal transfer or other malicious acts involving nuclear materials, other radioactive substances, or their associated facilities." IAEA (2019), defines nuclear safety as "the achievement of proper operating conditions, prevention of accidents, or mitigation of accident consequences, resulting in protection of workers, the public and the environment from undue radiation hazards." On the other hand, Healey (2016) argues that nuclear safety deal directly with the internal threats resulting from human failure, while nuclear security deals with external threats. Thus, safety and security are two sides of the same coin. Accordingly, it is very important to have a reliable protective and preventative safety and security system that includes but is not limited to site safety and security, transport safety and security, Information/cyber-security, and personnel safety and security.

From a nuclear security perspective, an effective and reliable security system must be able to dynamically identify and react with all threats spectrum related to the nuclear industry. Moreover, the durability and depth of security required also depend greatly on the importance of the risk. Therefore, Risk/Threat-Based Dynamic Design of Nuclear Facilities (Fig. 7.2) defines this within the particular context derived from a state's current assessment of national and international threats as well as the vulnerability assessment of the facilities. It is essential for any nuclear security system to consider a number of measures including insider threat mitigation for the nuclear sector, which include but not limited to (1) prevention and exclusion of insiders by identifying abnormal behavior or characteristics through identity verification, trustworthiness assessment, escort and surveillance of infrequent workers, contractors and visitors, enhancing security awareness; (2) minimizing opportunities for malicious and theft acts by limiting authority and knowledge access, confidentiality restrictions, work permits, process quality assurance, employee satisfaction, physical separation of areas, noncentralization of activities and authorities, finally law enforcement ; (3) the measures of threat/risk mitigation including insiders threat in any nuclear facility must be able to predict, detect, prevent, protect, delay, and respond to such malicious acts or

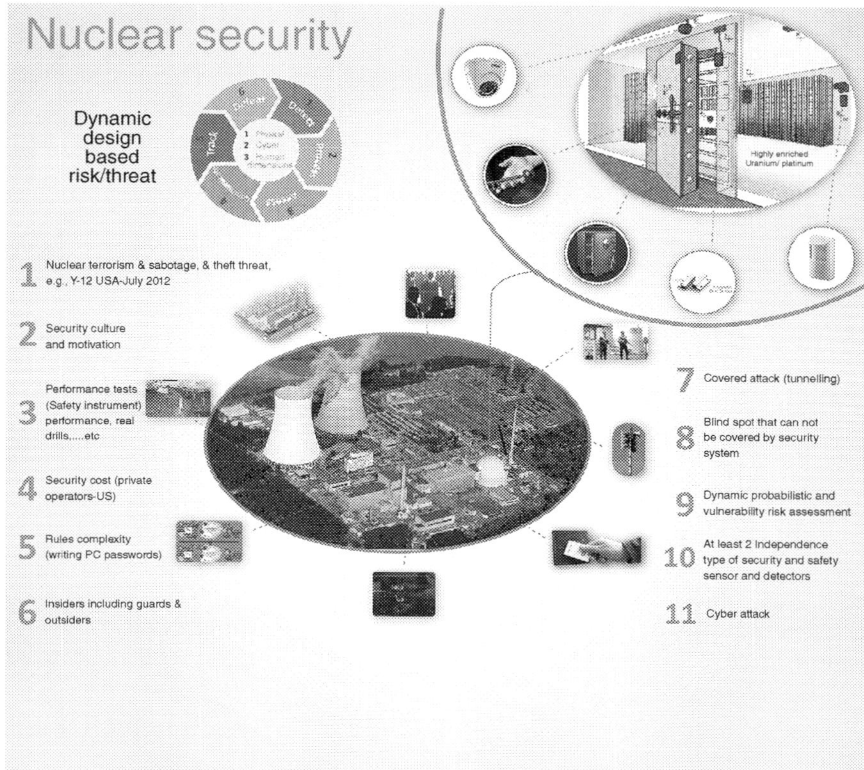

FIGURE 7.2 Risk/threat-based dynamic design of nuclear facilities.

abnormal behaviors. This can be done by the adoption of reliable security sensors, accompanying system (tow person rule), surveillance, operational processes monitoring, constant video surveillance of both vaults, and material interactions, guard patrol, secured guard checkpoint design, tracking the movement and location of personnel within the facility, detection of any contraband items using radiation detectors and scanners, routine investigation, inspections and audits for personals and materials, the introduction of multiple reliable layers of different physical or procedural barriers such as many layers of walls and barbed fences at borders around assets, mechanical and electronic locks on doors, intelligent software to detect attacks to force insiders to use more sophisticated tools, complicate the requirements for resources, logistics, training and skills, paying close attention at resignation/termination. All these measures are a result of available studies and assessments of previous incidents. They are strongly recommended to be implemented to bridge any gaps in the nuclear security system and to enhance its reliability and effectiveness.

7.6 Indicators of nuclear terrorism and illicit nuclear trafficking

The available literature on the danger of extremist terrorist groups that carry extremist political and religious ideologies indicates their desire to obtain power through the possession of chemical, biological, radiological, and nuclear weapons to increase their military strength. Therefore, nuclear terrorists seek to obtain radioactive or nuclear materials from nuclear facilities with the purpose of building a dirty bomb or obtaining a nuclear weapon. Accordingly, the terrorists' agenda seeks to recruit insiders by force using social engineering or spear-phishing. These insiders then can facilitate the terrorists 'mission to obtain radioactive or nuclear materials or to carry out sabotage actions to a nuclear facility.

According to the IAEA, there are several cases of smuggling, theft of nuclear materials, and illegal possession that has been reported over the years. That is why it is essential to take concrete steps to protect nuclear energy. Therefore, in view of the serious concerns, the ITDB has concentrated on additional security measures to develop a nuclear security system. According to the database of the IAEA's information system on incidents of illicit trafficking, statistics depict that there was a total of 3686 confirmed cases reported by 139 participating States from 1993 to December 31, 2019, as illustrated in Fig. 7.3 (ITDB, 2020).

Of these, 290 are cases involving nuclear and radioactive trafficking and misuse. Between 1993 and 2019, it has been reported that 12 trafficking and misuse cases of highly enriched uranium, two cases of plutonium, and five cases of plutonium beryllium neutron sources. It may appear that these cases represent only a small number, however, in a number of them, about kilograms of nuclear material that could be used in the manufacture of weapons were confiscated, while the majority included quantities of grams of seized material. These small quantities may be part of many frequent thefts, that will form larger, unsecured stocks. Some of these incidents included attempts to sell or transport these items across international borders. Incidents involved in attempts to sell nuclear and radioactive material indicate that the demand for these materials tremendous.

Out of 3686 reported cases, there are 1023 incidents that were difficult to determine whether they are related to smuggling or misuse but most of the incidents in this group are related to theft (ITDB, 2020). Such incidents could trigger illegal trafficking. Often, such incidents could be as a result of stealing nuclear or radioactive material during transportation. This large number of incidents indicates flaws in the security system for transporting nuclear and radioactive material inside or outside nuclear facilities; therefore, rigid steps need to be taken to secure the security

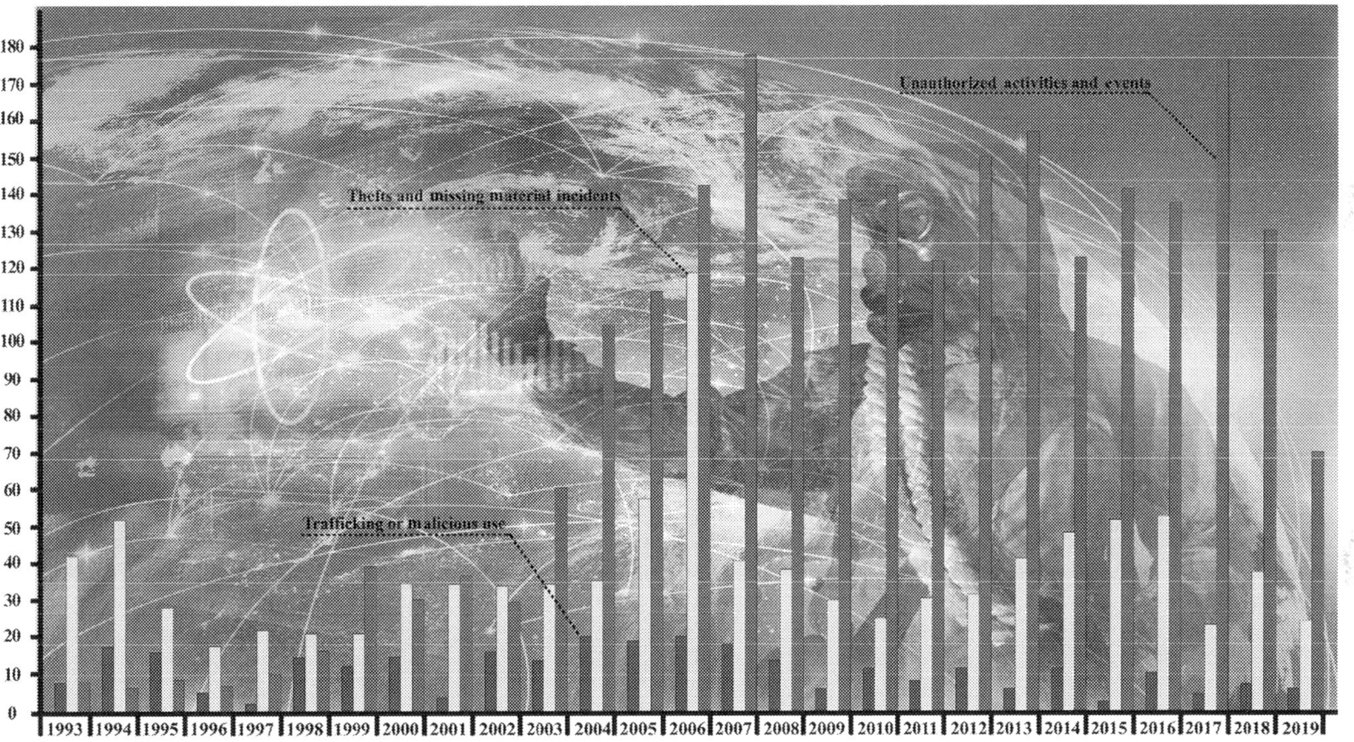

FIGURE 7.3 Incidents reported to the ITDB from 1993–2019 that are confirmed, or likely, to be connected with Illicit trafficking or malicious use of nuclear and other radioactive material. *ITDB*, incident and trafficking database.

system with relation to nuclear and radioactive materials transportations. According to ITDB reports, most of the stolen or lost nuclear and radioactive materials were those used by the industrial and medical sectors. Moreover, the majority of lost items in industrial sources are of equipment were those used in construction and mining. This depicts that nuclear security in the medical and industrial sector are vulnerable and need to be reviewed urgently. Devices containing radioactive materials from these industries can be easily stolen, thieves can get a large amount of money by selling these materials in the black market or to the scrap agents. Most of these devices contain long-lived isotopes such as iridium-192, cesium-137, and americium-241(ITDB, 2020). The ITDB classifies the activity of sealed radioactive sources according to IAEA safety standards based on their adverse health effects. Accordingly, they are classified into one to five categories from more dangerous and fatal to less dangerous.

Out of 3686, there are about 2373 cases that are not related to trafficking and misuse (ITDB, 2020). But they are divided into three main taxonomy.

- Unauthorized disposal (e.g., radioactive sources entering the scrap metal industry).
- Unauthorized shipment (e.g., international transportation of radioactively contaminated metal across borders).
- Discovery of radioactive material (e.g., uncontrolled radioactive sources).

This type of incident poses a serious concern to the public especially if enriched uranium is found beyond metal recycling and regulatory control such as the radiological accident that occurred in Goiânia, Brazil, on September 13, 1987. In this accident, it has been reported that more than 200 cases of radiation poisoning, four of them died, 28 suffered from radiation burns, $20 million is the estimated costs for cleaning up 1 km^2 of contaminated land (3500 m^3) (Sauer, 2016). Radiologically contaminated metal recycling way can inevitably create a huge health problem for consumers if in particular if household goods are made of them. About twenty cases of contaminated scrap metal shipments with high enriched uranium, two plutonium cases, and eight cases of plutonium neutron sources were reported between 1993 and 2019. Unfortunately, the incidence of illegal nuclear smuggling is on the rise and is a permanent problem for some countries. Such incidents properly identify flaws in the security system and emphasize the importance of the deployment of advanced monitoring systems at national borders such as X-rays or gamma scanners. There is also an urgent need to establish a rigid security system to protect the unauthorized disposal of radioactive materials and to enact deterrent laws and legislation. In addition to enhancing the public awareness of nuclear security and the dangers of such materials.

7.7 Detection of smuggled nuclear and radioactive materials

The possession of nuclear or radioactive materials by terrorist groups is extremely dangerous. Therefore, the detection of smuggled nuclear weapons and nuclear materials such as plutonium and enriched uranium in nuclear facilities or international borders is very important to thwarting nuclear proliferation and nuclear terrorism. Nuclear science and technologies have contributed to the detection of smuggled nuclear materials and have played an important role in enhancing national security, thwarting the proliferation of nuclear weapons, and preventing nuclear terrorism. Their use in national security dates back to the Manhattan Project.

Several nuclear countries tried to improve their nuclear protection systems necessary to defend their borders and nuclear facilities from any terrorist attack or smuggling particularly after the events of September 11. They also highlighted the fact that border patrols deter terrorists from importing or smuggling nuclear or nuclear weapons. Many countries have recently mounted radiological imaging scanners in their border checkpoints to find smuggled nuclear material, even though shielded by lead containers (Fig. 7.4). Usually, gamma rays and neutrons immitted from

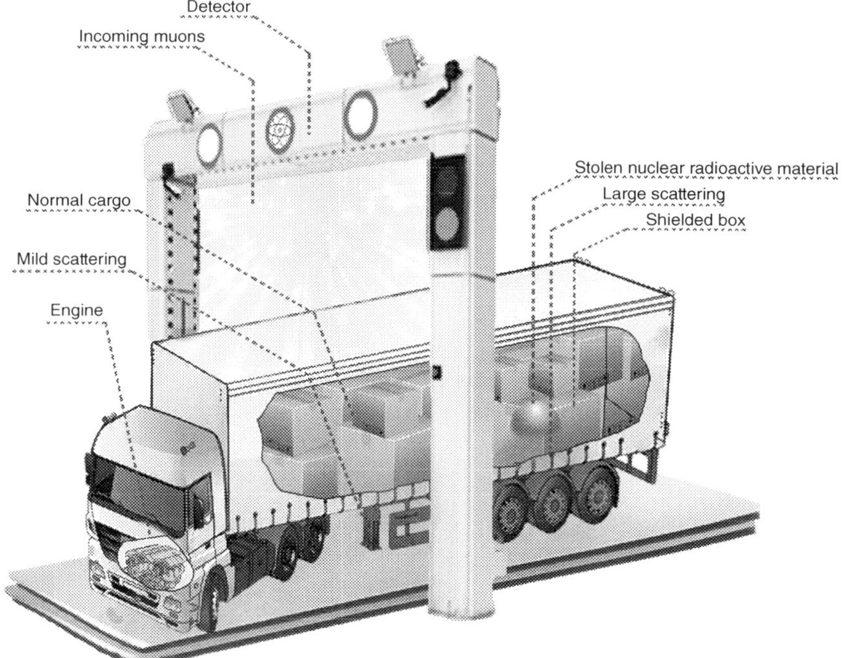

FIGURE 7.4 Radiological imaging scanners at the border inspection stations for identifying smuggled nuclear and radioactive materials hidden in vehicles.

radioactive materials that are not enclosed in a protected container are easily detected by the radiologically imaging scanners. However, well-educated terrorists can avoid such radiation from being detected by placing it in a lead shielded container especially when smuggled materials are of high atomic numbers (high-Z). Several new detection technologies introduced by Madelia (2010), such as the gamma detector response and analysis software, this technology comprises neutron detectors that are able to determine container content through the analysis of neutrons count rate that is analyzed using gamma-ray spectroscopy to identify their signatures of the smuggled materials. On the other hand, materials that contain high densities are detected using the Cargo Advanced Automated Radiographing System technology. This technique is built on the fact that detected material show more transparency to highly energized photons depending on the density and atomic number, it, therefore, produces a 3D image of the contents of a container, which have a major effect on the scattering of muons (a subatomic particle). Fluorescence Imaging of nuclear resonance determines materials based on the spectrum of gamma rays, which a nucleus releases when photons of particular energy are hit. Finally, unlike other devices working in a very close range, the threat assessment system and the photonuclear inspection can determine special nuclear materials up to 1 km away. High-energy photons can induce fission within nuclear materials with distant targets to produce characteristics signatures that can be detected.

Moreover, from the perspectives of the tomography of the gamma and muon. It has been found that gamma radiography core principle is that high-energy gamma radiation could penetrate any materials even in a sealed container, allowing materials to be identified within the containers without opening it. On the other hand, muon's radiograph almost same to gamma radiography in principle, but it depends on the use of cosmic-ray muons that have penetration ability and collide with atmospheric nuclei to make copious muon quantities. Due to the ability of the muons to penetrate, and not to interfere strongly with the atmosphere, therefore, it takes benefit of this penetrability also to identifying hidden nuclear contraband in vehicles passing over border inspection stations.

So, when a suspected vehicle is scanned using muons radiography, the muons scatter in various directions and interact with matter in two different ways:

(1) the atomic electrons leading to the continuous loss of energy, and
(2) the atomic nuclei, which lead to large angle shifts in the direction of the muon.

Both interactions produce radiographic signals, which classify the material in the truck. For instance, materials that include high-Z material, like uranium and plutonium cause muon rays to scattering more compared to normal materials such as steel or water (Fig. 7.4). Cosmic ray muons are used by the muon's detectors above and below a truck as an active interrogation probe of nuclear material.

7.8 Nuclear security is a national and international security issue—nuclear intelligence

Bunn and Sagan (2017) contend that the terrorist attack in 2011 was intended to target nuclear power plants and nuclear facilities in the United States. However, the terrorists observed that there would be massive destruction to the environment, human, and economic disruption. Thus, the terrorist attack plan for the September 11 attacks was altered to target World Trade Centres instead of nuclear facilities. The CIA received intelligence that terrorist groups were collaborating and forcing nuclear scientists in an attempt to acquire a nuclear bomb and weapons of mass destruction. A number of CIA and law enforcement agencies in several countries have therefore worked closely together in an unparalleled way to disfigure potential nuclear terrorist threats by exchanging information and evidence with a number of countries participating in the war on terror.

As nuclear terrorism is one of the world's main challenges in the globalized era, nuclear threats have since become increasing. Terrorist groups seek to obtain nuclear materials to strengthen their military power as well as they may attempt to target nuclear plants. Thus, if terrorists are able to procure smuggled or stolen nuclear material directly from nuclear facilities through the black market (a secret place to sell stolen nuclear materials that are sometimes identified as lost or whose fate in the original facility is uncertain), the situation will become more alarming. There are significant quantities and missing nuclear materials that have so far been found as per available studies; furthermore, it was indicated by records that nuclear material is found and confiscated by chance or by cooperation between law enforcement officials and intelligence organizations.

Another challenge that hinders the progress of the investigation of the missing nuclear materials is the reluctance of some nuclear countries to announce the missing nuclear material or because they secretly have supplied them to other governments or non-state actors for political purposes, including improving political alliances via the transfer of nuclear technology. Thus, the International Atomic Energy Agency's policy on cooperation between the Member States in the fields of intelligence through exchanging information to the IAEA to deter a nuclear terrorist attack is extremely necessary to add to this policy a mandate. This is in compliance with the International Convention of Acts of Nuclear Terrorism, which seeks to improve cooperation between countries by knowledge exchange and help in inquiries and extradition as well as it is in line with Convention on the Physical Protection of Nuclear Material (CPPNM) that outlined the issues of illicit trafficking or possession, or the misuse of nuclear, radioactive materials, and associated international nuclear material transportation (IAEA, 2020). This convention includes 115 signatories and 117 state parties according to the UN (2021), including the nuclear club, which are China, France, India, Russia, the United Kingdom, and the United States. Accordingly, this treaty provides a solid base of the necessity for Member States and involved parties to cooperate in the intelligence field to confront nuclear terrorism. Accordingly, there is an urgent need to expand the intelligence collaboration between the international intelligence agencies and the International Atomic Energy Agency in relation to confronting nuclear terrorism.

Since the IAEA is a multilateral organization explicitly established to prevent nuclear proliferation and promote nuclear peace applications, nuclear control activities are very necessary to track or disclose any missing nuclear material that may reach terrorist hands and eventually the nuclear black market. Moreover, nuclear intelligence cooperation will reinforce the IAEA's responsibility and potential to fulfill its mission to the mutual interests of all its Member States and to achieve global nuclear protection, which should be viewed as a national and international security issue in the fight against nuclear terrorism and the confrontation with the nuclear black market. There must be consciousness by the countries of the world of their inability to work alone to fight nuclear terrorism and the international cooperation on nuclear intelligence is the only hope of defeating nuclear terrorism. Accordingly, the international community should therefore initiate intelligence cooperation in the fight against nuclear terrorism, whether the IAEA Member States or not, and promote peaceful utilization of nuclear technology under the umbrella of the International Atomic Energy Agency, which serves all countries against international criminals. Such cooperation should be viewed from the same perspective of international cooperation with Interpol against international criminals. Similarly, as the International cooperation in regard to IAEA ITDB, International Radiation Monitoring Information System, The Unified System for Information Exchange in Incidents and Emergencies, which are all aimed at providing assistance to countries on nuclear or radiation emergencies, sharing information with the purpose to enhancing the nuclear security. Having international intelligence cooperation against nuclear terrorism and the nuclear black market will help to save the world from nuclear terrorist attacks.

7.9 Main nuclear security conventions and legal instruments

The international nuclear security conventions aim to limit criminal and other unauthorized acts against nuclear or radioactive materials, installations, or activities. Global commitment to these conventions and legal instruments contributes to the establishment of a reliable global nuclear security that contributes to combating nuclear terrorism and eliminating the nuclear black market. The following are the major international agreements on nuclear security.

7.9.1 Treaty on the nonproliferation of nuclear weapons

Treaty on the nonproliferation of nuclear weapons (NPT) is a major international treaty that became effective in 1970 (UN, 2020). It is considered as the key pillar of the nonproliferation system. Along with the five nuclear club states (France, China, the Russian Federation, the United Kingdom, and the United State), a total of 191 countries are parties to the treaty (UN, 2020). This treaty has three main pillars, which are the disarmament, nonproliferation, and peaceful uses of nuclear energy (UN, 2020). Nuclear weapon states are not to acquire nuclear weapons, and all countries can access nuclear technology for peaceful applications. The fundamental rule of the NPT is that recognized nuclear-weapon states can move toward disarmament. The NPT was extended indefinitely by the parties in 1995. Accordingly, States Parties meet every 5 years to review their successes and challenges in implementing the NPT at a review conference.

(Treaty on the nonproliferation of nuclear weapons (NPT) is attached in Appendix 5)

7.9.2 Convention on the physical protection of nuclear material

According to the IAEA (2020), the CPPNM is the only legally binding international agreement mandating standards of physical protection for nuclear materials. It entered into force on February 8, 1987, and about 155 States are the parties in this convention with 44 Signatories (NTI, 2020). The CPPNM obliges states, during international nuclear material transportation, to take physical security steps, and allows States to prosecute and extradite malicious use of nuclear material. The CPPNM modification, introduced in 2005 and came into force on May 8, 2016, aims to enforce more physical security measures in nuclear facilities, storage, and transportation of nuclear materials (Boon and Huq, 2011). Two new crimes are added to this convention, which is contraband of nuclear material and sabotage of nuclear installations. It obliges states to work together in order to respond to nuclear events more effectively, including materials loss or robbery or material theft or sabotage. To protect the security of information obtained from other countries on physical security, responsibilities have also been increased by it.

(CPPNM is attached in Appendix 6)

7.9.3 Convention on early notification of a nuclear accident & Convention on assistance in the case of a nuclear accident or radiological emergency

IAEA (2020) argues that after the accident of the Chernobyl Nuclear Power Plant in 1986, Convention on Early Notification of a Nuclear Accident and the Convention on Assistance in the Case of a Nuclear Accident or Radiological Emergency was passed. The main objective of the Early Nuclear Accident Notification Convention that was adopted on September 26, 1986, with the total parties of 127 States is to notify party countries of any potential nuclear or radiological accident or transborder release through a nuclear accident notification system that could be of radiological safety significance for another state (NEA, 2020). Under this treaty, the States Parties are required to report the time, location, releases of radiation, and other data necessary for the evaluation of the situation (IAEA, 2020).

On the other hand, the Convention on Assistance in the Case of a Nuclear Accident or Radiological Emergency (Convention on Assistance) was adopted on September 26, 1986, with a total parties of 127 States (NEA, 2020). This convention provides a mechanism for collaboration between States Parties, along with the IAEA timely assistance and help in the case of nuclear accidents or radiological emergencies (NEA, 2020). States Parties to this treaty must inform the IAEA of their available experts, equipment, and other support materials. In case of assistance is required by a state due to a nuclear accident or radiological emergency, the IAEA serves as a focal point and sends a request for assistance to other States Parties and the state party decides whether it can render the requested help and to identify its scope and terms.

(Convention on Early Notification of a Nuclear Accident & Convention on Assistance in the Case of a Nuclear Accident or Radiological Emergency are attached in Appendix 7 and 8)

7.9.4 Convention on nuclear safety

According to NEA (2020), convention on nuclear safety was adopted June 17, 1994, with the total parties of 127 States. This convention is considered an incentive agreement that aims to achieve a high level of nuclear safety around the world by strengthening national measures and international cooperation related to safety (IAEA, 1994). The contracting parties under this convention are obligated to maintain a high level of safety in civil nuclear power plants including the location, service, design, commissioning, construction, financial and human resources, safety evaluation and verification, quality assurance, and response to emergencies requirements. Under this convention, the parties engage in a peer review process, including the preparation and review of national reports, to be addressed every 3 years at the IAEA meetings held in Vienna, which aim to achieve a high degree of nuclear protection (IAEA, 1994).

(Convention on nuclear safety is attached in Appendix 8)

7.10 Conclusion

The literature refers that the world has changed after the terrorist events that occurred on September 11, 2001, but in fact, the world has changed after August 6, 1945, when they saw the horror of nuclear energy in World War II where the famous Albert Einstein equation has been abused in establishing the nuclear bomb that was intended initially as a weapon of deterrence to stop the spread of world war. However, people still look at the dark side of the nuclear energy since this accident, despite that Einstein requested the US President to stop Manhattan project who has ignored Einstein's and nuclear scientists' letter and continued with the building of nuclear bomb rather than restoring peace. The policymakers realized this big mistake and started to promote nuclear energy as the world's future development, US President Eisenhower was the first president who emphasized in his famous speech the development of atom for peace and peaceful applications. With the expansion of peaceful nuclear applications since World War II, and unfortunately with the adherence of some nuclear states to strengthening their vertical spread to be great powers, this makes it necessary that international nuclear security must be strengthened and developed against any threat in particular nuclear terrorism. This led to the dawn of a new era in the nuclear security system that is still continuing developing day by day as a result of the scientific revolution.

The need for inherent nuclear safety and security system was coupled with the increasing global demand for nuclear energy, as there was a continuous dependence on nuclear energy in order to generate sustainable and clean energy. Not only that, but that nuclear and radioisotopes technologies contributed to advancing development in a large number of sectors including a wide range of industries, agriculture, health, archaeology, space invasion and nuclear forensic and the detection of commercial fraud, and many other peaceful applications, which requires international guarantees and controls for protection of radioactive and nuclear materials used in these applications. It also requires a measure of attention and policy to ensure their safety and security in the long term for fear of falling into terrorist hands. Accordingly, strengthening the nuclear safety and security system through the use of modern technologies scientific methodologies plays an important role in bridging many gaps, such as human reliability and narrowing opportunities for penetration of nuclear facilities using advanced technologies such as sensor, unmanned aerial vehicles (drones) in routine nuclear security monitoring and emergency assessment, robots in the automation of the dangerous sites, which all prevent any nuclear or radiological unwanted or planned accidents. There is no doubt that expanding the scope of its use in the security of the infrastructure of nuclear facilities will constitute a major positive shift in the future of nuclear energy.

Future risks to national and global nuclear security will continue by terrorist groups or insiders whose aim is terrorism. Concurrently there will be an increased threat through more technologically advanced methods. While there is still a real threat from attacks on the nuclear infrastructure or from terrorist groups that may be able to obtain nuclear or radioactive materials. The possibility of an attack of this kind remains a source of concern according to ITBD reports of the increased reported accident related to trafficking or malicious use of nuclear and radioactive materials. Therefore, as a deterrent to prevent such attacks, strengthening the nuclear security system through enacting and adhering to international legislation, introducing artificial intelligence to the instrumented security system, and intensifying intelligence cooperation to confronting nuclear terrorism will remain a very important matter because nuclear security is an international security issue.

Despite the enormous benefits that these smart technological technologies have brought to society such as automation, including their usefulness to the nuclear sector, cyber terrorism, and the smart technologies that terrorists often try to use or get access through may be widely viewed as one of the biggest potential threats to nuclear safety and security. Therefore, an integrated physical safety, security, and prevention barriers with cybersecurity will work side by side in order to stem the threat from within or outside, which is necessary and urgent to confront the current potential threats to the global nuclear sector. One of the most important threats to nuclear security is the threat of nuclear terrorism, which may target a direct attack on nuclear facilities, and if this is not possible due to the strength of nuclear intelligence and the nuclear security system, they often resort to recruiting agents from within nuclear facility to steal important information or to do destructive acts of sabotage that may cause a nuclear or radiological disaster. Accordingly, the nuclear safety and security system requires reliable emergency and response plan to be in place to confront any nuclear and radiological emergency to ensure the safety and security of workers and the public.

Nuclear security receives great attention in States policies, as the increase in peaceful nuclear activities, the vertical proliferation of nuclear weapons and the increase in terrorist threats have taken the nuclear security dimension seriously within the defence strategies of countries. The nuclear terrorist threat is not the responsibility of superpowers only but rather includes all countries, due to the spread of nuclear technologies in most parts of the world, so States must join hands in the nuclear intelligence field with the enforcement of the international conventions s to confront the nuclear black market and nuclear terrorists. In short, nuclear security is based on three basic rules, the first: the nuclear security system based on effective and reliable techniques, and the second: the highly reliable and professional manpower, and the first and second rules will not work well without merging them with a third rule, which is fostering the nuclear security awareness and law enforcement.

References

Abaimov, S., Ingram, P., 2017. Hacking UK Trident. British American Security Information Council (BASIC), London, pp. 1–38.

Al Nabhani, K., Khan, F.I., Yang, M., 2017. Dynamic modeling of TENORM exposure risk using SMART approach. J. Pet. Exp. Prod. Technol. 3, 1–14.

Boon, K, Huq, A., 2011. Hacking UK Trident Security, 118. Oceana Publications, Oxford, pp. 1–425.

Bunn, M., Sagan, S., 2017. Insider Threats. Cornell University Press, USA, pp. 1–216.

Carr, J., 2010. Inside Cyber Warfare. Chapter 1: Assessing the Problem. O'Reilly Media, Inc. USA, 1–207.

Cyber Operations and Cyber Terrorism, 2005. U.S. Army Training and Doctrine Command, Handbook No. 1.02, http://stinet.dtic.mil/cgi-bin/GetTRDoc? [Accessed 4 Dec. 2020].

Du Bois, R., Richeldi, L., 2009. Interstitial Lung Diseases: European Respiratory Monograph. European Respiratory Society, UK (46), pp. 1–408.

Healey, A., 2016. The insider threat to nuclear safety and security. Securi. J. (29) 23–38.

Heinrich, H., 1931. Industrial Accident Prevention: A Scientific Approach. McGraw-Hill Book Company, New York, pp. 11–366.

Hunker, J., Probst, C., 2011. Insiders and insider threats: an overview of definitions and mitigation techniques. J. Wireless Mobile Net., Ubiq. Comput., Depend. Appl. 2 (1), 4–27.

IAEA, 2020. Convention on the physical protection of nuclear material. Available from https://www.iaea.org/publications/documents/conventions/convention-physical-protection-nuclear-material. [Accessed 8 Dec. 2020].

IAEA, 2020. Convention on Early Notification of a Nuclear Accident. https://www.iaea.org/topics/nuclear-safety-conventions/convention-early-notification-nuclear-accident.

IAEA, 1994. Convention on nuclear safety. https://www.iaea.org/sites/default/files/infcirc449.pdf. [Accessed 16 Dec. 2020].

IAEA, 2006. Dangerous quantities of radioactive material (D-values). IAEA, Vienna, pp. 1–144.

IAEA, 2019. IAEA Safety Glossary. Terminology Used in Nuclear Safety and Radiation Protection, 2018 Edition. IAEA, Vienna, pp. 1–261.

IAEA, 2011. Nuclear Security Recommendations on Physical Protection of Nuclear Material and Nuclear Facilities (INFCIRC/225/Revision 5). IAEA Nuclear Security Series No. 13, IAEA, Vienna, pp. 1–57.

IAEA, 2017. Operational intervention levels for reactor emergencies and methodology for their derivation- EPR-NPP-OILs. IAEA, Vienna, 1–149.

IAEA, 2015. Preparedness and Response for a Nuclear or Radiological Emergency: General Safety Requirements. GSR Part 7, IAEA, Vienna, pp. 1–102.

IAEA, 2008. Preventive and Protective Measures against Insider Threats: Implementing Guide. IAEA Nuclear Security Series ISSN 1816-9317. IAEA, Vienna, 1–36.

ITDB-IAEA Illicit trafficking database. Fact Sheet for 1993–2004. (2020). https://www.iaea.org/sites/default/files/16/05/fact_figures2004.pdf. [Accessed 6 Dec. 2020].

Johnson, A., 2011. Examining the foundation. In: Safety & Health. National Safety Council Congress & Expo. https://www.safetyandhealthmagazine.com/articles/print/6368-examining-the-foundation. [Accessed 2 Dec. 2020].

Leveson, N., 2020. Safety and security are two sides of the same coin. The Coupling of Safety and Security. SpringerBriefs in Applied Sciences and Technology Bieder C., Pettersen Gould K. Springer, Cham. https://doi.org/10.1007/978-3-030-47229-0_3.

Manuele, F., 2011. Advanced Safety Management Focusing on Z10 and Serious Injury Prevention. John Wiley & Sons, USA, pp. 1–432.

McKinnon, R., 2019. Cause, Effect, and Control of Accidental Loss with Accident Investigation Kit. CRC Press, Boca Raton, pp. 1–280.

Medalia, J., 2010. Detection of nuclear weapons and materials: science, technologies, observations. CRS Report for Congress. Prepared for Members and Committees of Congress, 1–106.

NEA, 2020. Convention on Assistance in the case of a nuclear accident or radiological emergency (Convention on Assistance). https://www.oecd-nea.org/jcms/pl_29131/convention-on-assistance-in-the-case-of-a-nuclear-accident-or-radiological-emergency-convention-on-assistance. [Accessed 17 Dec. 2020].

NEA, 2020. Convention on early notification of a nuclear accident (early notification convention). https://www.oecd-nea.org/jcms/pl_29135/convention-on-early-notification-of-a-nuclear-accident-early-notification-convention. [Accessed 14 Dec. 2020].

NEA, 2020. Convention on nuclear safety (CNS). https://www.oecd-nea.org/jcms/pl_29136/convention-on-nuclear-safety-cns. [Accessed 16 Dec. 2020].

NTI, 2020. Convention on the physical protection of nuclear material (CPPNM). https://www.nti.org/learn/treaties-and-regimes/convention-physical-protection-nuclear-material-cppnm/. [Accessed 12 Dec. 2020].

Ramsden, E., 1996. Chemistry of the Environment. Nelson Thornes, UK, pp. 1–160.

Rasmussen, J., 1993. Perspectives on the concept of human error. In: Society for Technology in Anesthesia Conference. New Orleans, LA.

Rathnayaka, S., Khan, F., Amayotte, P., 2013. Accident modelling and risk assessment framework for safety critical decision-making: application to deep-water drilling operation. J. Risk Reliab. 227 (1), 81–105.

Rathnayakaa, S, Khana, F, Amyotte, P, 2011. SHIPP methodology: predictive accident modeling approach. Part I: methodology and model description. Process Saf. Environ. Prot. 89, 151–164.

Sauer, T., 2016. Nuclear Terrorism: Countering the Threat. Taylor & Francis, New York, pp. 1–262.

Schonbeck, M., Rausand, M., Rouvroye, J., 2009. Human and organizational factors in the operational phase of safety instrumented systems: a new approach. Saf. Sci. 48, 310–318.

Senders, J., Moray, N. (Eds.), 1991. Human Errors: Their Causes, Prediction, and Reduction. Lawrence Erlbaum Associates Inc, Hillsdale, NJ.

UN, 2020. Treaty on the non-proliferation of nuclear weapons (NPT). https://www.un.org/disarmament/wmd/nuclear/npt/#:~:text=The%20Treaty%20represents%20the%20only,the%20Treaty%20was%20extended%20indefinitely. [Accessed 11 Dec. 2020].

UN, 2021. United Nations. Treaty collections. https://treaties.un.org/pages/ViewDetailsIII.aspx?src=TREATY&mtdsg_no=XVIII-15&chapter=18&Temp=mtdsg3&clang=_en. [Accessed 9 Dec. 2020].

CHAPTER

8

Conclusions and recommendations

OUTLINE

8.1 Conclusion

The birth of the peaceful nuclear age came more than half a century ago as a result of the treaty on the non-proliferation of nuclear weapons (NPT), which established a framework for increasing access to the peaceful uses of nuclear energy. Under Articles III and IV of the nonproliferation of nuclear weapons treaty (NPT), the International Atomic Energy Agency (IAEA) assists member states to adopt nuclear and theoretical technology in a wide range of peaceful nuclear applications and harness them to serve some sectors of sustainable three-dimensional development (economic, social, and environmental). Accordingly, a lot of studies, efforts, and experiments have been made to harness the atom for peace and development. Since then, the world has witnessed a shift in peaceful nuclear energy applications in various sectors of sustainable development, which has grown exponentially. It includes but is not limited to clean nuclear power, health, efficient plant and animal production, pest control, climate change reduction, improved water and soil quality, desertification control, wildlife improvement, sustainable environmental protection, nuclear forensic science in criminal investigations. In addition to many other industrial applications such as food irradiation to keep it fresh, oil spill tracking, characterization of the oil reservoir in the geological formation, and increasing the efficiency of oil production, sterilization of medical instruments, food imitation control,

Applications of Nuclear and Radioisotope Technology: The Atom for Peace and Sustainable Development
DOI: https://doi.org/10.1016/B978-0-12-821319-3.00004-X

water desalination, and many other applications. This book discusses all those nuclear technologies and radioisotopes and their peaceful applications in advancing sustainable development and the science behind them in seven chapters.

Chapter 1 presented an overview of the history of atomic theory, related discoveries, and explained the basic principles in nuclear science. Where a review of the literature revealed that Greek scholars are considered to be the first to establish atomic theory during the fifth century. Unfortunately, their theories were ignored at that time until the 16th and 17th centuries because they were materialistic and not based on religious foundations. However, modern science in the early 19th century revived the atomic theory using quantitative and experimental data. Scientists such as Fermi, Bohr, and Szilard proved to the world that these small atoms can create huge energy through their breakthrough discovery of self-sustaining fission-reaction and the fusion reaction. This significant discovery has changed the world but unfortunately, politicians abused nuclear energy during World War II. Scientists were not satisfied with the use of the power of atomic energy in wars and destructions. Accordingly, nuclear scientists devoted their efforts, studies, and scientific research to harnessing atomic and nuclear technology for peaceful applications. The birth of peaceful nuclear applications was announced by President Dwight D. Eisenhower on December 8, 1953, through his speech in front of the United Nations council under the title of "Atoms for Peace." Accordingly, the International Atomic Energy Agency was established by the United Nation in 1957. The main roles of the IAEA are to assist in the peaceful use of nuclear technology, and fostering nuclear safety.

Chapter 2 highlighted the role of international law via NPT in promoting nuclear peaceful applications. In this regard, Chapter 2 argued that nuclear weapons pose a major threat to humanity, assets, the environment, and international security. There have been speculations that a nuclear war is impending. This has followed the rising tension between some countries in the world as a result of the race to acquire nuclear weapons, which may result in a nuclear catastrophe. Nuclear weapons have caused terrible loss and devastation in the targeted regions during World War II. Hiroshima and Nagasaki are tragic examples that will not be forgotten by humanity, and it will be a terrible experience if the nuclear states are forced to use their nuclear weapons. These are some of the salient reasons why the use of atom for peace and sustainable development need to be and encouraged instead of racing to use the atom and nuclear technology for the weapons of mass destruction. Therefore, the reasons and motives for addressing the issue of reducing the spread of nuclear weapons have become an international necessity to put an end to the proliferation of nuclear and mass destruction weapons in the light of the global changes of contemporary international law as well as the nonproliferation of nuclear weapons treaty that both suffer from gaps and weaknesses, which both lead to the continuation of international violations and the fragile peace that threaten the security of the world. The fragile peace will continue if the privileges and rights granted by some essential international treaties are not revised as soon as possible. For example, the NPT grants and allows expressly the nuclear countries to develop their nuclear proliferation either vertically or horizontally, and similarly article 4 of the International Convention for the Suppression of Acts of Nuclear Terrorism expressly excludes the application of the convention to the use of nuclear devices during armed conflicts. As well as the importance of the ratification of the Comprehensive Nuclear-Test-Ban Treaty by the five nuclear countries. The increasing severity of this issue has been worsened by the frequent violations of the global security measures by prominent nuclear countries that have double standards stands and contributed to providing allied and friendly countries with nuclear technology for military use (such as Israel, India, and Pakistan) as well as they have rejected all resolutions on the establishment of a nuclear-weapon-free zone in the Middle East. The unexpected political and nuclear conflicts between the United States and North Korea and Iran may change the features of the world map by pushing a small button called the nuclear button. Accordingly, Chapter 2 underlined on more efforts are needed first from the nuclear states to abandon their nuclear program to become a real model strive for international peace and security so that NPT and Nuclear Weapons Tests Treaty becomes effective to stop and control any nuclear proliferation. Subsequently, other non-nuclear states such as Iran, North Korea, Israel, India, Pakistan will become convinced to be part of NPT and will ultimately help in avoiding any possible nuclear war in the coming future.

Arguably, nuclear technology plays a fundamental role in sustainable development in various sectors of the economy, energy, health, agriculture, sciences, industry, and the environment. In this context, the applications of nuclear science and radioisotope technology in power generation, hydrogen economics, and transportation have been discussed in Chapter 3, in more detail, whereas this chapter emphasized the importance of energy in its various forms as one of the main elements necessary to ensure the sustainability of the economic growth of any country in light of global population inflation and high global energy demand. Providing electric and thermal energy effectively and efficiently is vital for industrial production operations and providing services to all sectors of the economy and other development. Accordingly, the production of the necessary and sufficient energy at the lowest cost and with the least environmental damage becomes prerequisites for ensuring continued growth and development and ensuring the continued competitiveness of the economy including hydrogen economy and preserving the environment. Therefore, the production of electricity and hydrogen through the use of nuclear energy is one of the most effective, less costly, and environmentally friendly energy production methods compared to the methods of producing electric energy through the use of fossil fuels or renewable energies.

This chapter outlined that the history of nuclear science is related to the discovery of proton, electron, neutron, radioactivity, and fission reaction. Many noteworthy scientists played their crucial part in developing nuclear power as an alternative to fossil fuel generation of energy. The period from 1895 to 1945 marks the flourishing period of nuclear science. Several theories have been put forward to transform nuclear science as an energy source. The era from the 1950s onwards can be viewed as an application of all nuclear research on electric power generation. Numerous designs have been developed for nuclear power plants by taking advantage of the most important nuclear accidents in the history of the nuclear industry. These scenarios of these accidents have contributed to developing the modern nuclear designs of nuclear power plants to meet the highest safety standards established by the IAEA. IAEA has made guidelines and safety measures and standards for all types of nuclear power plants starting from the construction, operation, safety, nuclear fuel transport, and used fuel management. According to the available literature, about 16% of the world's electricity is supplied by nuclear energy, and the leading countries in nuclear power generation are the developed economies including the United States, Russia, Japan, United Kingdom, France, Korea, and Canada. The first nuclear power plant was inaugurated in Russia in 1954. There are different types of nuclear power plants used in the world. The design of nuclear power plants varies according to the type of nuclear reactor used. The most common types of nuclear reactors include pressure tube graphite-moderated reactors, boiling water reactors, graphite-moderated gas-cooled reactors, pressurized water reactors, and pressure tube heavy water-moderated reactors. The design of these nuclear reactors depends on a number of factors such as nuclear fuel type, nuclear reaction type, coolant type, steam cycle, and moderator type. According to the scientific facts, 1 kg fission reaction of U-235 in a nuclear power plant yields 8.2×10^{13} J of heat energy that can be converted into electrical energy, which is sufficient to operate a 1000 MW power plant.

Chapter 3 asserted that nuclear power plants are the only power plants that generate an enormous amount of electricity at very cost-effective in the long run and without greenhouse gas emissions. The main cost of nuclear power generation is the construction of a power plant. However, once this cost is overcome, they produce electricity at a stable and low price as compared to fossil fuel-run power plants. The electricity can be produced uninterrupted for a long time in a nuclear power plant. The fuel used in a nuclear power plant is either natural uranium or enriched uranium and the refueling of a nuclear power plant usually is done every18–24 months. All these depend on the reactor type. However, strict safety measures are required for operating a nuclear power plant safely and securely. Since the 1990s, the record of the safety of nuclear power plants is outstanding. The remarkable improvement in nuclear safety and security records is based on continuous evaluation and lessons learned from rare nuclear accidents. The scarcity of nuclear accidents, in and of itself, is a landmark achievement for this industry.

Moreover, this chapter revealed that the huge thermal energy produced by nuclear power plants can also be exploited to produce hydrogen, which functions as an energy carrier and not as an energy source like fossil fuels. This hydrogen is the base of the new hydrogen economy that will establish a new concept of modern energy infrastructures. Hydrogen is now researched as an effective means of storing the surplus energy being generated by renewable energy sources like wind or solar power and nuclear power. The supporters of the hydrogen economy argue that the future energy needs of the world would rely temporarily on the production of hydrogen by using fossil fuels in the short run and production by renewable and nuclear resources in the long-term. The arbiters of this concept claim that hydrogen is unable to compete with electricity generated by fossil fuels or directly from other renewables. Hydrogen has several uses in transport, industry, power generation using fuel cells. Many prototypes of different vehicles and ships that have been designed to run on hydrogen proved their success. If the vehicles, marine transport, and aviation are shifted to run on hydrogen, many million tons of greenhouse gas emissions will be reduced.

The industry is considered one of the most common fields that use nuclear and radioisotope technologies. There is currently no developed industry that does not use nuclear and radioisotope technologies in their production process, technical inspecting process, quality and defects monitoring process. Chapter 4 provided a brief overview of a number of applications of nuclear science and radioisotope technology in the industries, and in environmental sustainability, which is just the tip of the iceberg.

Chapter 4 shed light on a number of nuclear and radioisotope technologies applications in the industry. The most important of which is petroleum and minerals explorations and production optimization using different approaches, such as neutron activation analysis method. This approach reveals the type of elements and the proportion of their presence in the ground. The Interwell radiotracer approach is another nuclear application used in the oil and gas industry to enhance the secondary and tertiary recovery of oil from the reservoirs. Not only that but, we can calculate the velocity of oil flow inside the pipelines by injecting the labeled radioactive isotopes into the petroleum pipeline and tracking the passage of the isotope in the pipeline, and also detecting if there is a leak in the oil pipelines. Industrial nondestructive radiography such as X-ray or gamma-ray computed tomography or neutron radiography, which are another significant nuclear application in the industry, utilizes ionizing rays, to detect faults in the flowlines, examine weld strength, and detect any deposits, therefore these technologies are widely used in examining oil and gas pipelines, as well as examining water pipes in major projects. Moreover, they can be used in examining and detection of cracks in the buildings, structures and the bodies of aircraft, cars, and ships. Nucleonic gauges are another vital nuclear application used widely in many industries due to their ability to measure accurately different parameters including density, weight, levels of any matter in any harsh or risky environment depending on the interaction between ionizing radiation and matter. Accordingly, industry worldwide started to employ this technology for refining and revamping procedures and materials to monitor and enhance production efficiency in many industries including but not limited to mining and mineral ore processing, environmental monitoring, the paper and plastics industries, the cement industries, civil engineering, petrochemical industries, the oil and gas industries, and many other industries.

This chapter also argued that soon after the invention of the crosslinking of olefin polymers using ionizing radiation in the mid of 20th century, radiation crosslinking polymerization technique has begun to be implemented in a wide range of processing industries, such as oil and gas pipelines, vehicle tires plies, electrical cables and wire, polymeric foams, and food packaging, medical devices. Radiation crosslinking polymerization technique is a green technology that does not require any external chemicals or thermal treatment and instead, it relies on gamma radiation or electron beam as a source of the irradiation process. Processing radiation improves the generation of intermolecular bonds between polymeric chains, such as mechanical properties, resistance to corrosive liquids, thermal stability, withstanding high temperature, suppressing of flame propagation, and exhibiting improved abrasion tolerance, fluid resistance. As a result, the use of processing radiation in the industry improved the quality and reliability of products.

Chapter 4 concluded that nuclear technology has proven itself as a competent tool with enormous benefits in a variety of applications ranging from power generation to improving industrial efficiency and optimization of natural resource exploration and extraction including oil, gas, minerals, and water, to the environment protection from greenhouse gases and marine life from ocean acidification. Not only that but, nuclear and radioisotope technologies applications in the industry play a significant role in combating food fraud and verifying the authenticity of imported food products, which often can be easily forged. However, it became very easy to verify the quality of food items and react more quickly at vital checkpoints to prevent the flow of contaminated products or adulterated using Isotope analysis that helps to establish the geological sources of foodstuffs and the authenticity of their composites. Moreover, nuclear technology is credited for saving the life of tens of thousands of people from landmines and improving the quality of materials for safe and reliable use in addition to many other industrial applications. The power of nuclear technology has transformed many industries in the past decades. Although some challenges remain, workers in this field remain optimistic that they will be overcome in the near future to provide better solutions to a large number of problems for the benefit of the humankind and advancing the wheel of progress and prosperity through the sustainable development for the resources and the environment.

Chapter 5 shed light on the importance of nuclear and radioisotope technology and their vital role in promoting sustainable development, protecting the environment, and enhancing the welfare of peoples in providing competitive solutions to the sustainable development of agriculture to securing food and fighting hunger. Whereas this chapter elucidated the significant role of nuclear and radioisotope technology has played in combating hunger and reducing malnutrition in a large number of poor and densely populated countries, improving environmental sustainability, ensuring food safety and health, addressing soil erosion and irrigation water scarcity problems, and pest control.

This chapter addressed the dilemma of food security and the global challenges in fighting hunger due to global population growth. This chapter indicated that such problem arises especially in the developing world, where the population growth rate is high in light of the severe shortage of food and water resources due to the impact of climate change and global warming. Reports indicate that the most important challenges in the agricultural sector lie in the efficiency of fertilization, loss of irrigation water, optimal absorption of nutrients, soil quality, and associated erosion problems. In this regard, Chapter 5 has cast light on the advancement of nuclear and radioisotope technology in the agriculture sector that contributed to securing food in sufficient quantities and distinct types to meet this shortfall. Where nuclear scientists have been able to use the nuclear properties of labeled radioisotopes in measuring and monitoring soil quality, soil erosion, measuring absorption rates, rationalizing the use of fertilizers, and improving the quality and abundance of agricultural and animal production. For instance, the technique of using the labeled nitrogen probe helps in measuring the soil moisture and therefore determining the amount of water required, rationalizing irrigation, and saving water. This process is often combined with the technique of measuring absorption rates that help to know exactly the required quantities of water and fertilizers for the plants. Fertilizers that farmers use mostly contain nitrogen or phosphorous, which are very expensive and often used in large and random quantities, amounting to millions of tons and dollars annually.

Excessive use of these fertilizers has direct and indirect effects on human health and the environment that also play a significant role in increasing the emission of greenhouse gases. On the opposite, nuclear scientists have been able to harness atomic energy and radioisotope technology by using labeled phosphorous or labeled nitrogen radioisotopes to determine the exact amount of fertilizer needed for the plants. This has contributed to reducing the impact of the global warming issue as well as contributed to saving a lot of money and fertilizers. Not only that but, nuclear technology and radioisotope technology helped in the early detection of the factors that cause disease. This helped a lot in changing the genetic structure of plants and animals through the use of nuclear mutation technology in such a way that enhances their immunity system, increases the speed of plant and animal growth, and abundance of production. This approach relies on DNA markers in establishing a genomic map, by causing genetic changes in plant or animal genes as a result of exposing them to certain radioactive doses of gamma rays or other ionizing radiation.

In addition to its prominent role in protecting the environment by reducing the level of chemical pesticides to control pests. Scientists have tended to think of an environmentally clean alternative, which is the use of the nuclear sterile technique for insects. The basic principle of this technique is that exposing the males of that pest in the late cocoon phase to specific doses of radiation to make them sterile. This is usually achieved by using gamma rays Issued by Cesium-137 and Cobalt-60. This method has been successful in a number of countries to eliminate the Mediterranean fly and many other pests that cause significant losses in crops according to many international reports This chapter revealed the important roles of peaceful nuclear applications and radioisotope technology in providing competitive solutions to the sustainable development of agriculture and food. This includes but is not limited to fighting hunger, reduce malnutrition, improve environmental sustainability, ensure food safety and authenticity, address soil erosion and irrigation water scarcity problems, and pest control. Therefore, nuclear applications are effective and low-cost solutions to many of the dilemmas facing the agriculture sector in the world to improve the agricultural level, supporting national economies, and supporting global food security.

Nuclear science and its peaceful applications witnessed great importance after World War II, atomic energy scientists have focused their research on developing the use of nuclear science and its peaceful applications in advanced science and scientific research such as space exploration, nuclear medicine, nuclear forensics, archaeology, geology, palaeontology, hydrology, and many more, which have been discussed in Chapter 6.

This chapter highlighted the prominent role nuclear and isotopic technology has played in predicting and treating millions of patients around the world who suffer from serious and incurable diseases such as cancer, heart, and brain diseases that traditional medicine has been unable to treat. It argued that diagnostic nuclear imaging techniques, such as Positron Emission Tomography-PET and Single Emission Photon Tomography-SPECT, the use of radiotracers, and internal or external irradiation techniques examine precisely the efficiency of the work of human body organs and identify the presence of any malignant tumors of cancer cells and eliminating them without harming healthy cells. Accordingly, nuclear and isotopic technology also boosts the progress of medicine in predicting many neurological diseases, such as dementia and Alzheimer's. It also discussed that ionizing radiation is a crucial application of nuclear science and radioisotope technology that are used widely in the medical sector to sterilize medical equipment and products.

Chapter 6 also explained how nuclear and radioisotope technology helped in solving the mystery of many cases in the courts and provided important and absolute information that could be used as conclusive evidence during the trial. The use of atomic sciences in criminal investigations has witnessed a remarkable development in light of the development of crime's tools, as criminals try their best to leave no evidence at the crime scene, but some microscopic and macroscopic evidence is always left behind. Nuclear sciences assist forensics medicine in gaining access to this evidence and uncovering clues of crimes by tracking this evidence using nondestructive quantitative and qualitative radiological analysis methods of samples obtained from the crime scene. For instance, analysis of nuclear activation showed that Oswald's rifle was used to assassinate President John F. Kennedy in 1977 when the assassination accident of President John F. Kennedy was reinvestigated

Moreover, Chapter 6 elaborated how radioisotopes dating methods are a vital branch of nuclear science due to the fact that they are widely used in many applications such as archaeology, climatology, hydrology, geology, meteorology, and nuclear forensics science. Nuclear dating methods are very useful and precise tools for the determination of the age of rocks, the ancient artworks, cave paintings, and engravings, and the death date of the organic matters. Accordingly, geochemists, geologists, geophysicists, and archaeologists have joined hands with nuclear scientists to widen the research and maximizing the use of nuclear peaceful applications in these fields. In archaeology, for instance, the history of art paintings, cave engravings, and skeletons has always been a controversial topic until the emergence of radiometric dating method that helped in extrapolating the history of the old nations with high precision such as the Mayan civilization.

Not only that, but this chapter indicated the significant role that nuclear science and radioisotope technology play in water resources management and its sustainability. It argued that isotope hydrology is one of the modern scientific research methods used by experts in many countries around the world to trace the effect of freshwater movement according to the water cycle. This helps to provide a better understanding in assessing and identifying ground and surface water sources, the age, recharging mechanism, pollutants, movement, and interactions of this water within different geological formations. Isotope hydrology relies on the analysis of various labeled isotopes in the natural water to generate a detailed hydrological map. This helps scientists to make evidence-based decisions on sustainable resource management for sustainable development. For instance, this chapter explained from the chemistry point of view that the water molecule is composed of two hydrogen atoms and one atom of oxygen. Oxygen, as well as hydrogen, have different isotopes, such as ^{16}O, ^{17}O, ^{18}O, ^{1}H, ^{2}H, ^{3}H, respectively. Scientifically, this proves that water is not the same and is different from one place to another based on the scientific fact that isotopes have the same number of electrons and protons but different numbers of neutrons and therefore different atomic mass. Accordingly, some water may carry heavier atoms while some others carry lighter atoms. Scientific studies reveal that the presence of certain naturally occurring radioactive isotopes tracers assists in knowing the age of water that may range from a few months to a million years. For instance, the presence of ^{3}H indicates that water is less than 100 years old. While the presence of ^{14}C indicates that the water age is around 40,000 years old. On the other hand, the water containing ^{81}Kr, ^{4}He, ^{129}I, and ^{36}Cl indicates that age range from 40,000 years up to million years. Moreover, the isotopes composition of groundwater in hydrological environments provides significant information about the regional climatic conditions at the time of recharging. Stable and radioactive isotopes, including ^{18}O, ^{14}C, ^{2}H, and ^{3}H, can give vital information about physical and chemical processes acting on groundwater and hydrological features of aquifers that include the rate and mechanism of recharge, aquifer interconnections time, and its origin. These parameters are difficult to be measured using any technique other than isotopic technique. This information is vital for the sustainable management and assessment of water resources, especially in the dry regions and in developing countries.

Finally, Chapter 6 argued that the peaceful application of nuclear science and radioisotope technology was not limited to the Earth, but also reached the moon and neighboring planets. During the last 70 years, the exploration of space was considered to be very expensive and one of the most complex jobs for humans due to the harsh conditions and complex challenges in space. However, this is seen as a significant advancement in human history. This chapter argued that nuclear science has been playing significant roles in making space exploration more successful, reliable, and cheaper since the launch of the first nuclear-powered spacecraft using radioisotope thermoelectric generator that has been launched on June 29, 1961, by the American Navy and named "Transit 4A navigation satellite."

Nuclear fission energy is the best option for high thrust propulsion mission's in terms of technical maturity and energy density required for long space missions. There are three types of space propulsions that are powered by nuclear fission energy, namely nuclear electric propulsion, nuclear pulse propulsion, and nuclear thermal propulsion. Nuclear thermal-based rockets utilize nuclear fission reactions and work on the principle similar to nuclear power plants and propulsion ships. This chapter outlined how the fission reaction provides the necessary propulsion power to penetrate the gravity forces that also help in reducing the exposure time of space radiation that astronauts may be exposed to during their travel between Earth and the red planet. The capability to provide reliable and sustainable long-term operational electricity required for space missions that conventional systems such as solar power, fuel cells are unable to provide as compared to nuclear-based systems. This is achieved by converting the energy produced by the radioactive decay of plutonium or americium-241into electrical energy via radioisotope thermoelectric generator, and this energy is used to power the satellites orbiting around the Earth for many years, space equipment, and international space stations.

Chapter 7 discussed a very important topic, which is nuclear safety and security. Therefore, the need for an inherent system of nuclear safety and security has been coupled with the increasing global demand for nuclear energy, as there has been a continuous reliance on nuclear energy in order to generate sustainable and clean energy. Not only that but nuclear and radioisotope technologies have contributed to advancing development in a large number of sectors including a wide range of industries, agriculture, health, archaeology, space invasion, nuclear forensics, commercial fraud detection, and many other peaceful applications, which requires international guarantees and controls for protection of radioactive and nuclear materials used in these applications. It also requires a measure of attention and policy to ensure their safety and security in the long-term for fear of falling into terrorist hands. Accordingly, strengthening the nuclear safety and security system through the use of modern technologies scientific methodologies plays an important role in bridging many gaps, such as human reliability and narrowing opportunities for penetration of nuclear facilities through the adoption of advanced technologies in nuclear security layers, such as sensor, drones, artificial intelligence, quantitative threat assessment, and dynamic risk modeling, automation of the dangerous operation. All these options help to prevent any nuclear or radiological unwanted or planned threat. There is no doubt that expanding the scope of its use in the security of the infrastructure of nuclear facilities will constitute a major positive shift in the future of nuclear energy and its security.

According to the IAEA's Incident and Trafficking Database, the reports indicate that the number of confirmed incidents reported by 139 countries from 1993 until December 31, 2019, are 3638 incidents, which are troubling numbers. These incidents are related to malicious acts targeting theft, unauthorized possess, smuggling of nuclear and radiological materials, or nuclear installations sabotage. These figures are really worrying, and investigation reports indicated the involvement of insiders in many of these incidents. Insiders pose a real threat to the security and safety of nuclear facilities. Therefore, this chapter argued that it is very important to develop a strong and reliable nuclear security system in particular after the events of September 11, 2001, that were also targeting nuclear facilities. As a result of the increasing nuclear illicit trafficking and fears of terrorist threats to nuclear and radiological facilities and materials, the issue of nuclear security has received great international attention since the past few decades. This is evident in strengthening the collaboration between the international institutions and safeguard conventions such as the Convention on the Physical Protection of Nuclear Material was adopted in 1979 and the International Convention for the Suppression of Acts of Nuclear Terrorism that was adopted in 2005 with a significant aim, which is to foster nuclear security in nuclear facilities to ensure the safety and security of nuclear facilities, public, and to confronting any terrorist or sabotage acts that may cause a nuclear disaster by recruiting insiders in these facilities to meet to terrorist agendas. In addition to that, there is an urgent need to further intensify international cooperation in the field of nuclear security intelligence to combat nuclear terrorism and eliminate the nuclear black market.

8.2 Recommendations

1. The international institutions and treaties designed for the nonproliferation of nuclear weapons can only be successful with the cooperation of all the member states as emphasized by the liberalism ideology for achieving global peace. Accordingly, it is important that all forms of threats should be eliminated to make all existing states feel safe through trust-building and serious cooperation to get rid of all nuclear weapons. Therefore, the five main nuclear powers are primarily responsible for cooperation to ensure that nuclear energy is safely used for positive purposes. The cooperation of the five nuclear states will enhance the authority of the international treaties and institutions such as the NPT, the Security Council, and the International Court of Justice to make them more effective in their duties. Nuclear states should take the initiative to utilized enriched uranium in nuclear warheads into reactor fuel to generate electricity that benefits people rather than killing them. In 1994, the Megatons to Megawatts treaty with Russia to reuse nuclear warheads into reactor fuel is a good example of how to harness the atom for peace and sustainable development. In addition to Obama's

initiatives in the US nuclear plans, where 10% of US electricity comes from dismantled nuclear weapons. Thus, more efforts are needed, first from the nuclear states to abandon their nuclear program to become a real model strive for international peace and security so that NPT and Nuclear Weapons Tests Treaty becomes effective to stop and control any nuclear proliferation. Subsequently, other non-nuclear states such as Iran, North Korea, Israel, India, Pakistan will become convinced to be part of NPT. This will ultimately help in avoiding any possible nuclear war in the coming future and having a global zone free of weapons of mass destruction.

2. The energy in its various forms is one of the main elements necessary to ensure the sustainability of the economic growth of any country in light of global population inflation and high global energy demand. Providing electric and thermal energy, and water effectively and efficiently is vital for industrial production operations and providing services to all sectors of the economy and other development. Accordingly, the production of the necessary and sufficient energy at the lowest cost and with the least environmental damage becomes prerequisites for ensuring continued growth and development and ensuring the continued competitiveness of the economy including hydrogen economy and preserving the environment. In this context, nuclear energy is the only energy that is capable of producing electricity, hydrogen, and water simultaneously because it is the best way to produce energy effectively, sustainably, at the lowest cost, and environmentally friendly, compared to the conventional methods of producing energy through the use of fossil fuels or renewable energies. Thus, serious developing countries should adopt nuclear energy as a source of three-dimensional energy production (electricity, water, and hydrogen) without greenhouse gas emissions and as a radical solution to the issue of global warming.
3. Countries interested in advancing their development, sustainability, and prosperity must start to adopt nuclear technologies and their peaceful applications in their development plans due to the remarkable successes they have presented in various sectors, such as energy, health, industry, agriculture, environmental conservation, the sustainability of water resources, archaeology, space exploration, crime detection, and many more.
4. There is an urgent need to enhance global awareness of nuclear sciences, their peaceful applications, and their contribution to advancing sustainable development and global prosperity.
5. There is an urgent need to enhance nuclear security system to be able at high certainty and reliability to combat and confront any nuclear or cyber terrorism threat. This can be achieved through the adoption of aggressive national security strategies, law enforcement, strengthening nuclear security systems using artificial intelligence techniques that are

impenetrable by cyber-terrorist, and strengthening international intelligence cooperation to confront terrorists, insiders, and outsiders, the nuclear black market.

6. Always to remember that nuclear safety and security are two sides of the same coin since they deal with the internal and external threats, respectively. Accordingly, there is an urgent need to couple dynamic and quantitative risk analysis techniques that consider a series of failure events as the cause of the accident occurring in the system or subsystems with reliability theory that precedes the cause of electromechanical or human failure.

Glossary of terms

Accident An event that happens unintentionally and unexpectedly, typically causing damage to humans, property, or the environment.

Accident Modelling A technique used to analyze why and how an accident occurs by modeling it in a scenario. It is used to predict and characterize accidents.

Atom The smallest unit of ordinary matter and comprises of the nucleus. The nucleus is comprised of positively charged protons and, typically, a similar number of neutrons, which have no electrical charge that prevents the repulsive forces between protons. Protons and neutrons are referred to as "nucleons." There are electrons orbiting around the nucleus and they are negatively charged. When the number of protons and electrons are equal, the atom is said to be electrically neutral, whereas if it has a greater or lesser number of protons, it is said to be positively or negatively charged, accordingly.

Bayesian Updating Theorem A mathematical inference used to update the posterior probability of a hypothesis that was based on prior knowledge as more evidence or information becomes available.

Cyberterrorism According to Cyber Operations and Cyber Terrorism (2005) Kevin Colman defined cyberterrorism as "the premeditated use of disruptive activities, or the threat thereof, against computers and/or networks, with the intention to cause harm or further social, ideological, religious, political or similar objectives, or to intimidate any person in furtherance of such objectives."

Decay Chain A series of radioactive decays of different radioactive decay products, referred to as a sequential series of radioactive transformations of unstable atoms to become stable ones.

Deductive Reasoning A process of formal reasoning wherein a concordance of premises leads to a specific and logically certain conclusion. If the premises are true, and the principles of deductive reasoning are used, the conclusion must also be true.

Design-Based Threat The introduction of different security layers based on threat assessment for a nuclear facility within the particular context derived from a state's current assessment of national and international threats as well as the vulnerability assessment of the facility.

Dynamic Modeling Is a dynamical process that describes the dynamic change in the behavior and interaction of system components. It describes the behavior of a system over time.

Enhanced Oil Recovery Technology Technologies are used to increase the amount of oil that can be extracted from a reservoir. Usually, this entails injecting a substance into an injection well in order to increase depleted pressure and reduce the oil's viscosity in the reservoir.

Exploration Well A borehole that is drilled in order to determine the presence of oil or gas.

Failure Probability The likelihood that a system or system component will fail at a given time.

Fat Man Was the codename used for the second atomic bomb, which was a plutonium implosion-type bomb. This bomb was developed by the United States and used during the Second World War. It was dropped in the city of Nagasaki, Japan on August 9, 1945. A hundred thousand of people died as a result of this bomb.

Fission Is a nuclear reaction process by which atoms split up when they are bombarded with neutrons. The process of splitting an atom is accompanied by a massive release of heat that can be converted into energy. It is argued that nuclear fission was first discovered by Hahn and Strassmann in 1938.

Food Irradiation Is an advanced method used to make food healthier and fresher with an extended shelf life by reducing the risk of foodborne illness and delaying or eliminating sprouting or ripening. This is achieved through exposing food and food packaging to a precalculated dosage of ionizing radiation, such as from gamma rays, X-rays, or electron beams.

Fusion Is a nuclear reaction process by which two small, light nuclei join together to make one heavy nucleus. The fusion process of atoms is accompanied by a massive release of heat that can be converted into energy.

Gamma Radiation Electromagnetic energy (photons) emitted by some radionuclides as a product of radioactive decay. Gamma photons constitute the most energetic photons on the electromagnetic spectrum.

Gaseous Diffusion Methodology Is a process used for separating and enriching isotopes based on the molecular diffusion of a gaseous isotopic mixture through porous membranes.

Gas Centrifugation Methodology Is a process that is based on the mass differential between the two isotopes that are required to be separated from each other. Accordingly, this process is commonly used to separate Uranium-235 isotope from Uranium-238 isotope because Uranium 238U possesses three more neutrons in its nucleus, which means that it has a higher mass compared to 235U. This process requires first to convert the solid mixture into gaseous form (Uranium hexafluoride (UF6)), then UF6 gas is entered into a centrifuge cylinder and rotated at high speed. Accordingly, a strong centrifugal force is created that forces more of the denser gas molecules comprising the 238U toward the wall of the cylinder, while the lighter gas molecules of 235U gathered near the center. The slightly enriched stream in 235U is removed and fed into the subsequent higher stage, while the slightly depleted stream gets recycled back into a lower stage.

Half-life The time taken for half of the atoms comprising a radioactive material to disintegrate during radiological decay.

Human error An action, either intentional or unintentional, that does not adhere correctly to policy or procedure, and that may lead to consequences such as injury, harm, or loss. It is considered the foremost contributing factor in industrial disasters and accidents. The terminologies "human error" or "human factor" are mostly considered interchangeable in the safety and risk assessment literature and can be distinguished as the underlying causes of accidents (human factors) and immediate causes (human errors).

Induced Nuclear Mutation Is the procedure of exposing plants or animals to artificial radiation in an attempt to come up with desirable new genes to breed with other plants or animals.

Inductive Reasoning A process of reasoning wherein multiple premises that are viewed as true are combined to supply evidence for a conclusion. The conclusion of an inductive argument is typically general and probable rather than certain.

In-Situ Recovery A process in mining that entails drilling boreholes in a formation and injecting a lixiviant solution in order to dissolve minerals that naturally occur in a solid-state in order to recover other minerals, such as uranium.

Ionizing Radiation The process through which an atom is charged or ionized. This occurs when radiation has enough energy to remove tightly bound electrons from the orbit of an atom.

Isotopes Atoms that have an equal number of protons and electrons and, hence, the same atomic number, but a different number of neutrons. This means that isotopes have different atomic mass and physical properties but will have the same chemical properties.

Laser Isotope Separation Is a process used for isotope enrichment purposes based on the fact that each isotope has a unique signature of spectroscopic that their atoms can absorb certain electromagnetic radiation (light) at an individual and well-defined wavelength specific to each atomic or molecular species. This causes atoms to become excited to higher vibrational, rotational, or electronic energy levels if they are exposed to a particular wavelength. Their state will change and may enter preferentially into chemical reactions. The excited positively charged atoms are ultimately collected electrostatically on a cathode and this method is called "Photochemical Isotope Separation." This approach can be used for Uranium-235 separation from Uranium-238 (uranium enrichment) and there are currently two standard Laser Isotope Separation processes used in the industry. These include Atomic Vapor Laser Isotope Separation AVLIS and Molecular Laser Isotope Separation

Legislation The process of enacting laws or a collective body of laws.

Little Boy Was the codename used for the first atomic bomb, which was a uranium gun-type bomb. This bomb was developed by the United States and used during the Second World War. It was dropped in the city of Hiroshima, Japan on August 6, 1945. A hundred thousand people died as a result of this bomb.

Lixiviant A liquid medium, such as a groundwater solution that is mixed with oxygen, is used to extract the desired metal from a formation. It is used in hydrometallurgy.

Neutron Activation Is a process that takes place when neutrons hit the targeted atomic nuclei where some of the neutrons are captured by the targeted nuclei, which becomes heavier than before, unstable, and entering excitation states. Eventually, the unstable radioactive nucleus will try to become more stable by decaying its excess energy in the form of emitting gamma rays, or particles such as alpha or beta.

Nuclear The energy is produced by the nucleus of an atom when it is divided, decays, or joins another nucleus.

Nuclear Energy for Space Applications Is the use of nuclear energy for the purposes of exploring outer space. it is usually either large fission systems to launch spacecraft at very high speed, small fission systems, or radioactive decay, through the radioactive decay of materials (radioisotope battery) to generate electricity or heat needed to heat delicate devices. Small fission reactors are used to run Earth observation satellites. The production of electrical or thermal nuclear power could last for decades.

Nuclear Forensics Is a technique used for the examination of forensic evidence or nuclear material or other radioactive materials, or evidence contaminated with radionuclides using nuclear analysis and radiometric techniques such as isotope ratio mass spectrometry, gamma-ray spectrometer analysis, particle-induced X-ray emission or proton-induced X-ray emission, neutron activation analysis, and many other techniques to determine what materials are present at the crime scene.

Nuclear Medicine Is a branch of medicine that uses radioactive materials, or radiopharmaceuticals, nuclear imaging techniques for medical research, or medical diagnosis, and treatment such as examination of organs' function and structure.

Nuclear Proliferation The spread of nuclear weapons, nuclear weapons technology, or fissile material to countries that do not already own them.

Nuclear Reactor Is a device in which a nuclear fission chain reaction is initiated, sustained, and controlled that generates energy in a form of heat or radiation, which can be converted into electricity or used for research purposes.

Nuclear Safety IAEA defines it as "the achievement of proper operating conditions, prevention of accidents, or mitigation of accident consequences, resulting in protection of workers, the public and the environment from undue radiation hazards."

Nuclear Security IAEA defines it as "the process of prevention and detection of and response to the theft, sabotage, unauthorized access, illegal transfer or other malicious acts involving nuclear materials, other radioactive substances, or their associated facilities."

Nucleonic Gauges Instruments are also known as nucleonic control systems or referred to as Ionizing Radiation Gauging Devices, which are used for measuring and assessing the interaction between ionizing radiation and matter.

Political System The formal and legal institutions that constitute a state or government.

Pollution The introduction of contaminants into an environment, often leading to adverse changes, harm, damage, or loss to the people, assets, and the environment.

Posteriori Knowledge Knowledge requires evidence to be proven.

Posterior Probability A revised calculation of the likelihood of an event occurrence. It is calculated by using Bayesian updating theorem to update a prior probability.

Prior Knowledge Knowledge that already available based on experience or historical data. Prior knowledge may be accurate or inaccurate. Such knowledge may be self-evident and therefore not require proof.

Prior Probability A term used in Bayesian statistical inference, which is the likelihood of an event's occurrence, based on previously available data (historical data).

Probability A branch of mathematics that entails calculating the likelihood that an event will occur. Probability is expressed as a number between 0 (will never occur) and 1 (will always occur).

Prompt Fission Neutrons Logging A process that entails a pulsed source of neutrons flux, emitting about 108 neutrons per second. This accelerates deuterium ions into tritium. The neutrons flux targets geological formations and collide with U-253, which leads to the slow-neutron-induced fission of U-235 into the formation. Epithermal neutrons and thermal neutrons that return from the formation following fission are counted separately in an epithermal/thermal neutron detector. This detector indicates the percentage ratio of U-235, the ratio of epithermal to thermal neutrons being directly proportional to that percentage.

Radiation Crosslinking Polymerization Is a process that uses the effect of high energy induced ionized radiation such as beta- or gamma-rays in which the material absorbs radiation energy, and its chemical bonds are broken, and free radicals are formed that react to new chemical bonds to be extremely resistant as per required property.

Radiotracer It is also known as a radiotracer, or a radioactive label, which is a chemical compound through which one or more atoms have been replaced by a labeled radioisotope. radiotracer is used to explore the mechanism of chemical reactions by tracing the path that the radioisotope follows from reactants to products.

Red Tide A phenomena that refer to the harmful algal bloom that causes ocean acidification. it is caused as a result of a reduction in the pH of the ocean over time, caused primarily by the absorption of carbon dioxide.

Qualitative Risk Assessment A process that qualitatively characterizes the level of risk associated with a particular hazard or activity by assessing the probability of injury and the severity of the associated consequences, typically by drawing from historical data. The risk matrix is an example of this type of assessment.

Quantitative Risk Assessment A process that entails a numerical estimate of the probability of the risk or the defined risk that will result from a particular hazard. It is sometimes referred to as probabilistic risk assessment.

Radiation The emission of energy as electromagnetic waves such as gamma radiation or as particles, such as alpha a beta particle, through a material medium or through space.

Radiation Protection A practice that has been defined by the International Atomic Energy Agency (IAEA Safety Glossary—draft 2016 revision) as "The protection of people from harmful effects of exposure to ionizing radiation, and the means for achieving this." The IAEA also states. "The accepted understanding of the term radiation protection is restricted to protection of people," while some organizations such as ICRP extends the definition to include the protection of nonhuman species or the protection of the environment.

Radioactive Decay (Nuclear Decay or Nuclear Radiation Radioactivity) The process wherein an unstable atomic nucleus stabilizes by emitting excess energy in the form of radiation, such as alpha and beta particles, or electromagnetic waves, such as gamma radiations.

Radioactive Waste Any material whether it is liquid, gas, or solid, that contains a radioactive nuclear substance and is produced via nuclear power generation, nuclear fission, or nuclear technology or from other applications such as the oil and gas industry, mining, and research and medicine.

Radioactivity Concentration The amount of activity per unit mass or volume of material wherein radionuclides are essentially distributed uniformly. In the SI system, it is measured in becquerel per gram (Bq/g), where Bq is the number of radioactive transformations that occur in a particular radioactive isotope per second.

Radiometric Dating Is a scientific method of dating specimens in different fields such as archaeology, climatology, hydrology, geology, meteorology, and nuclear forensic sciences by determining the relative proportions of particular radioactive isotopes present in a sample using different nuclear methods such as rubidium–strontium dating method, uranium–thorium dating method, potassium–argon dating method, radiocarbon dating method.

Radionuclide An atom with excess nuclear energy, making it unstable. When a radionuclide decays it emits nuclear radiation.

Radioisotope Is the unstable form of an element that attempts to become more stable by releasing the excess energy in form of radiation or particles or energy such as alpha or beta or gamma radiation.

Rational Reasoning Theory A systemic process that entails logical, inductive, and probabilistic analysis. It is used in risk management and to rationally appraises active and passive factors within a system and subsystems that can contribute to an accident by investigating the performance of identified safety prevention barriers.

Regulations Rules or orders issued by a government, a regulatory agency, or an executive authority have the force of law.

Risk The likelihood of occurrence of unwanted events such as injury or harm, often as a result of exposure to hazards that may result in various levels of consequences.

Risk Assessment A systematic process that entails hazard identification and an evaluation of the risks involved in a certain activity, alongside an appraisal of potential consequences. This process can be either qualitative or quantitative.

Safety Reasonable protection from the risk of harm or injury, or loss or damage that can inflict on property or people or environment, whether it be accidental or deliberate. It judges risk acceptability.

Safety and Prevention Barriers Physical or nonphysical barriers are used in the prevention, mitigation, and control of accidents or other undesirable events.

Soil Erosion Is a naturally occurring process causing the removal of the topsoil removal, which is the most fertile top layer of the soil. This leads to the accumulation of minerals and nutrients in the soil in other places, often degrading traditional ecosystems.

Sterile Insect Technology Is a nuclear technology used to mitigate and control pest hazards in humans, plants, livestock, and crops. It facilitates the loss of fertilization by exposing millions of male larvae to specific radiation doses and after that, releasing sterile insect males in the affected areas where they naturally and successfully mate with the wild females but rather inhibiting egg production.

Uncertainty Refers to situations that involve imperfect or incomplete knowledge, sometimes due to unknown variables. Uncertainty can arise for subjective or objective reasons.

Uranium A chemical element with atomic number 92. Its most common isotopes are U-238, with 146 neutrons, and which accounts for about 99.3% of uranium, and U-235, which is the only naturally occurring fissile isotope and has 145 neutrons. it accounts for about 0.7% of uranium. U-238 can be used to produce a fissile isotope of plutonium and has a half-life of 4.5 billion years.

APPENDIX

1

List of countries that have cleared irradiated food for human consumption (World Health Organization, 1988)

Applications of Nuclear and Radioisotope Technology: The Atom for Peace and Sustainable Development
DOI: https://doi.org/10.1016/B978-0-12-821319-3.00061-0

TABLE A1 List of countries that have cleared irradiated food for human consumption (Updated March 22, 1988).

Country	Product	Purpose of irradiation	Type of clearance	Dose permitted (kGy)	Date of approval
Argentina	Strawberries	Shelf-life extension	Unconditional	2.5 max.	April 30, 1987
	Potatoes	Sprout inhibition	Unconditional	0.03–0.15	April 30, 1987
	Onions	Sprout inhibition	Unconditional	0.02–0.15	April 30, 1987
	Garlic	Sprout inhibition	Unconditional	0.02–0.15	April 30, 1987
Bangladesh	Chicken	Shelf-life extension/ decontamination	Unconditional	Up to 8	December 28, 1983
	Papaya	Insect disinfestation/ control of ripening	Unconditional	Up to 1	December 28, 1983
	Potatoes wheat and ground	Sprout inhibition	Unconditional	Up to 0.15	December 28, 1983
	Wheat products	Insect disinfestation	Unconditional	Up to 1	December 28, 1983
	Fish	Shelf-life extension/ decontamination/ Insect disinfestation	Unconditional	Up to 2.2	December 28, 1983
	Onions	Sprout inhibition	Unconditional	Up to 0.15	December 28, 1983
	Rice	Insect disinfestation	Unconditional	Up to 1	December 28, 1983
	Frog legs	Decontamination	Provisional		
	Shrimp	Shelf-life extension/ decontamination	Provisional		
	Mangoes	Shelf-life extension/Insect disinfestation/control ripening	Unconditional	Up to 1	December 28, 1983
	Pulses	Insect disinfestation	Unconditional	Up to 1	December 28, 1983
	Spices	Decontamination/Insect disinfestation	Unconditional	Up to 10	December 28, 1983

(*continued on next page*)

TABLE A1 *(continued)*

Country	Product	Purpose of irradiation	Type of clearance	Dose permitted (kGy)	Date of approval
Belgium	Potatoes	Sprout inhibition	Provisional	Up to 0.15	July 16, 1980
	Strawberries	Shelf-life extension	Provisional	Up to 3	July 16,1980
	Onions	Sprout inhibition	Provisional	Up to 0.15	October 16, 1980
	Garlic	Sprout inhibition	Provisional	Up to 0.15	October 16, 1980
	Shallots	Sprout inhibition	Provisional	Up to 0.15	October 16, 1980
	Black/white pepper	Decontamination	Provisional	Up to 10	October 16, 1980
	Paprika powder	Decontamination	Provisional	Up to 10	October 16, 1980
	Arabic gum	Decontamination	Provisional	Up to 10	September 29, 1983
	Spices (78 different products) (semi)-dried vegetables	Decontamination	Provisional	Up to 10	September 29, 1983
	(Seven different products)	Decontamination	Provisional	Up to 10	September 29, 1983
Brazil	Rice	Insect disinfestation	Unconditional	Up to 1	March 7, 1985
	Potatoes	Sprout inhibition	Unconditional	Up to 0.15	March 7, 1985
	Onions	Sprout inhibition	Unconditional	Up to 0.15	March 7, 1985
	Beans	Insect disinfestation	Unconditional	Up to 1	March 7, 1985
	Maize	Insect disinfestation	Unconditional	Up to 0.5	March 7, 1985
	Wheat	Insect disinfestation	Unconditional	Up to 1	March 7, 1985
	Wheat flour	Insect disinfestation	Unconditional	Up to 1	March 7, 1985
	Spices (13 different products)	Decontamination/Insect disinfestation	Unconditional	Up to 10	March 7, 1985
	Papaya	Insect disinfestation/control of ripening	Unconditional	Up to 1	March 7, 1985
	Strawberries	Shelf-life extension	Unconditional	Up to 3	March 7, 1985

(continued on next page)

TABLE A1 *(continued)*

Country	Product	Purpose of irradiation	Type of clearance	Dose permitted (kGy)	Date of approval
Brazil *(contd)*	Fish and fish products (fillets, salted, smoked dried, dehydrated)	Shelf-life extension/ decontamination/ Insect disinfestation	Unconditional	Up to 2.2	8 March 1985
	Poultry	Shelf-life extension/ decontamination	Unconditional	Up to 7	8 March 1985
Bulgaria	Potatoes	Sprout inhibition	experimental batches	0.1	April 30, 1972
	Onions	Sprout inhibition	experimental batches	0.1	April 30, 1972
	Garlic	Sprout inhibition	experimental batches	0.1	April 30, 1972
	Grain	Insect disinfestation	experimental batches	0.3	April 30, 1972
	Dry food concentrates	Insect disinfestation	experimental batches	1	April 30, 1972
	Dried fruits	Insect disinfestation	experimental batches	1	April 30, 1972
	Fresh fruits (tomatoes, peaches, apricots, cherries, raspberries, grapes)	Shelf-life extension	experimental batches	2.5	April 30, 1972
Canada	Potatoes	Sprout inhibition	Unconditional	Up to 0.1	November 9, 1960 June 14, 1963
	Onions	Sprout inhibition	Unconditional	Up to 0.15	March 25, 1965
	Wheat, flour, wholewheat	Insect disinfestation	Unconditional	Up to 0.75	February 25, 1969

(continued on next page)

TABLE A1 (*continued*)

Country	Product	Purpose of irradiation	Type of clearance	Dose permitted (kGy)	Date of approval
	Poultry	Decontamination	Test marketing	Up to 7	June 20, 1973
	Cod and haddock fillets	Shelf-life extension	Test marketing	Up to 1.5	October 2, 1973
	Spices and certain dried vegetables' seasonings	Decontamination	Unconditional	Up to 10	October 3, 1984
	Onion powder	Decontamination	Unconditional	Up to 10	December 12, 1983
Chile	Potatoes	Sprout inhibition	experimental batches		October 31, 1974
			Test marketing		December 29, 1982
			Unconditional	Up to 0.15	
	Papaya	Insect disinfestation	Unconditional	Up to 1	December 29, 1982
	Wheat and ground wheat products	Insect disinfestation	Unconditional	Up to 1	December 29, 1982
	Strawberries	Shelf-life extension	Unconditional	Up to 3	December 29, 1982
	Chicken	Decontamination	Unconditional	Up to 7	December 29, 1982
	Onions	Sprout inhibition	Unconditional	Up to 0.15	December 29, 1982
	Rice	Insect disinfestation	Unconditional	Up to 1	December 29, 1982
	Teleost fish and fish products	Shelf-life extension/decontamination/Insect disinfestation	Unconditional	Up to 2.2	December 29, 1982
	Cocoa beans	Decontamination/Insect disinfestation	Unconditional	Up to 5	December 29, 1982
	Dates	Insect disinfestation	Unconditional	Up to 1	December 29, 1982

(*continued on next page*)

TABLE A1 *(continued)*

Country	Product	Purpose of irradiation	Type of clearance	Dose permitted (kGy)	Date of approval
Chile *(contd)*	Mangoes	Shelf-life extension/Insect disinfestation/control of ripening	Unconditional	Up to 1	December 29, 1982
	Pulses	Insect disinfestation	Unconditional	Up to 1	December 29, 1982
	Spices and condiments	Decontamination/Insect disinfestation	Unconditional	Up to 10	December 29, 1982
China	Potatoes	Sprout inhibition	Unconditional	Up to 0.20	November 30, 1984
	Onions	Sprout inhibition	Unconditional	Up to 0.15	November 30, 1984
	Garlic	Sprout inhibition	Unconditional	Up to 0.10	November 30, 1984
	Peanuts	Insect disinfestation	Unconditional	Up to 0.40	November 30, 1984
	Grain	Insect disinfestation	Unconditional	Up to 0.45	November 30, 1984
	Mushrooms	Growth inhibition	Unconditional	Up to 1	November 30, 1984
	Sausage	Decontamination	Unconditional	Up to 8	November 30, 1984
Czechoslovakia	Potatoes	Sprout inhibition	experimental batches	Up to 0.1	November 26, 1976
	Onions	Sprout inhibition	experimental batches	Up to 0.08	November 26, 1976
	Mushrooms	Growth inhibition	experimental	Up to 2	November 26, 1976
Denmark	Spices and herbs	Decontamination	Unconditional	Up to 15 max. Up to 10 average	December 23, 1985
Finland	Dry and dehydrated spices and herbs	Decontamination	Unconditional	Up to 10 average	November 13, 1987
	All foods for patients requiring a sterile diet	Sterilization	Unconditional	unlimited	November 13, 1987

(continued on next page)

TABLE A1 (*continued*)

Country	Product	Purpose of irradiation	Type of clearance	Dose permitted (kGy)	Date of approval
France	Potatoes	Sprout inhibition	Provisional	0.075-0.15	November 8, 1972
	Onions	Sprout inhibition	Provisional	0.075-0.15	August 9, 1977
	Garlic	Sprout inhibition	Provisional	0.075-0.15	August 9, 1977
	Shallots spices and aromatic substances	Sprout inhibition	Provisional	0.075-0.15	August 9, 1977
	172 products including powdered onion and garlic!	Decontamination	Unconditional	Up to 11	February 10, 1983
	gum arabic	Decontamination	Unconditional	Up to 9	June 16, 1985
	Muesli-like cereal	Decontamination	Unconditional	Up to 10	June 16, 1985
	Dehydrated vegetables	Decontamination	Unconditional	Up to 10	June 16, 1985
	mechanically deboned Poultry meat	Decontamination	Unconditional	Up to 5	February 16, 1985
	Dried fruits	Insect disinfestation	Unconditional	1 max.	January 6, 1988
	Dried vegetables	Insect disinfestation	Unconditional	1 max.	January 6, 1988
German Democratic Republic	Onions	Sprout inhibition	Test marketing	50	1981
	Onions	Sprout inhibition	Unconditional	20	January 30, 1984
	Enzyme solutions	Decontamination	Unconditional	10	June 7, 1983
	Spices	Decontamination	Provisional	Up to 10	December 29, 1982

(*continued on next page*)

TABLE A1 (*continued*)

Country	Product	Purpose of irradiation	Type of clearance	Dose permitted (kGy)	Date of approval
Hungary	Potatoes	Sprout inhibition	Test marketing	0.1	December 23, 1969
	Potatoes	Sprout inhibition	Test marketing	0.15 max.	January 10, 1972
	Potatoes	Sprout inhibition	Test marketing	0.15 max.	March 5, 1973
	Onions	Sprout inhibition	Test marketing		March 5, 1973
	Strawberries mixed	Shelf-life extension	Test marketing		March 5, 1973
	spices (black pepper, cumin, paprika, dried garlic: for use in sausages)	Decontamination	experimental batches	5	April 2, 1974
	Onions	Sprout inhibition	Test marketing	0.06	August 6, 1975
	ONIONS mixed dry ingredients for	Sprout inhibition	experimental batches	0.06	September 6, 1976
	Canned hashed meat	Decontamination	experimental batches	5	November 20, 1976
	Potatoes	Sprout inhibition	Test marketing	0.10	May 4, 1980
	Onions onions (for dehydrated	Sprout inhibition	experimental batches	0.05	September 15, 1980
	flakes processing)	Sprout inhibition	Test marketing	0.05	November 18, 1980
	Mushrooms l*Agaricus)*	Growth inhibition	Test marketing	2.5	20 June 1981
	Strawberries	Shelf-life extension	Test marketing	2.5	June 20, 1981
	Potatoes	Sprout inhibition	Test marketing	0.1	October 13, 1981
	Potatoes	Sprout inhibition	Test marketing	0.10	December 2, 1981
	Spices for sausage production	Decontamination	Test marketing	5	January 4, 1982
	Strawberries	Shelf-life extension	Test marketing	2.5	April 15, 1982

(*continued on next page*)

TABLE A1 (*continued*)

Country	Product	Purpose of irradiation	Type of clearance	Dose permitted (kGy)	Date of approval
	Mushrooms {*Agaricus*}	Growth inhibition	Test marketing	2.5	April 15, 1982
	Mushrooms (*Pleurotus*}	growth inhibition	Test marketing	3	April 15, 1982
	Grapes	Shelf-life extension	Test marketing	2.5	April 15, 1982
	Cherries	Shelf-life extension	Test marketing	2.5	April 15, 1982
	Sour cherries	Shelf-life extension	Test marketing	2.5	April 15, 1982
	Red currants	Shelf-life extension	Test marketing	2.5	April 15, 1982
	Onions	Sprout inhibition	Unconditional	0.05±0.02	June 23, 1982
	Spices for sausage	Decontamination	Test marketing	5	June 28, 1982
	Pears	Shelf-life extension	Test marketing	2.5	December 7, 1982
	Pears	Shelf-life extension	Test marketing	l .u + Uau 12 treatment	January 24, 1983
	Spices potatoes (for processing into flakes)	Decontamination	Test marketing	5	1983
		Sprout inhibition	Test marketing	0.1	January 28, 1983
	Frozen chicken	Decontamination	Test marketing	4	October 3, 1983
	Sour cherries (canned)		conditional	0.2 average	February 20, 1984 April 23, 1985
	Black pepper	Decontamination	conditional	6 minimum	May 1985
	Spices	Decontamination	conditional	5—6 minimum	April 25, 1986
	Spices	Decontamination	Unconditional	8, 6 average	August 19, 1986
India	Potatoes	Sprout inhibition	Unconditional	Codex Standard	January 1986
	Onions	Sprout inhibition	Unconditional	Codex Standard	January 1986
	Spices	Disinfection	for export only	Codex Standard	January 1986
	Frozen shrimps and frog legs	Disinfection	for export only	Codex Standard	January 1986

(*continued on next page*)

TABLE A1 (*continued*)

Country	Product	Purpose of irradiation	Type of clearance	Dose permitted (kGy)	Date of approval
Indonesia	Dried spices tuber and root crops (potatoes, shallots, garlic	Decontamination	Unconditional	10 max.	December 29, 1987
	and rhizomes)	Sprout inhibition	Unconditional	0.15 max.	December 29, 1987
	Cereals	Disinfestation	Unconditional	1 max.	December 29, 1987
Israel	Potatoes	Sprout inhibition	Unconditional	0.15 max.	July 5, 1967
	Onions	Sprout inhibition	Unconditional	0.10 max.	July 25, 1968
	Poultry and poultry sections	Shelf-life extension/decontamination	Unconditional	7 max.	April 23, 1982
	Onions	Sprout inhibition	Unconditional	0.15	March 6, 1985
	Garlic	Sprout inhibition	Unconditional	0.15	March 6, 1985
	Shallots	Sprout inhibition	Unconditional	0.15	March 6, 1985
	spices (36 different products)	Decontamination	Unconditional	10	March 6, 1985
	fresh fruits and vegetables grains, cereals, pulses, cocoa & coffee beans, nuts,	Disinfestation	Unconditional	1 average	January 1987
	edible seeds	Disinfestation	Unconditional	1 average	January 1987
	Mushrooms, strawberries	Shelf-life extension	Unconditional	3 average	January 1987

(*continued on next page*)

TABLE A1 (*continued*)

Country	Product	Purpose of irradiation	Type of clearance	Dose permitted (kGy)	Date of approval
	poultry and poultry sections	Decontamination	Unconditional	7 average	January 1987
	spices & condiments, dehydrated & dried vegetables, edible herbs	Decontamination	Unconditional	10 average	January 1987
	poultry feeds	Decontamination	Unconditional	15 average	January 1987
Italy	Potatoes	Sprout inhibition	Unconditional	0.075—0.15	August 30, 1973
	Onions	Sprout inhibition	Unconditional	0.075—0.-15	August 30, 1973
	Garlic	Sprout inhibition	Unconditional	0.075-0.15	August 30, 1973
Japan	Potatoes	Sprout inhibition	Unconditional	0.15 max.	August 30, 1972
The Netherlands	Asparagus	Shelf-life extension/growth inhibition	experimental batches	2 max.	May 7, 1969
	Cocoa beans	Insect disinfestation	experimental batches	0.7 max.	May 7, 1969
	Strawberries	Shelf-life extension	experimental batches	2.5 max.	May 7, 1969
	Mushrooms	Growth inhibition	Unconditional	2.5 max.	October 23, 1969
	Deep frozen meals	Sterilization	Hospital patients	25 min.	November 27, 1969
	Potatoes	Sprout inhibition	Unconditional	0.15 max.	March 23, 1970
	Shrimps	Shelf-life extension	experimental batches	0.5-1	November 13, 1970
	Onions	Sprout inhibition	experimental batches	0.15	February 5, 1971
	Spices and condiments	Decontamination	experimental batches	8-10	September 13, 1971

(*continued on next page*)

TABLE A1 (*continued*)

Country	Product	Purpose of irradiation	Type of clearance	Dose permitted (kGy)	Date of approval
The Netherlands (*contd*)	poultry, eviscerated (in plastic bags)	Shelf-life extension	experimental batches	3 max.	December 31, 1971
	Chicken	Shelf-life extension/decontamination	Unconditional	3 max.	May 10, 1976
	Fresh, tinned and liquid foodstuffs	Sterilization	Hospital patients	25 min.	March 8, 1972
	Spices	Decontamination	Provisional	10	October 4, 1974
	Powdered batter mix	Decontamination	Test marketing	1.5	October 4, 1974
	vegetable filling	Decontamination	Test marketing	0.75	October 4, 1974
	endive (prepared, cut)	Shelf-life extension	Test marketing	1	January 14, 1975
	onions	Sprout inhibition	Unconditional	0.05 max.	June 9, 1975
	spices	Decontamination	Provisional	10	June 26, 1975
	peeled potatoes	Shelf-life extension	Test marketing	0.5	May 12, 1976
	chicken	Shelf-life extension/decontamination	Unconditional	3 max.	May 10, 1976
	shrimps	Shelf-life extension	Test marketing	1	June 15, 1976
	fillets of haddock, coal-fish, whiting	Shelf-life extension	Test marketing	1	September 6, 1976
	fillets of cod and plaice	Shelf-life extension	Test marketing	1	September 7, 1976
	fresh vegetables (prepared, cut, soup greens)	Shelf-life extension	Test marketing	1	September 6, 1977

(*continued on next page*)

TABLE A1 (*continued*)

Country	Product	Purpose of irradiation	Type of clearance	Dose permitted (kGy)	Date of approval
	Spices	Decontamination	Provisional	10	April 4, 1978
	frozen frog legs	Decontamination	Provisional	5	September 25, 1978
	rice and ground rice products	Insect disinfestation	Provisional	1	March 15, 1979
	rye bread	Shelf-life extension	Provisional	5 max.	February 12, 1980
	spices	Decontamination	Provisional	7 max.	April 15, 1980
	frozen shrimp	Decontamination	Provisional	7 max.	May 9, 1980
	malt	Decontamination	Provisional	10 max.	February 8, 1983
	boiled and cooled shrimp	Shelf-life extension	Provisional	1 max.	February 8, 1983
	frozen shrimp	Decontamination	Provisional	7 max.	February 8, 1983
	frozen fish	Decontamination	Provisional	6 max.	August 24, 1983
	egg powder	Decontamination	Provisional	6 max.	August 25, 1983
	dry blood protein	Decontamination	Provisional	7 max.	August 25, 1983
	dehydrated vegetables	Decontamination	Provisional	10 max.	October 27, 1983
	refrigerated snacks of minced meat	Shelf-life extension	Test marketing	2	July 12, 1984
New Zealand	Herbs and spices (one batch)	Decontamination	Provisional	8	March 1985
Norway	Spices	Decontamination	Unconditional	Up to 10	
The Philippines	Potatoes	Sprout inhibition	Provisional	0.15 max.	September 13, 1972
	Onions	Sprout inhibition	Provisional	0.07	1981
	Garlic	Sprout inhibition	Provisional	0.07	1981
	Onions and garlic	Sprout inhibition	Test marketing		July 9, 1984
					September 29,1 986

(*continued on next page*)

TABLE A1 *(continued)*

Country	Product	Purpose of irradiation	Type of clearance	Dose permitted (kGy)	Date of approval
Poland	Potatoes	Sprout inhibition	Provisional	Up to 0.1 5	1982
	Onions	Sprout inhibition	Provisional		March 1983
Republic of Korea	Potatoes	Sprout inhibition	unconditiona	0.15 max.	September 28, 1987
	Onions	Sprout inhibition	Unconditional	0.15 max.	September 28, 1987
	Garlic	Sprout inhibition	Unconditional	0.15 max.	September 28, 1987
	Chestnuts fresh and	Sprout inhibition	Unconditional	0.25 max.	September 28, 1987
	dried mushrooms	growth inhibition/Insect disinfestation	Unconditional	1.00 max.	September 28, 1987
South Africa	Potatoes	Sprout inhibition	Unconditional	0.12-0.24	January 19, 1977
	Dried bananas	Insect disinfestation	Provisional	0.5 max.	July 28, 1977
	Avocados	Insect disinfestation	Provisional	0.1 max.	July 28, 1977
	Onions	Sprout inhibition	Unconditional	0.05-0.15	August 25, 1978
	Garlic	Sprout inhibition	Unconditional	0.1-0.20	August 25, 1978
	Chicken	Shelf-life extension/decontamination	Unconditional	2-7	August 25, 1978
	Papaya	Shelf-life extension	Unconditional	0.5-1.5	August 25, 1978
	Mango	Shelf-life extension	Unconditional	0.5-1.5	August 25, 1978
	Strawberries	Shelf-life extension	Unconditional	1-4	August 25, 1978
	Bananas	Shelf-life extension	Unconditional		1982
	Litchis	Shelf-life extension	Unconditional		1982
	Pickled mango (achar)	Shelf-life extension	Unconditional		1982
	Avocados	Shelf-life extension	Unconditional		1982
	Frozen fruit juices	Shelf-life extension	Unconditional		
	Green beans		Unconditional		
	Tomatoes	Control of ripening	Unconditional		

(continued on next page)

TABLE A1 (*continued*)

Country	Product	Purpose of irradiation	Type of clearance	Dose permitted (kGy)	Date of approval
	Brinjals		Unconditional		
	Soya pickle products		Unconditional		
	Ginger		Unconditional		
	Vegetable paste		Unconditional		
	Bananas (dried)	Insect disinfestation	Unconditional		
	Almonds	Insect disinfestation	Unconditional		
	cheese powder	Insect disinfestation	Unconditional		
	yeast powder		Unconditional		
	herbal tea		Unconditional		
	various spices		Unconditional		
	various dehydrated vegetables		Unconditional		
Spain	Potatoes	Sprout inhibition	Unconditional	0.05-0.15	November 4 1969
	Onions	Sprout inhibition	Unconditional	0.08 max.	1971
Thailand	Onions	Sprout inhibition	Unconditional	0.1 max.	March 20, 1973
	Potatoes, onions, garlic	Sprout inhibition	Unconditional	0.15	December 4, 1986
	Dates	Disinfestation	Unconditional	1	December 4, 1986
	Mangoes, papaya	Disinfestation/delay of ripening	Unconditional	1	December 4, 1986
	Wheat, rice, pulses	Disinfestation	Unconditional	1	December 4, 1986

(*continued on next page*)

TABLE A1 (*continued*)

Country	Product	Purpose of irradiation	Type of clearance	Dose permitted (kGy)	Date of approval
Thailand (*contd*)	Cocoa beans	Disinfestation	Unconditional	1	December 4, 1986
	Fish and fishery products	Disinfestation	Unconditional	1	December 4, 1986
	Fish and fishery products	Reduce microbial load	Unconditional	2.2	December 4, 1986
	Strawberries	Shelf-life extension	Unconditional	3	December 4, 1986.
	nam	decontamination	Unconditional	4	December 4, 1986
	moo yor	Decontamination	Unconditional	5	December 4, 1986
	Sausage	Decontamination	Unconditional	5	December 4, 1986
	Frozen shrimps	Decontamination	Unconditional	5	December 4, 1986
	Cocoa beans	Reduce microbial load	Unconditional	5	December 4, 1986
	Chicken	Decontamination/Shelf-life extension	Unconditional	7	December 4, 1986
	spices & condiments, dehydrated	Insect disinfestation	Unconditional	1	December 4, 1986
	onions and onion powder	Decontamination	Unconditional	10	December 4, 1986
Union of Soviet Socialist Republics	Potatoes	Sprout inhibition	Unconditional	0.1 max.	March 14, 1958
	Potatoes	Sprout inhibition	Unconditional	0.3 (1 MeV-	July 17, 1973
	grain	Insect disinfestation	Unconditional	electrons) 0.3	1959
	Fresh fruits and vegetables	Shelf-life extension	experimental batches	2—4	July 11, 1964

(*continued on next page*)

TABLE A1 (*continued*)

Country	Product	Purpose of irradiation	Type of clearance	Dose permitted (kGy)	Date of approval
	Semiprepared raw beef, pork & rabbit products (in plastic bags)	Shelf-life extension	experimental batches	6-8	July 11, 1964
	Dried fruits	Insect disinfestation	Unconditional	1	February 15, 1966
	Dry food concentrates (buckwheat mush, gruel, rice, pudding)	Insect disinfestation	Unconditional	0.7	June 6, 1966
	Poultry, eviscerated (in plastic bags)	Shelf-life extension	experimental batches	6	July 4, 1966
	Culinary prepared meat products (fried meat, entrecote) (in plastic bags)	Shelf-life extension	Test marketing	8	February 1, 1967
	Onions	Sprout inhibition	Test marketing	0.06	February 25, 1967
	Onions	Sprout inhibition	Unconditional	0.06	July 17, 1973
United Kingdom	Any food for consumption by patients who require a sterile diet as an essential factor in their treatment	Sterilization	Hospital patients	0.2-0.5	December 1 1969

(*continued on next page*)

TABLE A1 (*continued*)

Country	Product	Purpose of irradiation	Type of clearance	Dose permitted (kGy)	Date of approval
United States	Wheat and wheat flour	Insect disinfestation	Unconditional		August 21 1963
	White potatoes	Shelf-life extension	Unconditional	0.05-0.1	June 30, 1964
	White potatoes	Shelf-life extension	Unconditional	0.05-0.15	November 1, 1965
	spices and dry vegetable seasonings (38 commodities)	Decontamination/Insect disinfestation	Unconditional	30 max.	July 5, 1983
	dry or dehydrated enzyme preparations (including immobilized enzyme preparations)	Control of insects and/or micro-organisms	Unconditional	10 kGy max.	June 10, 1985
	pork carcasses or fresh, nonheat processed cuts of pork carcasses	Control of *Trichineiia spiralis*	Unconditional	0.3 min.—1.0 max.	July 22, 1985
	fresh foods	Delay or maturation	Unconditional	1	April 18, 1986
	Food	Disinfestation	Unconditional	1	April 18, 1986

(*continued on next page*)

TABLE A1 (*continued*)

Country	Product	Purpose of irradiation	Type of clearance	Dose permitted (kGy)	Date of approval
	Dry or dehydrated enzyme preparations	Decontamination	Unconditional	10	April 18, 1986
	dry or dehydrated aromatic vegetable substances	Decontamination	Unconditional	30	April 18, 1986
Uruguay	Potatoes	Sprout inhibition	Unconditional		June 23, 1970
Yugoslavia	Cereals	Insect disinfestation	Unconditional	Up to 10	December 17, 1984
	Legumes	Insect disinfestation	Unconditional	Up to 10	December 17, 1984
	Onions	Sprout inhibition	Unconditional	Up to 10	December 17, 1984
	Garlic	Sprout inhibition	Unconditional	Up to 10	December 17, 1984
	Potatoes	Sprout inhibition	Unconditional	Up to 10	December 17, 1984
	Dehydrated fruits & vegetables	Sprout inhibition	Unconditional	Up to 10	December 17, 1984
	Dried mushrooms		Unconditional	Up to 10	December 17, 1984
	Egg powder	decontamination	Unconditional	Up to 10	December 17, 1984
	Herbal teas, tea extracts	Decontamination	Unconditional	Up to 10	December 17, 1984
	Fresh poultry	Shelf-life extension/decontamination	Unconditional	Up to 10	December 17, 1984

(*continued on next page*)

TABLE A1 (*continued*)

Country	Product	Purpose of irradiation	Type of clearance	Dose permitted (kGy)	Date of approval
Recommendations published by international organizations					
FAO/IAEA/WHO	Potatoes	Sprout inhibition	Provisional	0.15 max.	April 12, 1969
Expert Committee 1969	Wheat and ground wheat products	Insect disinfestation	Provisional	0.75 max.	April 12, 1969
FAO/IAEA/WHO	Potatoes	Sprout inhibition	Unconditional	0.03-0.15	September 7, 1976
Expert Committee 1976	Onions	Sprout inhibition	Provisional	0.02—0.15	September 7, 1976
	Papaya	Insect disinfestation	Unconditional	0.5-1	September 7, 1976
	Strawberries wheat and ground	Shelf-life extension	Unconditional	1-3	September 7, 1976
	wheat products	Insect disinfestation	Unconditional	0.15-1	September 7, 1976
	Rice	Insect disinfestation	Provisional	0.1-1	September 7, 1976
	Chicken	Shelf-life extension/decontamination	Unconditional	2-7	September 7, 1976
	Cod & redfish	Shelf-life extension/ decontamination	Provisional	2-2.2	September 7, 1976
FAO/IAEA/WHO Expert Committee 1980	Any food product	Sprout inhibition /shelf-life extension/decontamination insect disinfestation/control of ripening/growth inhibition	Unconditional	Up to 10	November 3, 1980

APPENDIX

2

General Safety Requirements Nuclear Security Guidelines No. GSR Part 7: Preparedness and Response for a Nuclear or Radiological Emergency

Applications of Nuclear and Radioisotope Technology: The Atom for Peace and Sustainable Development
DOI: https://doi.org/10.1016/B978-0-12-821319-3.00068-3

IAEA Safety Standards

for protecting people and the environment

Preparedness and Response for a Nuclear or Radiological Emergency

Jointly sponsored by the
FAO, IAEA, ICAO, ILO, IMO, INTERPOL,
OECD/NEA, PAHO, CTBTO, UNEP, OCHA, WHO, WMO

General Safety Requirements

No. GSR Part 7

CONTENTS

1. INTRODUCTION

BACKGROUND

1.1. This IAEA Safety Requirements publication is governed by the fundamental safety objective and the fundamental safety principles established in the IAEA Safety Standards publication Fundamental Safety Principles (SF-1) [1]. In particular, this publication addresses Principle 9, which is concerned with the arrangements that must be made for preparedness and response for a nuclear or radiological emergency [1].

1.2. This publication also allows for consistency with Essential Element No. 11 of the IAEA Nuclear Security Fundamentals [2], which is concerned with the planning for, preparedness for and response to a nuclear security event. It therefore addresses the emergency arrangements that must be in place irrespective of the initiator of the emergency, which could be a natural event, a human error, a mechanical or other failure, or a nuclear security event.

1.3. In 2002, the IAEA published the Safety Requirements publication, Preparedness and Response for a Nuclear or Radiological Emergency (GS-R-2)[1], jointly sponsored by seven international organizations (the Food and Agriculture Organization of the United Nations (FAO), the IAEA, the International Labour Organization (ILO), the OECD Nuclear Energy Agency (OECD/NEA), the Pan American Health Organization (PAHO), the United Nations Office for the Coordination of Humanitarian Affairs (OCHA) and the World Health Organization (WHO)). This Safety Requirements publication is a revised edition of IAEA Safety Standards Series No. GS-R-2, updated to take into account developments and experience gained since 2002. In the revision process, due consideration has been given to — but was not limited to — experience gained from the response to the accident at the Fukushima Daiichi nuclear power plant and to recommendations of the International Commission on Radiological Protection (ICRP) [3]. The IAEA Safety Guides Criteria for Use in Preparedness

---

[1] FOOD AND AGRICULTURE ORGANIZATION OF THE UNITED NATIONS, INTERNATIONAL ATOMIC ENERGY AGENCY, INTERNATIONAL LABOUR ORGANIZATION, OECD NUCLEAR ENERGY AGENCY, PAN AMERICAN HEALTH ORGANIZATION, UNITED NATIONS OFFICE FOR THE COORDINATION OF HUMANITARIAN AFFAIRS, WORLD HEALTH ORGANIZATION, Preparedness and Response for a Nuclear or Radiological Emergency, IAEA Safety Standards Series No. GS-R-2, IAEA, Vienna (2002).

and Response for a Nuclear or Radiological Emergency (GSG-2) [4] and Arrangements for Preparedness for a Nuclear or Radiological Emergency (GS-G-2.1) [5] elaborate on the requirements established in GS-R-2 and provide recommendations and guidance on their implementation. In addition, Planning and Preparing for Emergency Response to Transport Accidents Involving Radioactive Material (TS-G-1.2 (ST-3)) [6] provides guidance on planning and preparing for emergency response to transport accidents involving radioactive material.

1.4. This Safety Requirements publication addresses the requirements for preparedness and response for a nuclear or radiological emergency (including requirements for the transition to an existing exposure situation). Other Safety Requirements publications refer to and are consistent with these requirements in relation to emergency preparedness and response.

1.5. The response to a nuclear or radiological emergency may involve many national organizations (e.g. the operating organization and response organizations at the local, regional and national levels) as well as international organizations. The functions of many of these organizations may be the same for the response to a nuclear or radiological emergency as for the response to a conventional emergency. However, the response to a nuclear or radiological emergency might also involve specialized agencies and technical experts. Therefore, in order to be effective, the response to a nuclear or radiological emergency has to be well coordinated, and emergency arrangements have to be appropriately integrated with arrangements for the response to a conventional emergency and with the response measures for a nuclear security event.

1.6. Safety measures and security measures have in common the aim of protecting human life and health and protecting the environment. Paragraph 1.10 of Ref. [1] states that "Safety measures and security measures must be designed and implemented in an integrated manner so that security measures do not compromise safety and safety measures do not compromise security." This emphasizes the importance of effective coordination between safety measures and security measures in relation to the response to a nuclear or radiological emergency.

1.7. This publication also provides guidance for (1) preparedness and response for a nuclear or radiological emergency by the relevant international organizations and (2) the inter-agency coordination performed through the Inter-Agency Committee on Radiological and Nuclear Emergencies (IACRNE).

1.8. It is assumed that States applying these requirements have in place an infrastructure for the purpose of regulating the safety of facilities and activities that could pose radiation risks. This includes laws and regulations governing the safe operation of facilities and the safe conduct of activities, and an independent regulatory body with responsibilities for establishing and enforcing rules for safe operation and safe conduct. In this context, the IAEA has issued General Safety Requirements publications on the Governmental, Legal and Regulatory Framework for Safety (GSR Part 1) [7] and on Radiation Protection and Safety of Radiation Sources: International Basic Safety Standards (GSR Part 3) [8].

1.9. In addition, it is assumed that States applying these requirements have in place an infrastructure for the purpose of regulating the nuclear security of nuclear material and other radioactive material, associated facilities and associated activities, as well as nuclear security measures for nuclear material and other radioactive material out of regulatory control. This also includes an independent regulatory body as well as other competent authorities with responsibilities for regulating nuclear security. In this context, IAEA Nuclear Security Series publications [9–11] provide recommendations.

OBJECTIVE

1.10. The present publication establishes the requirements for an adequate level of preparedness and response for a nuclear or radiological emergency. The application of these requirements is also intended to mitigate the consequences of a nuclear or radiological emergency if such an emergency arises despite all efforts made to prevent it.

1.11. The fulfilment of these requirements will contribute to the harmonization worldwide of arrangements for preparedness and response for a nuclear or radiological emergency.

1.12. These requirements are intended to be applied by the government at the national level by means of adopting legislation and establishing regulations, and by making other arrangements, including assigning responsibilities (e.g. to the operating organization or the operating personnel of a facility or an activity, local or national officials, response organizations or the regulatory body) and verifying their effective fulfilment.

1.13. The requirements are also intended for use by response organizations, operating organizations and the regulatory body in respect of preparedness and response for a nuclear or radiological emergency, as well as by authorities with responsibilities for emergency preparedness and response at the local and regional level and, as appropriate, by relevant international organizations at the international level.

SCOPE

1.14. The requirements apply for preparedness and response for a nuclear or radiological emergency in relation to all those facilities and activities, as well as sources, with the potential for causing radiation exposure, environmental contamination or concern on the part of the public warranting protective actions and other response actions.

1.15. The requirements also apply to preparedness and response for a nuclear or radiological emergency in relation to off-site jurisdictions that may need to take protective actions and other response actions.

1.16. The requirements apply for preparedness and response for a nuclear or radiological emergency irrespective of the initiator of the emergency, whether the emergency follows a natural event, a human error, a mechanical or other failure, or a nuclear security event[2]. The requirements do not cover preparedness for, or response measures that are specific to, nuclear security events, for which recommendations are provided in Refs [9–11]. Such response measures include activities for the identification, collection, packaging and transport of evidence contaminated with radionuclides, nuclear forensics and related actions in the context of investigation into the circumstances surrounding a nuclear security event. The requirements established here do provide for a coordinated and integrated approach to preparedness and response for a nuclear or radiological emergency arising from a nuclear security event that necessitates protective actions and other response actions to be taken for protection of members of the public, workers and emergency workers, helpers in an emergency and patients.

[2] A 'nuclear security event' is an event that has potential or actual implications for nuclear security that must be addressed. Such events include criminal or intentional unauthorized acts involving or directed at nuclear material, other radioactive material, associated facilities or associated activities. A nuclear security event, for example, sabotage of a nuclear facility or detonation of a radiological dispersal device, may give rise to a nuclear or radiological emergency.

STRUCTURE

1.17. This publication comprises six sections. Section 2 provides for the interpretation and entry into force of the requirements. Section 3 establishes the goals of emergency preparedness and response. Section 4 establishes the general requirements that are to be met before effective emergency arrangements can be made, defines by using a graded approach the emergency preparedness categories for which the requirements have been established and elaborates on the development of a protection strategy on the basis of the hazards assessed. Section 5 establishes the requirements to be met for performing the functions critical for an effective emergency response. Section 6 establishes requirements for the infrastructure necessary to develop and maintain adequate arrangements for preparedness. Guidance values for restricting exposure of emergency workers in a nuclear or radiological emergency are provided in Appendix I. Generic criteria for use in emergency preparedness and response are provided in Appendix II. Annex I presents the applicability of paragraphs in the text for each emergency preparedness category.

2. INTERPRETATION, RESOLUTION OF CONFLICTS AND ENTRY INTO FORCE

DEFINITIONS

2.1. Terms used in this publication have the meanings given under 'Definitions' on page 79. If not otherwise defined under Definitions, terms are used as defined in the IAEA Safety Glossary, 2007 Edition [12].

INTERPRETATION

2.2. Except as specifically authorized by the statutory governing body of a Sponsoring Organization, no interpretation of this standard by any officer or employee of the Sponsoring Organization other than a written interpretation by the Director General of the Sponsoring Organization shall be binding on the Sponsoring Organization.

RESOLUTION OF CONFLICTS

2.3. The requirements of this standard are established in addition to and not in place of other applicable requirements, such as those of relevant binding conventions and national laws and regulations.

2.4. In cases of conflict between the requirements of this standard and other applicable requirements, the government or the regulatory body, as appropriate, shall determine which requirements are to be enforced.

2.5. Nothing in this standard shall be construed as restricting any actions that may otherwise be necessary for protection and safety or as relieving the parties referred to in this standard from complying with applicable laws and regulations.

ENTRY INTO FORCE

2.6. The Secretariat envisages that, for the IAEA's own operations and for those operations assisted by the IAEA, arrangements will be made to meet these requirements within a period of no more than one year from the date of publication of this standard.

2.7. This standard shall come into force within a period of no more than one year from the date of publication of this standard for all the Sponsoring Organizations in accordance with their respective mandates.

2.8. If a State decides to adopt this standard, this standard shall come into force at the time indicated in the formal adoption by that State, and preferably within a period of no more than one year from the date of its publication.

3. GOALS OF EMERGENCY PREPAREDNESS AND RESPONSE

GOAL OF EMERGENCY PREPAREDNESS

3.1. The goal of emergency preparedness is to ensure that an adequate capability is in place within the operating organization and at local, regional and national levels and, where appropriate, at the international level, for an effective response in a nuclear or radiological emergency. This capability relates to an integrated set of infrastructural elements that include, but are not limited to: authority and responsibilities; organization and staffing; coordination; plans and procedures; tools, equipment and facilities; training, drills and exercises; and a management system.

GOALS OF EMERGENCY RESPONSE

3.2. In a nuclear or radiological emergency, the goals of emergency response are:

(a) To regain control of the situation and to mitigate consequences;
(b) To save lives;
(c) To avoid or to minimize severe deterministic effects;
(d) To render first aid, to provide critical medical treatment and to manage the treatment of radiation injuries;
(e) To reduce the risk of stochastic effects;
(f) To keep the public informed and to maintain public trust;
(g) To mitigate, to the extent practicable, non-radiological consequences;
(h) To protect, to the extent practicable, property and the environment;
(i) To prepare, to the extent practicable, for the resumption of normal social and economic activity.

4. GENERAL REQUIREMENTS

Requirement 1: The emergency management system

The government shall ensure that an integrated and coordinated emergency management system for preparedness and response for a nuclear or radiological emergency is established and maintained.

4.1. The government shall ensure that an emergency management system is established and maintained on the territories of and within the jurisdiction of the State for the purposes of emergency response to protect human life, health, property and the environment in the event of a nuclear or radiological emergency.

4.2. The emergency management system shall be designed to be commensurate with the results of the hazard assessment (see paras 4.18–4.26) and shall enable an effective emergency response to reasonably foreseeable events (including very low probability events).

4.3. The emergency management system shall be integrated, to the extent practicable, into an all-hazards emergency management system (see paras 5.6 and 5.7).

4.4. The government shall ensure the coordination of and consistency of national emergency arrangements with the relevant international emergency arrangements[3].

Requirement 2: Roles and responsibilities in emergency preparedness and response

The government shall make provisions to ensure that roles and responsibilities for preparedness and response for a nuclear or radiological emergency are clearly specified and clearly assigned.

General

4.5. The government shall make adequate preparations to anticipate, prepare for, respond to and recover from a nuclear or radiological emergency at the operating organization, local, regional and national levels, and also, as appropriate, at the international level. These preparations shall include adopting legislation and establishing regulations for effectively governing the preparedness and response for a nuclear or radiological emergency at all levels (see para. 1.12).

4.6. The government shall ensure that arrangements are in place for effectively governing the provision of prompt and adequate compensation of victims for damage due to a nuclear or radiological emergency.

[3] Arrangements set under the Assistance Convention and under the Early Notification Convention [13] are examples of international emergency arrangements that are relevant for States Parties to these Conventions.

4.7. The government shall ensure that all roles and responsibilities for preparedness and response for a nuclear or radiological emergency are clearly allocated in advance among operating organizations, the regulatory body and response organizations[4].

4.8. The government shall ensure that response organizations, operating organizations and the regulatory body have the necessary human, financial and other resources, in view of their expected roles and responsibilities and the assessed hazards, to prepare for and to deal with both radiological and non-radiological consequences of a nuclear or radiological emergency, whether the emergency occurs within or beyond national borders.

4.9. The government shall ensure that operating organizations, response organizations and the regulatory body establish, maintain and demonstrate leadership in relation to preparedness and response for a nuclear or radiological emergency [14].

Coordinating mechanism

4.10. The government shall establish a national coordinating mechanism[5] to be functional at the preparedness stage, consistent with its emergency management system, with the following functions:

(a) To ensure that roles and responsibilities are clearly specified and are understood by operating organizations, response organizations and the regulatory body (see para. 4.7);
(b) To coordinate the hazard assessment within the State (see paras 4.18–4.26) and periodic reviews of the assessed hazards (see para. 4.25);
(c) To coordinate and ensure consistency between the emergency arrangements of the various response organizations, operating organizations and the regulatory body at local, regional and national levels under the all-hazards approach, including those arrangements for response to relevant nuclear security events, and, as appropriate, those arrangements of other States and of international organizations;

---

[4] This also includes the allocation of roles and responsibilities, as appropriate, among members of the government.

[5] The mechanism for ensuring coordination may differ for different tasks. It may involve an existing body or a newly established body (e.g. a committee consisting of representatives from different organizations and bodies) that has been given the authority to ensure the necessary coordination.

(d) To ensure consistency among requirements for emergency arrangements, contingency plans and security plans of operating organizations specified by the regulatory body and by other competent authorities with responsibilities for regulating nuclear security, as relevant, and to ensure that these arrangements and plans are integrated (see para. 4.14(b));
(e) To ensure that appropriate emergency arrangements are in place, both on the site and off the site, as appropriate, in relation to facilities and activities under regulatory control, both within the State and, as relevant, beyond its borders, and also for sources that are not under regulatory control[6];
(f) To coordinate arrangements made for enforcing compliance with the national requirements for emergency preparedness and response as established by legislation and regulations (see paras 1.12, 4.5 and 4.12);
(g) To coordinate a subsequent analysis of an emergency, including analysis of the emergency response (see Requirement 19);
(h) To ensure that appropriate and coordinated programmes of training and exercises are in place and implemented, and that training and exercises are systematically evaluated;
(i) To coordinate effective communication with the public in preparedness for a nuclear or radiological emergency.

Regulatory body

4.11. The government shall ensure that arrangements for preparedness and response to a nuclear or radiological emergency for facilities and activities under the responsibility of the operating organization are dealt with through the regulatory process.

4.12. The regulatory body is required to establish or adopt regulations and guides to specify the principles, requirements and associated criteria for safety upon which its regulatory judgements, decisions and actions are based [7]. These regulations and guides shall include principles, requirements and associated criteria for emergency preparedness and response for the operating organization (see also paras 1.12 and 4.5).

---

[6] Examples of sources not under regulatory control are sources that have been abandoned, lost or stolen and sources under governmental control but not under regulatory control. Examples also include radioactive material that is out of regulatory control as discussed in Ref. [11].

4.13. The regulatory body shall require that arrangements for preparedness and response for a nuclear or radiological emergency be in place for the on-site area for any regulated facility or activity that could necessitate emergency response actions. Appropriate emergency arrangements shall be established by the time the source is brought to the site, and complete emergency arrangements shall be in place before the commencement of operation of the facility or commencement of the activity. The regulatory body shall verify compliance with the requirements for such arrangements.

4.14. Before commencement of operation of the facility or commencement of the activity, the regulatory body shall ensure, for all facilities and activities under regulatory control that could necessitate emergency response actions, that the on-site emergency arrangements:

(a) Are integrated with those of other response organizations, as appropriate;
(b) Are integrated with contingency plans in the context of Ref. [9] and with security plans in the context of Ref. [10];
(c) Provide, to the extent practicable, assurance of an effective response to a nuclear or radiological emergency.

4.15. The regulatory body shall ensure that the operating organization is given sufficient authority to promptly take necessary protective actions on the site in response to a nuclear or radiological emergency that could result in off-site consequences.

Operating organization

4.16. The operating organization shall establish and maintain arrangements for on-site preparedness and response for a nuclear or radiological emergency for facilities or activities under its responsibility, in accordance with the applicable requirements (see paras 1.12, 4.5 and 4.12).

4.17. The operating organization shall demonstrate that, and shall provide the regulatory body with an assurance that, emergency arrangements are in place for an effective response on the site to a nuclear or radiological emergency in relation to a facility or an activity under its responsibility.

Requirement 3: Responsibilities of international organizations in emergency preparedness and response

Relevant international organizations shall coordinate their arrangements in preparedness for a nuclear or radiological emergency and their emergency response actions.[7]

Requirement 4: Hazard assessment

The government shall ensure that a hazard assessment is performed to provide a basis for a graded approach in preparedness and response for a nuclear or radiological emergency.

4.18. Hazards shall be identified and potential consequences of an emergency shall be assessed to provide a basis for establishing arrangements for preparedness and response for a nuclear or radiological emergency. These arrangements shall be commensurate with the hazards identified and the potential consequences of an emergency.

4.19. For the purposes of these safety requirements, assessed hazards are grouped in accordance with the emergency preparedness categories shown in Table 1. The five emergency preparedness categories (hereinafter referred to as 'categories') in Table 1 establish the basis for a graded approach to the application of these requirements and for developing generically justified and optimized arrangements for preparedness and response for a nuclear or radiological emergency.

[7] The Inter-Agency Committee on Radiological and Nuclear Emergencies and its Joint Radiation Emergency Management Plan of the International Organizations are examples of such coordination.

TABLE 1. EMERGENCY PREPAREDNESS CATEGORIES

Category	Description
I	Facilities, such as nuclear power plants, for which on-site events[a, b] (including those not considered in the design[c]) are postulated that could give rise to severe deterministic effects[d] off the site that would warrant precautionary urgent protective actions, urgent protective actions or early protective actions, and other response actions to achieve the goals of emergency response in accordance with international standards[e], or for which such events have occurred in similar facilities.
II	Facilities, such as some types of research reactor and nuclear reactors used to provide power for the propulsion of vessels (e.g. ships and submarines), for which on-site events[a, b] are postulated that could give rise to doses to people off the site that would warrant urgent protective actions or early protective actions and other response actions to achieve the goals of emergency response in accordance with international standards[e], or for which such events have occurred in similar facilities. Category II (as opposed to category I) does not include facilities for which on-site events (including those not considered in the design) are postulated that could give rise to severe deterministic effects off the site, or for which such events have occurred in similar facilities.
III	Facilities, such as industrial irradiation facilities or some hospitals, for which on-site events[b] are postulated that could warrant protective actions and other response actions on the site to achieve the goals of emergency response in accordance with international standards[e], or for which such events have occurred in similar facilities. Category III (as opposed to category II) does not include facilities for which events are postulated that could warrant urgent protective actions or early protective actions off the site, or for which such events have occurred in similar facilities.
IV	Activities and acts that could give rise to a nuclear or radiological emergency that could warrant protective actions and other response actions to achieve the goals of emergency response in accordance with international standards[e] in an unforeseen location. These activities and acts include: (a) transport of nuclear or radioactive material and other authorized activities involving mobile dangerous sources such as industrial radiography sources, nuclear powered satellites or radioisotope thermoelectric generators; and (b) theft of a dangerous source and use of a radiological dispersal device or radiological exposure device[f]. This category also includes: (i) detection of elevated radiation levels of unknown origin or of commodities with contamination; (ii) identification of clinical symptoms due to exposure to radiation; and (iii) a transnational emergency that is not in category V arising from a nuclear or radiological emergency in another State. Category IV represents a level of hazard that applies for all States and jurisdictions.

Please see table notes on following page

TABLE 1. EMERGENCY PREPAREDNESS CATEGORIES (cont.)

Category	Description
V	Areas within emergency planning zones and emergency planning distances[g] in a State for a facility in category I or II located in another State.

[a] That is, on-site events involving an atmospheric or aquatic release of radioactive material, or external exposure (due, for example, to a loss of shielding or a criticality event), that originates from a location on the site.

[b] Such events include nuclear security events.

[c] This includes events that are beyond the design basis accidents and, as appropriate, conditions that are beyond design extension conditions.

[d] See 'deterministic effect' under Definitions.

[e] See the goals of emergency response in para. 3.2 and the generic criteria in Appendix II.

[f] A radiological dispersal device is a device to spread radioactive material using conventional explosives or other means. A radiation exposure device is a device with radioactive material designed to intentionally expose members of the public to radiation. They could be fabricated, modified or improvised devices.

[g] See para. 5.38.

4.20. The government shall ensure that for facilities and activities, a hazard assessment on the basis of a graded approach is performed. The hazard assessment shall include consideration of:

(a) Events that could affect the facility or activity, including events of very low probability and events not considered in the design;
(b) Events involving a combination of a nuclear or radiological emergency with a conventional emergency such as an emergency following an earthquake, a volcanic eruption, a tropical cyclone, severe weather, a tsunami, an aircraft crash or civil disturbances that could affect wide areas and/or could impair capabilities to provide support in the emergency response;
(c) Events that could affect several facilities and activities concurrently, as well as consideration of the interactions between the facilities and activities affected;
(d) Events at facilities in other States or events involving activities in other States.

4.21. The government shall ensure that the hazard assessment identifies those facilities and locations at which there is a significant likelihood of encountering a dangerous source that is not under control.[8]

4.22. The government shall ensure that the hazard assessment includes consideration of the results of threat assessments made for nuclear security purposes [9–11].[9]

4.23. In the hazard assessment, facilities and activities, on-site areas, off-site areas and locations shall be identified for which a nuclear or radiological emergency could — with account taken of the uncertainties in and limitations of the information available — warrant any of the following:

(a) Precautionary urgent protective actions to avoid or to minimize severe deterministic effects by keeping doses below levels approaching the generic criteria at which urgent protective actions and other response actions are required to be undertaken under any circumstances, with account taken of Appendix II;
(b) Urgent protective actions and other response actions to avoid or to minimize severe deterministic effects and to reduce the risk of stochastic effects, with account taken of Appendix II;
(c) Early protective actions and other response actions, with account taken of Appendix II;
(d) Other emergency response actions such as longer term medical actions, with account taken of Appendix II, and emergency response actions aimed at enabling the termination of the emergency (see Requirement 18); or
(e) Protection of emergency workers in accordance with Requirement 11 and with account taken of Appendix I.

---

[8] Examples of such facilities and locations are: scrap metal processing facilities, border crossing points, seaports, airports and abandoned military facilities or other facilities where dangerous sources might have been used in the past.

[9] This includes consideration of 'strategic locations', i.e. locations of high security interest in the State which are potential targets for attacks using nuclear and other radioactive material and locations for detection of nuclear and other radioactive material that is out of regulatory control, in line with Ref. [11].

4.24. The government shall ensure that the hazard assessment also identifies non-radiation-related hazards[10] to people on the site and off the site that are associated with the facility or activity and that may impair the effectiveness of the response actions to be taken.

4.25. The government shall ensure that a review of the hazard assessment is performed periodically with the aims of: (a) ensuring that all facilities and activities, on-site areas, off-site areas and locations where events could occur that would necessitate protective actions and other response actions are identified, and (b) taking into account any changes in the hazards within the State and beyond its borders, any changes in assessments of threats for nuclear security purposes, the experience and lessons from research, operation and emergency exercises, and technological developments (see paras 6.30, 6.36 and 6.38). The results of this review shall be used to revise the emergency arrangements as necessary.

4.26. The government through the regulatory body shall ensure that operating organizations review appropriately and, as necessary, revise the emergency arrangements (a) prior to any changes in the facility or activity that affect the existing hazard assessment and (b) when new information becomes available that provides insights into the adequacy of the existing arrangements.[11]

Requirement 5: Protection strategy for a nuclear or radiological emergency

The government shall ensure that protection strategies are developed, justified and optimized at the preparedness stage for taking protective actions and other response actions effectively in a nuclear or radiological emergency.

4.27. The government shall ensure that, on the basis of the hazards identified and the potential consequences of a nuclear or radiological emergency, protection strategies are developed, justified and optimized at the preparedness stage for taking protective actions and other response actions effectively in a nuclear or radiological emergency to achieve the goals of emergency response.

---

[10] Examples of non-radiation-related hazards are the release of toxic chemicals, e.g. uranium hexafluoride (UF_6), fires, explosions and floods.

[11] Examples of such changes and available information include the movement of irradiated nuclear fuel to a new location, projected flooding, and information on storms or other meteorological hazards.

4.28. Development of a protection strategy shall include, but shall not be limited to, the following:

(1) Consideration shall be given to actions to be taken to avoid or to minimize severe deterministic effects and to reduce the risk of stochastic effects. Deterministic effects shall be evaluated on the basis of relative biological effectiveness (RBE) weighted absorbed dose to a tissue or organ. Stochastic effects in a tissue or organ shall be evaluated on the basis of equivalent dose to the tissue or organ. The detriment associated with the occurrence of stochastic effects in individuals in an exposed population shall be evaluated on the basis of the effective dose.
(2) A reference level expressed in terms of residual dose shall be set, typically as an effective dose in the range 20–100 mSv, acute or annual, that includes dose contributions via all exposure pathways. This reference level shall be used in conjunction with the goals of emergency response (see para. 3.2) and the specific time frame in which particular goals are to be achieved.[12]
(3) On the basis of the outcome of the justification and the optimization of the protection strategy, national generic criteria for taking protective actions and other response actions, expressed in terms of projected dose or of dose that has been received, shall be developed with account taken of the generic criteria in Appendix II. If the national generic criteria for projected dose or received dose are exceeded, protective actions and other response actions, either individually or in combination, shall be implemented.

[12] The application solely of the reference level for effective dose would not be sufficient to develop the protection strategy. Consideration needs to be given to the particular goal to be met in the response, the time to allow for actions to be taken effectively, and the appropriate dose quantity to be used to ensure that organ doses will be kept below those at which protective actions and other response actions are justified (see para. 4.28 (1)). For example, actions to avoid or to minimize severe deterministic effects are to be taken urgently when projected doses expected to be received within a short period of time exceed those given in Table II.1 of Appendix II for the RBE weighted absorbed dose to a tissue or organ. In this case, if such doses are received, then prompt and appropriate medical actions are necessary. Moreover, selection of a particular value (to be used for optimization purposes and for retrospective assessment of the effectiveness of actions and strategy taken) within the proposed range of 20–100 mSv acute or annual effective dose would depend on the phase of the emergency, the practicality of reducing or preventing exposures, and other factors. In the urgent phase of an emergency, an effective dose of 100 mSv, acute or annual, might be justified as one of the dosimetric bases for implementing and optimizing a protection strategy. In the later phases, such as during the transition, an effective dose of 20 mSv per year may be justified as one of the dosimetric bases for implementing and optimizing a protection strategy to enable the transition to an existing exposure situation to be made.

(4) Once the protection strategy has been justified and optimized and a set of national generic criteria has been developed, pre-established operational criteria (conditions on the site, emergency action levels (EALs) and operational intervention levels (OILs)) for initiating the different parts of an emergency plan and for taking protective actions and other response actions shall be derived from the generic criteria[13]. Arrangements shall be established in advance to revise these operational criteria, as appropriate, in the course of a nuclear or radiological emergency, with account taken of the prevailing conditions as they evolve.

4.29. Each protective action, in the context of the protection strategy, and the protection strategy itself shall be demonstrated to be justified (i.e. to do more good than harm), with account taken not only of those detriments that are associated with radiation exposure but also of those detriments associated with impacts of the actions taken on public health[14], the economy, society and the environment.

4.30. The government shall ensure that interested parties are involved and are consulted, as appropriate, in the development of the protection strategy.

4.31. The government shall ensure that the protection strategy is implemented safely and effectively in an emergency response through the implementation of emergency arrangements, including but not limited to:

(a) Promptly taking urgent protective actions and other response actions with account taken of Appendix II to avoid or to minimize severe deterministic effects, if possible, on the basis of observed conditions and before any exposure occurs;
(b) Taking early protective actions and other response actions to reduce the risk of stochastic effects with account taken of Appendix II;
(c) Providing for registration, health screening and longer term medical follow-up, as appropriate, with account taken of Appendix II;
(d) Taking actions to protect emergency workers, with account taken of guidance values provided in Appendix I;

[13] The operational criteria (i.e. operational intervention levels) need to be derived for a representative person with account taken of those members of the public that are most vulnerable to radiation exposure (i.e. pregnant women and children).

[14] Examples of such impacts include possible deaths among patients evacuated without the necessary medical care and possible reduced life expectancy due to resettlement.

(e) Taking actions to mitigate non-radiological consequences, with account taken of Appendix II;
(f) Assessing the effectiveness of the actions taken and adjusting them as appropriate on the basis of prevailing conditions and available information as well as the reference level expressed in terms of residual dose;
(g) Revising the protection strategy as necessary and its further implementation;
(h) Discontinuing protective actions and other response actions when they are no longer justified.

5. FUNCTIONAL REQUIREMENTS

GENERAL

5.1. The requirements established in this section address the functions that are essential for the emergency response in a nuclear or radiological emergency to be effective and for achieving the goals of emergency response (see para. 3.2).

Requirement 6: Managing operations in an emergency response

The government shall ensure that arrangements are in place for operations in response to a nuclear or radiological emergency to be appropriately managed.

5.2. For facilities in categories I, II and III, arrangements shall be made for the on-site emergency response to be promptly executed and managed without impairing the performance of the continuing operational safety and security functions both at the facility and at any other facilities on the same site. The transition from normal operations to operations under emergency conditions on the site shall be clearly specified and shall be effectively made. The responsibilities of all personnel who would be on the site in an emergency shall be designated as part of the arrangements for this transition. It shall be ensured that the transition to the emergency response and the performance of initial response actions do not impair the ability of operating personnel (such as operating personnel in the control room) to ensure safe and secure operation while taking mitigatory actions.

5.3. For facilities in categories I, II and III, and, where appropriate, for activities in category IV, arrangements shall be made for an off-site emergency response to be promptly executed, effectively managed and coordinated with an on-site emergency response.

5.4. For a site where several facilities in categories I and II are collocated, adequate arrangements shall be made to manage the emergency response at all the facilities if each of them is under emergency conditions simultaneously. This shall include arrangements to manage the deployment of and the protection of personnel responding on and off the site (see Requirement 11).

5.5. For facilities and activities in categories I, II, III and IV, arrangements have to be made, as far as practicable, so that the facility or activity has a nuclear security system or systems [9–11] that would be functional in a nuclear or radiological emergency.

5.6. Arrangements for response to a nuclear or radiological emergency shall be coordinated and integrated with arrangements at the local, regional and national levels for response to a conventional emergency and to a nuclear security event.[15] These arrangements shall take into consideration the fact that the initiator of the nuclear or radiological emergency may not be known early in the response.

5.7. Arrangements shall be made for the establishment and use of a clearly specified and unified command and control system for emergency response under the all-hazards approach as part of the emergency management system (see paras 4.1–4.3). The command and control system shall provide sufficient assurance for effective coordination of the on-site and off-site response. The authority and responsibility for directing the emergency response and for making decisions on emergency response actions to be taken shall be clearly assigned. The responsibility for directing the emergency response and for decision making on emergency response actions to be taken shall be promptly discharged following a notification of an emergency.

---

[15] The coordination and integration of arrangements for response to a nuclear or radiological emergency with arrangements for response to a nuclear security event includes coordination with and integration of arrangements for response measures such as identification, collection, packaging and transport of evidence contaminated with radionuclides, nuclear forensics and related activities in the context of an investigation into the circumstances surrounding a nuclear security event.

5.8. Arrangements shall be made for obtaining and assessing the information necessary for making decisions on the allocation of resources for all response organizations throughout a nuclear or radiological emergency.

5.9. For facilities in category I or II and areas in category V, arrangements shall be made for coordinating the emergency response between response organizations (including those of other States) within the emergency planning zones and emergency planning distances (see para. 5.38) and for providing mutual support.

5.10. Arrangements shall be made with other States, as appropriate, for coordinated response to a radiological emergency.

Requirement 7: Identifying and notifying a nuclear or radiological emergency and activating an emergency response

The government shall ensure that arrangements are in place for the prompt identification and notification of a nuclear or radiological emergency and for the activation of an emergency response.

5.11. An off-site notification point[16], or more than one, shall be established to receive notification of an actual or potential nuclear or radiological emergency. The notification point(s) shall be maintained in a state of continuous availability to receive any notification or request for support and to respond promptly, or to initiate a preplanned and coordinated off-site emergency response appropriate to the emergency class or the level of emergency response. The notification point(s) shall be able to initiate immediate communication by suitable, reliable and diverse means with the response organizations that are providing support.

5.12. For facilities in categories I and II and for areas in category V, the notification point shall be able to initiate immediate communication with the authority that has been assigned the responsibility to decide on and to initiate precautionary urgent protective actions and urgent protective actions off the site (see also para. 5.7).

5.13. For facilities and locations at which there is a significant likelihood of encountering a dangerous source that is not under control (see para. 4.21), arrangements shall be made to ensure that the on-site managers of operations and

[16] This may be the notification point used to receive notification of and to initiate an off-site emergency response to an emergency of any type (conventional, or nuclear or radiological).

other personnel are aware of the indicators of a potential radiological emergency, the appropriate notification, and protective actions and other response actions that are immediately warranted in an emergency. For facilities and locations for which there is a significant likelihood of encountering a dangerous source that is not under control and for an emergency at an unforeseen location, arrangements shall be made to ensure that the local officials responsible for the response and first responders are aware of the indicators of a potential radiological emergency, the appropriate notification, and protective actions and other response actions that are warranted to be taken immediately in an emergency.

5.14. The operating organization of a facility or activity in category I, II, III or IV shall make arrangements for promptly classifying, on the basis of the hazard assessment, a nuclear or radiological emergency warranting protective actions and other response actions to protect workers, emergency workers, members of the public and, as relevant, patients and helpers in an emergency, in accordance with the protection strategy (see Requirement 5). This shall include a system for classifying all types of nuclear or radiological emergency[17] as follows:

(a) *General emergency* at facilities in category I or II for an emergency that warrants taking precautionary urgent protective actions, urgent protective actions, and early protective actions and other response actions on the site and off the site. Upon declaration of this emergency class, appropriate actions shall promptly be taken, on the basis of the available information relating to the emergency, to mitigate the consequences of the emergency on the site and to protect people on the site and off the site.
(b) *Site area emergency* at facilities in category I or II for an emergency that warrants taking protective actions and other response actions on the site and in the vicinity of the site. Upon declaration of this emergency class, actions shall promptly be taken: (i) to mitigate the consequences of the emergency on the site and to protect people on the site; (ii) to increase the readiness to take protective actions and other response actions off the site if this becomes necessary on the basis of observable conditions, reliable assessments and/or results of monitoring; and (iii) to conduct off-site monitoring, sampling and analysis.
(c) *Facility emergency* at facilities in category I, II or III for an emergency that warrants taking protective actions and other response actions at the facility and on the site but does not warrant taking protective actions off the site.

[17] The emergency classes may differ from those specified in (a)–(e) provided that emergencies of all these types are included.

Upon declaration of this emergency class, actions shall promptly be taken to mitigate the consequences of the emergency and to protect people at the facility and on the site. Emergencies in this class do not present an off-site hazard.

(d) *Alert* at facilities in category I, II or III for an event that warrants taking actions to assess and to mitigate the potential consequences at the facility. Upon declaration of this emergency class, actions shall promptly be taken to assess and to mitigate the potential consequences of the event and to increase the readiness of the on-site response organizations.

(e) *Other nuclear or radiological emergency*[18] for an emergency in category IV that warrants taking protective actions and other response actions at any location. Upon declaration of this emergency class and the level of emergency response, actions shall promptly be taken to mitigate the consequences of the emergency on the site, to protect those in the vicinity (e.g. workers and emergency workers and the public) and to determine where and for whom protective actions and other response actions are warranted.

5.15. For facilities in category I, II or III and for category IV, arrangements shall be made to review the declared emergency class in the light of any new information and, as appropriate, to revise it.

5.16. The emergency classification system for facilities and activities in categories I, II, III and IV shall take into account all postulated emergencies, including those arising from events of very low probability. The operational criteria for classification shall include emergency action levels and other observable conditions (i.e. 'observables') and indicators of the conditions at the facility and/or on the site or off the site. The emergency classification system shall be established with the aim of allowing for the prompt initiation of an effective response in recognition of the uncertainty of the available information. It shall be ensured that any process for rating an event on the International Nuclear and Radiological Event Scale (INES) [15] does not delay the emergency classification or emergency response actions.[19]

---

[18] This class covers broad types of emergency (see Table 1 and paras 4.21 and 4.22). A graded approach may need to be taken when postulating emergencies and expected consequences within this class in order to determine the level of emergency response warranted.

[19] The emergency classification system is not to be confused with the INES. The INES is a scale developed for use by States solely for the purpose of communicating with the public on the safety significance of events associated with sources of radiation. The INES is not to be used as a basis for emergency response actions.

5.17. For facilities and activities in categories I, II and III, and for category IV, arrangements shall be made: (1) to promptly recognize and classify a nuclear or radiological emergency; (2) upon classification, to promptly declare the emergency class and to initiate a coordinated and preplanned on-site response; (3) to notify the appropriate notification point (see para. 5.11) and to provide sufficient information for an effective off-site response; and (4) upon notification, to initiate a coordinated and preplanned off-site response, as appropriate, in accordance with the protection strategy. These arrangements shall include suitable, reliable and diverse means of warning persons on the site, of notifying the notification point (see paras 5.41–5.43, 6.22 and 6.34) and of communication between response organizations.

5.18. In the event of a transnational emergency, the notifying State shall promptly notify[20,21] the IAEA of the emergency and, either directly or through the IAEA, those States that could be affected by it. The notifying State shall provide information on the nature of the emergency and on its potential transnational consequences, and shall respond to requests from other States and from the IAEA for information for the purposes of mitigating any consequences.

5.19. The State shall make known to the IAEA and to other States, directly or through the IAEA, its single warning point responsible for receiving emergency notifications and information from other States and information from the IAEA. This warning point shall be maintained in a state of continuous availability to receive any notification, request for assistance or request for verification and to promptly initiate a response or verification. The State shall promptly inform the IAEA and shall inform other States, directly or through the IAEA, of any changes that occur in respect of the warning point. The State shall make arrangements for promptly notifying and for providing relevant information, directly or through the IAEA, to those States that could be affected by a transnational emergency.

5.20. The notifying State shall have arrangements in place for promptly responding to requests from other States or from the IAEA for information in respect of a transnational emergency, in particular with regard to minimizing

[20] Such a notification is in accordance with the State's obligations under the general principles and rules of international law and, for the case of a transboundary release that could be of radiological safety significance for another State, it is in accordance with the Early Notification Convention [13].

[21] A transnational emergency that is considered to represent a public health emergency of international concern may also be expected to be notified in accordance with the International Health Regulations [16].

any consequences. These arrangements shall include making known to the IAEA and to other States, directly or through the IAEA, the notifying State's designated organization(s) for so doing.

5.21. Arrangements shall be made for promptly and directly notifying any State within the emergency planning zones and emergency planning distances (see para. 5.38) within which urgent protective actions and early protective actions and other response actions could be required to be taken.

5.22. Appropriate emergency response actions shall be initiated in a timely manner upon the receipt of a notification from another State or of information from the IAEA on a notification relating to an actual or potential transnational emergency that could have impacts on the State or its nationals.

Requirement 8: Taking mitigatory actions

The government shall ensure that arrangements are in place for taking mitigatory actions in a nuclear or radiological emergency.

5.23. The operating organization of a facility or activity in category I, II, III or IV shall promptly decide on and take actions[22] on the site that are necessary to mitigate the consequences of a nuclear or radiological emergency involving a facility or an activity under its responsibility.

5.24. Off-site emergency services shall be made available for the purpose of, and shall be capable of, supporting the on-site emergency response at facilities and activities in category I, II, III or IV.[23]

5.25. For facilities in category I, II or III, arrangements shall be made for mitigatory actions to be taken by the operating personnel, in particular:

(a) To prevent escalation of an emergency;
(b) To return the facility to a safe and stable state;

---

[22] Such actions may include actions with off-site consequences such as discharge of radioactive material to the environment, provided that the appropriate off-site organizations are notified in advance.

[23] This is not to be understood as diminishing the responsibility of the operating organization to have adequate capabilities to respond to an emergency arising in the facility or activity under its responsibility.

(c) To reduce the potential for, and to mitigate the consequences of, radioactive releases or exposures.

These arrangements shall take into account the full range of possible conditions affecting the emergency response, including those resulting from conditions in the facility and those resulting from impacts of postulated natural, human induced or other events and affecting regional infrastructure or affecting several facilities simultaneously. Arrangements shall include emergency operating procedures and guidance for operating personnel on mitigatory actions for severe conditions (for a nuclear power plant, as part of the accident management programme [17]) and for the full range of postulated emergencies, including accidents that are not considered in the design and associated conditions. As far as practicable, the continued functionality of nuclear security system(s) (see Refs [9–11]) needs to be considered in these arrangements.

5.26. The operating organization of a facility or activity in category I, II, III or IV shall assess and determine, at the preparedness stage, when and under what conditions assistance from off-site emergency services may need to be provided on the site, consistent with the hazard assessment and the protection strategy.[23]

5.27. For facilities in category I, II or III, arrangements shall be made, in particular by the operating organization, to provide technical assistance to the operating personnel. On-site teams for mitigating the consequences of an emergency (e.g. damage control, firefighting) shall be available and shall be prepared to perform actions at the facility. Paragraph 5.15 of Safety of Nuclear Power Plants: Design (SSR-2/1) [18] states that:

> "Any equipment that is necessary for actions to be taken in manual response and recovery processes shall be placed at the most suitable location to ensure its availability at the time of need and to allow safe access to it under the environmental conditions anticipated."

The operating personnel directing mitigatory actions shall be provided with information and technical assistance to allow them to take actions effectively to mitigate the consequences of the emergency. Arrangements shall be made to obtain support promptly from the emergency services (e.g. law enforcement agencies, medical services and firefighting services) off the site. Off-site emergency services shall be afforded prompt access to the facility, and shall be informed of on-site conditions and provided with instructions and with means for protecting themselves as emergency workers.

5.28. Arrangements shall be made for the operating organization of an activity in category IV, first responders in an emergency at an unforeseen location, and those personnel at locations where there is a significant likelihood of encountering a dangerous source that is not under control (see para. 4.21) to take promptly all practicable and appropriate actions to mitigate the consequences of a nuclear or radiological emergency. These arrangements shall include providing basic instructions and training in the means of mitigating the potential consequences of a nuclear or radiological emergency (see para. 5.44).

5.29. Arrangements shall be made to provide expertise and services in radiation protection promptly to local officials, first responders in an emergency at an unforeseen location and specialized services (e.g. law enforcement agencies) responding to emergencies involving activities and acts in category IV, and to those personnel at locations where there is a significant likelihood of encountering a dangerous source that is not under control (see para. 4.21). This shall include arrangements for on-call advice or other appropriate mechanisms and arrangements to dispatch to the site an emergency team capable of assessing radiation hazards, mitigating radiological consequences and managing the exposure of emergency workers. In addition, arrangements shall be made to determine whether and when additional assistance is necessary and to determine how to obtain such assistance (see paras 5.24 and 5.94).

5.30. Arrangements shall be made to initiate a prompt search in the event that a dangerous source could possibly be in the public domain as a result of its loss or unauthorized removal (see para. 5.47).

Requirement 9: Taking urgent protective actions and other response actions

The government shall ensure that arrangements are in place to assess emergency conditions and to take urgent protective actions and other response actions effectively in a nuclear or radiological emergency.

5.31. Arrangements shall be made so that the magnitudes of hazards and the possible development of hazardous conditions are assessed initially and throughout a nuclear or radiological emergency in order to promptly identify, characterize or anticipate, as appropriate, new hazards or the extent of hazards and to revise the protection strategy.

5.32. The operating organization of a facility in category I, II or III shall make arrangements to promptly assess and anticipate:

(a) Abnormal conditions at the facility;
(b) Exposures and radioactive releases and releases of other hazardous material;
(c) Radiological conditions on the site and, as appropriate, off the site;
(d) Any exposures or potential exposures of workers and emergency workers, the public and, as relevant, patients and helpers in an emergency.

5.33. These assessments as stated in para. 5.32 shall be used:

(a) For deciding on mitigatory actions to be taken by the operating personnel;
(b) As a basis for emergency classification (see para. 5.14);
(c) For deciding on protective actions and other response actions to be taken on the site, including those for the protection of workers and emergency workers;
(d) For deciding on protective actions and other response actions to be taken off the site;
(e) Where appropriate, to identify those individuals who could potentially have been exposed on the site at levels requiring appropriate medical attention in accordance with Appendix II.

5.34. These arrangements as stated in para. 5.32 shall include the use of pre-established operational criteria in accordance with the protection strategy (see para. 4.28(4)) and provision for access to instruments displaying or measuring those parameters that can readily be measured or observed in a nuclear or radiological emergency. In these arrangements, the expected response of instrumentation and of structures, systems and components at the facility under emergency conditions shall be taken into account.

5.35. The operating organization for activities in category IV shall make arrangements to assess promptly the extent and/or the significance of any abnormal conditions on the site, any exposures or any contamination. These assessments shall be used:

(a) For initiating the mitigatory actions;
(b) As a basis for protective actions and other response actions to be taken on the site;
(c) For determining the level for emergency response and for communicating the extent of the hazards to the appropriate off-site response organizations.

These arrangements shall include the use of pre-established operational criteria in accordance with the protection strategy (see para. 4.28(4)).

5.36. Arrangements shall be made such that information on emergency conditions, assessments and protective actions and other response actions that have been recommended and have been taken is promptly made available, as appropriate, to all relevant response organizations and to the IAEA throughout the emergency.

5.37. Arrangements shall be made for actions to save human life or to prevent serious injury to be taken without any delay on the grounds of the possible presence of radioactive material (see paras 5.39 and 5.64). These arrangements shall include providing first responders in an emergency at an unforeseen location with information on the precautions to take in giving first aid or in transporting an individual with possible contamination.

5.38. For facilities in category I or II, arrangements shall be made for effectively making decisions on and taking urgent protective actions, early protective actions and other response actions[24] off the site in order to achieve the goals of emergency response, on the basis of a graded approach and in accordance with the protection strategy. The arrangements shall be made with account taken of the uncertainties in and limitations of the information available when protective actions and other response actions have to be taken to be effective, and shall include the following:

(a) The specification of off-site emergency planning zones and emergency planning distances[25] for which arrangements shall be made at the preparedness stage for taking protective actions and other response actions effectively. These emergency planning zones and emergency planning

[24] Although defined under this overarching requirement, emergency planning zones and emergency planning distances are applicable for both urgent protective actions and early protective actions and other response actions. Within emergency planning zones, the main focus is on taking precautionary urgent protective actions, urgent protective actions and other response actions. However, within emergency planning distances, urgent decisions may be warranted, as a precaution, to prevent inadvertent ingestion and to restrict the consumption of food, milk and drinking water that could be directly contaminated following a significant release of radioactive material to the environment and then consumed.

[25] The off-site emergency planning zones and emergency planning distances may differ from those specified provided that, at the preparedness stage, such areas and distances are designated and arrangements are made to effectively take precautionary urgent protective actions, urgent protective actions and early protective actions and other response actions within these areas and distances in order to achieve the goals of emergency response.

distances shall be contiguous across national borders, where appropriate, and shall include:

(i) A precautionary action zone (PAZ), for facilities in category I, for which arrangements shall be made for taking urgent protective actions and other response actions, before any significant release[26] of radioactive material occurs, on the basis of conditions at the facility (i.e. conditions leading to the declaration of a general emergency; see para. 5.14), in order to avoid or to minimize severe deterministic effects.

(ii) An urgent protective action planning zone (UPZ), for facilities in category I or II, for which arrangements shall be made to initiate urgent protective actions and other response actions, if possible before any significant release of radioactive material occurs, on the basis of conditions at the facility (i.e. conditions leading to the declaration of a general emergency; see para. 5.14), and after a release occurs, on the basis of monitoring and assessment of the radiological situation off the site, in order to reduce the risk of stochastic effects.[27] Any such actions shall be taken in such a way as not to delay the implementation of precautionary urgent protective actions and other response actions within the precautionary action zone.

(iii) An extended planning distance (EPD) from the facility, for facilities in category I or II (beyond the urgent protective action planning zone), for which arrangements shall be made to conduct monitoring and assessment of the radiological situation off the site in order to identify areas, within a period of time that would allow the risk of stochastic effects in the areas to be effectively reduced by taking protective actions and other response actions within a day to a week or to a few weeks following a significant radioactive release.

(iv) An ingestion and commodities planning distance (ICPD) from the facility, for facilities in category I or II (beyond the extended planning distance), for which arrangements shall be made to take response

---

[26] A significant release of radioactive material is a radioactive release that could lead to severe deterministic effects off the site and thus warrants taking protective actions or other response actions off the site.

[27] Taking actions within the urgent protective action planning zone in order to reduce the risk of stochastic effects would not mean that no severe deterministic effects could possibly be observed within the urgent protective action planning zone. However, any severe deterministic effects are most likely to occur within the precautionary action zone.

actions (1) for protecting the food chain and water supply[28] as well as for protecting commodities other than food from contamination following a significant radioactive release and (2) for protecting the public from the ingestion of food, milk and drinking water and from the use of commodities other than food with possible contamination following a significant radioactive release.

(b) Criteria, based on the emergency classification and on conditions at the facility and off the site (see paras 4.28(3), 4.28(4), 5.14 and 5.15), for initiating and for adjusting urgent protective actions and other response actions within the emergency planning zones and emergency planning distances, in accordance with the protection strategy.

(c) Authority and responsibility to provide sufficient and updated information to the notification point at any time to allow for an effective off-site emergency response.

5.39. Within the emergency planning zones and emergency planning distances, arrangements shall be made for taking appropriate protective actions and other response actions effectively, as necessary, promptly upon the notification of a nuclear or radiological emergency. These arrangements shall include arrangements for:

(a) Prompt exercise of authority and discharge of responsibility for making decisions to initiate protective actions and other response actions upon notification of an emergency (see para. 5.12);

(b) Warning the permanent population, transient population groups and special population groups or those responsible for them and warning special facilities;

(c) Taking urgent protective actions and other response actions such as evacuation, restrictions on the food chain and on water supply, prevention of inadvertent ingestion, restrictions on the consumption of food, milk and drinking water and on the use of commodities, decontamination of evacuees, control of access and traffic restrictions;

(d) Protection of emergency workers and helpers in an emergency.

The arrangements shall be coordinated with all jurisdictions (including, to the extent practicable, jurisdictions beyond national borders, where relevant) within any emergency planning zone or distance. These arrangements shall ensure

[28] 'Water supply' refers to water supplies that use rainwater or other untreated surface water.

that services necessary for ensuring public safety (e.g. rescue services and health services for the care of critically ill patients) are provided continuously throughout the emergency, including during the period when protective actions and other response actions are being taken.

5.40. Within emergency planning zones and emergency planning distances, arrangements shall be made for the timely monitoring and assessment of contamination, radioactive releases and exposures for the purpose of deciding on or adjusting the protective actions and other response actions that have to be taken or that are being taken. These arrangements shall include the use of pre-established operational criteria in accordance with the protection strategy (see para. 4.28(4)).

5.41. The operating organization of a facility in category I, II or III shall make arrangements to ensure protection and safety for all persons on the site in a nuclear or radiological emergency. These shall include arrangements to do the following:

(a) To notify all persons on the site of an emergency on the site;
(b) For all persons on the site to take appropriate actions immediately upon notification of an emergency;
(c) To account for those persons on the site and to locate and recover those persons unaccounted for;
(d) To provide immediate first aid;
(e) To take urgent protective actions.

5.42. Arrangements as stated in para. 5.41 shall also include ensuring the provision, for all persons present in the facility and on the site, of:

(a) Suitable assembly points, provided with continuous radiation monitoring;
(b) A sufficient number of suitable escape routes;
(c) Suitable and reliable alarm systems and other means for warning and instructing all persons present under the full range of emergency conditions.

5.43. The operating organization of a facility in category I, II or III shall ensure that suitable, reliable and diverse means of communication are available at all times, under the full range of emergency conditions, for use in taking protective actions and other response actions on the site and for communication with off-site officials responsible for taking protective actions and other response actions off the site or within any emergency planning zones or emergency planning distances.

5.44. Operating personnel for activities in category IV, first responders in an emergency at an unforeseen location and those personnel at locations where there is a significant likelihood of encountering a dangerous source that is not under control (see para. 4.21) shall be provided with guidance and training on taking urgent protective actions and other response actions. This shall include guidance and training on the approximate radius of the inner cordoned off area in which urgent protective actions and other response actions would initially be taken and on the adjustment of this area on the basis of observed or assessed conditions on the site.

Requirement 10: Providing instructions, warnings and relevant information to the public for emergency preparedness and response

The government shall ensure that arrangements are in place to provide the public who are affected or are potentially affected by a nuclear or radiological emergency with information that is necessary for their protection, to warn them promptly and to instruct them on actions to be taken.

5.45. For facilities in category I or II and areas in category V, arrangements shall be made to provide the permanent population, transient population groups and special population groups or those responsible for them and special facilities within the emergency planning zones and emergency planning distances (see para. 5.38), before operation and throughout the lifetime of the facility, with information on the response to a nuclear or radiological emergency. This information shall include information on the potential for a nuclear or radiological emergency, on the nature of the hazards, on how people would be warned or notified, and on the actions to be taken in such an emergency. The information shall be provided in the languages mainly spoken by the population residing within the emergency planning zones and emergency planning distances. The effectiveness of these arrangements for public information shall be periodically assessed.

5.46. For facilities in category I or II and in areas in category V, arrangements shall be made to register those members of the public in special population groups and, as appropriate, those responsible for them, and to promptly issue them and the permanent population and transient population groups, as well as special facilities in the emergency planning zones and emergency planning distances, with a warning and with instructions to be followed upon declaration of a general emergency (see para. 5.14). This shall include providing instructions on the actions to be taken in the languages mainly spoken by the population residing within these emergency planning zones and emergency planning distances (see para. 5.38).

5.47. For facilities in category III and category IV, arrangements shall be made to provide the public with information and instructions in order to identify and locate people who may have been affected by a nuclear or radiological emergency and who may need response actions such as decontamination, medical examination or health screening. These arrangements shall include arrangements for issuing a warning to the public and providing information in the event that a dangerous source could be in the public domain as a consequence of its loss or unauthorized removal.

5.48. Arrangements shall be made by response organizations in a State to promptly provide information and advice to its nationals and to those people with interests in other States[29] in the event of a nuclear or radiological emergency declared beyond national borders, with due account taken of the response actions recommended in the State in which the emergency occurs as well as in the State(s) affected by that emergency (see paras 5.73 and 6.14).

Requirement 11: Protecting emergency workers and helpers in an emergency

The government shall ensure that arrangements are in place to protect emergency workers and to protect helpers in a nuclear or radiological emergency.

5.49. Arrangements shall be made to ensure that emergency workers are, to the extent practicable, designated in advance and are fit for the intended duty. These arrangements shall include health surveillance for emergency workers for the purpose of assessing their initial fitness and continuing fitness for their intended duties (see also GSR Part 3 [8]).

5.50. Arrangements shall be made to register and to integrate into operations in an emergency response those emergency workers who were not designated as such in advance of a nuclear or radiological emergency and helpers in an emergency. This shall include designation of the response organization(s) responsible for ensuring protection of emergency workers and protection of helpers in an emergency.

---

[29] Examples of people with interests in other States include people travelling, people working and/or living abroad, importers and exporters, and people working in companies operating abroad.

5.51. The operating organization and response organizations shall determine the anticipated hazardous conditions, both on the site and off the site, in which emergency workers might have to perform response functions in a nuclear or radiological emergency in accordance with the hazard assessment and the protection strategy.

5.52. The operating organization and response organizations shall ensure that arrangements are in place for the protection of emergency workers and protection of helpers in an emergency for the range of anticipated hazardous conditions in which they might have to perform response functions. These arrangements, as a minimum, shall include:

(a) Training those emergency workers designated as such in advance;
(b) Providing emergency workers not designated in advance and helpers in an emergency immediately before the conduct of their specified duties with instructions on how to perform the duties under emergency conditions ('just in time' training);
(c) Managing, controlling and recording the doses received;
(d) Provision of appropriate specialized protective equipment and monitoring equipment;
(e) Provision of iodine thyroid blocking, as appropriate, if exposure due to radioactive iodine is possible;
(f) Obtaining informed consent to perform specified duties, when appropriate;
(g) Medical examination, longer term medical actions and psychological counselling, as appropriate.

5.53. The operating organization and response organizations shall ensure that all practicable means are used to minimize exposures of emergency workers and helpers in an emergency in the response to a nuclear or radiological emergency (see para. I.2 of Appendix I), and to optimize their protection.

5.54. In a nuclear or radiological emergency, the relevant requirements for occupational exposure in planned exposure situations established in GSR Part 3 [8] shall be applied, on the basis of a graded approach, for emergency workers, except as required in para. 5.55.

5.55. The operating organization and response organizations shall ensure that no emergency worker is subject to an exposure in an emergency that could give rise to an effective dose in excess of 50 mSv other than:

(1) For the purposes of saving human life or preventing serious injury;
(2) When taking actions to prevent severe deterministic effects or actions to prevent the development of catastrophic conditions that could significantly affect people and the environment;
(3) When taking actions to avert a large collective dose.

5.56. For the exceptional circumstances of para. 5.55, national guidance values shall be established for restricting the exposures of emergency workers, in accordance with Appendix I.

5.57. The operating organization and response organizations shall ensure that emergency workers who undertake emergency response actions in which doses received might exceed an effective dose of 50 mSv do so voluntarily[30]; that they have been clearly and comprehensively informed in advance of associated health risks as well as of available protective measures; and that they are, to the extent possible, trained in the actions that they might be required to take. Emergency workers not designated as such in advance shall not be the first emergency workers chosen for taking actions that could result in their doses exceeding the guidance values of dose for lifesaving actions, as given in Appendix I. Helpers in an emergency shall not be allowed to take actions that could result in their receiving doses in excess of an effective dose of 50 mSv.

5.58. Arrangements shall be made to assess as soon as practicable the individual doses received in a response to a nuclear or radiological emergency by emergency workers and helpers in an emergency and, as appropriate, to restrict further exposures in the response to the emergency (see Appendix I).

5.59. Emergency workers and helpers in an emergency shall be given appropriate medical attention for doses received in a response to a nuclear or radiological emergency (see Appendix II) or at their request.

---

[30] The voluntary basis for response actions by emergency workers is usually covered in the emergency arrangements.

5.60. Emergency workers who receive doses in a response to a nuclear or radiological emergency shall normally not be precluded from incurring further occupational exposure. However, qualified medical advice[31] shall be obtained before any further occupational exposure occurs if an emergency worker has received an effective dose exceeding 200 mSv, or at the request of the emergency worker.

5.61. Information on the doses received in the response to a nuclear or radiological emergency and information on any consequent health risks shall be communicated, as soon as practicable, to emergency workers and to helpers in an emergency.

Requirement 12: Managing the medical response in a nuclear or radiological emergency

The government shall ensure that arrangements are in place for the provision of appropriate medical screening and triage, medical treatment and longer term medical actions for those people who could be affected in a nuclear or radiological emergency.

5.62. On the presentation by an individual of clinical symptoms of radiation exposure or other indications associated with a possible nuclear or radiological emergency, the medical personnel or other responsible parties who identify the clinical symptoms or other indications shall notify the appropriate local or national officials and shall take response actions as appropriate.

5.63. Arrangements shall be made for medical personnel, both general practitioners and emergency medical staff, to be made aware of the clinical symptoms of radiation exposure, and of the appropriate notification procedures and other emergency response actions to be taken if a nuclear or radiological emergency arises or is suspected.

[31] Such qualified medical advice is intended for assessing the continuing fitness of workers for their intended tasks involving occupational exposure in line with GSR Part 3 [8]. In line with para. 5.59 of this Safety Requirements publication, any emergency worker is to be given appropriate medical attention for doses received. To illustrate this, the generic criterion for dose that is received (100 mSv effective dose in a month), as provided in Table II.2 of Appendix II, will indicate that an emergency worker receiving such a dose needs to be registered and subjected to health screening and that the emergency worker will then need appropriate longer term medical follow-up in order to detect radiation induced health effects early and to treat them effectively.

5.64. Arrangements shall be made so that, in a nuclear or radiological emergency, individuals with possible contamination can promptly be given appropriate medical attention. These arrangements shall include ensuring that transport services are provided where needed and providing instructions[32] to medical personnel on the precautions to take.

5.65. For facilities in categories I, II and III, arrangements shall be made to manage an adequate number of any individuals with contamination or of any individuals who have been overexposed to radiation, including arrangements for first aid, the estimation of doses, medical transport and initial medical treatment in predesignated medical facilities.

5.66. For areas within emergency planning zones (see para. 5.38), arrangements shall be made for performing medical screening and triage and for assigning to a predesignated medical facility any individual exposed at levels exceeding the criteria in Table II.1 of Appendix II. These arrangements shall include the use of pre-established operational criteria in accordance with the protection strategy (see para. 4.28(4)).

5.67. Arrangements shall be made to identify individuals with possible contamination and individuals who have possibly been sufficiently exposed for radiation induced health effects to result, and to provide them with appropriate medical attention, including longer term medical follow-up. These arrangements shall include:

(a) Guidelines for effective diagnosis and treatment;
(b) Designation of medical personnel trained in clinical management of radiation injuries;
(c) Designation of institutions for evaluating radiation exposure (external and internal), for providing specialized medical treatment and for longer term medical actions.

These arrangements shall also include the use of pre-established operational criteria in accordance with the protection strategy (see para. 4.28(4)) and arrangements for medical consultation on treatment following any exposure

[32] These instructions include advice that universal precautions in health care against infection (e.g. surgical masks and gloves) generally provide medical personnel with adequate protection when treating individuals with possible contamination.

that could result in severe deterministic effects (see Appendix II) with medical personnel experienced in dealing with such injuries.[33]

5.68. Arrangements shall be made for the identification of individuals who are in those population groups that are at risk of sustaining increases in the incidence of cancers as a result of radiation exposure in a nuclear or radiological emergency. Arrangements shall be made to take longer term medical actions to detect radiation induced health effects among such population groups in time to allow for their effective treatment. These arrangements shall include the use of pre-established operational criteria in accordance with the protection strategy (see para. 4.28(4)).

Requirement 13: Communicating with the public throughout a nuclear or radiological emergency

The government shall ensure that arrangements are in place for communication with the public throughout a nuclear or radiological emergency.

5.69. Arrangements shall be made for providing useful, timely, true, clear and appropriate information to the public in a nuclear or radiological emergency, with account taken of the possibility that the usual means of communication might be damaged in the emergency or by its initiating event (e.g. by an earthquake or by flooding) or overburdened by demand for its use. These arrangements shall also include arrangements for keeping the international community informed, as appropriate. These arrangements shall take into account the need to protect sensitive information in circumstances where a nuclear or radiological emergency is initiated by a nuclear security event. Communication with the public in a nuclear or radiological emergency shall be carried out on the basis of a strategy to be developed at the preparedness stage as part of the protection strategy. Arrangements shall be made to adjust this strategy in the emergency response on the basis of prevailing conditions.

5.70. Arrangements shall be made to ensure that information provided to the public by response organizations, operating organizations, the regulatory body, international organizations and others in a nuclear or radiological emergency

[33] Such arrangements for medical consultation on treatment could include international assistance to be provided through or to be coordinated by the IAEA and by WHO; for example, under the Assistance Convention [13].

is coordinated and consistent, with due recognition of the evolutionary nature of an emergency.

5.71. Arrangements shall be made so that in a nuclear or radiological emergency information is provided to the public in plain and understandable language.

5.72. The government shall ensure that a system for putting radiological health hazards in perspective in a nuclear or radiological emergency is developed and implemented with the following aim:

— To support informed decision making concerning protective actions and other response actions to be taken;
— To help in ensuring that actions taken do more good than harm;
— To address public concerns regarding potential health effects.

In the development of such a system, due consideration shall be given to pregnant women and children as the individuals who are most vulnerable with regard to radiation exposure.

5.73. Arrangements shall be made to explain to the public any changes in the protective actions and other response actions recommended in the State and any differences from those recommended in other States (see paras 6.13–6.15).

5.74. Arrangements shall be made to identify and address, to the extent practicable, misconceptions, rumours and incorrect and misleading information that might be circulating widely in a nuclear or radiological emergency, in particular those that might result in actions being taken beyond those emergency response actions that are warranted[34] (see Requirement 16).

5.75. Arrangements shall be made to respond to enquiries from the public and from news media, both national and international, including enquiries received from or through the IAEA. These arrangements shall recognize the evolutionary

[34] Actions beyond those emergency response actions that are warranted include, but are not limited to: actions that interfere with prompt implementation of protective actions, such as self-evacuation both from within and from outside areas from which evacuation is ordered; actions that unnecessarily burden the health care system; actions that shun or otherwise discriminate against people or products from an area affected by a nuclear or radiological emergency; elective terminations of pregnancy that are not radiologically informed; and cancellations of commercial flights that are not radiologically informed.

nature of emergencies and the need to respond in a timely manner to enquiries even when the information requested is not yet available.

Requirement 14: Taking early protective actions and other response actions

The government shall ensure that arrangements are in place to take early protective actions and other response actions effectively in a nuclear or radiological emergency.

5.76. Within the extended planning distance (see para. 5.38), arrangements shall be made for effective relocation that may be required following a significant radioactive release and for the prevention of inadvertent ingestion, in accordance with the protection strategy (see Requirement 5). These arrangements shall include:

(a) Provision of instructions and advice to prevent inadvertent ingestion;
(b) Prompt monitoring and assessment;
(c) Use of pre-established operational criteria in accordance with the protection strategy (see para. 4.28(4));
(d) The means for accomplishing relocation and for assisting those persons who have been relocated;
(e) Provisions to extend monitoring and assessment and actions beyond the extended planning distance if necessary.

5.77. For areas within the ingestion and commodities planning distance (see para. 5.38), arrangements shall be made for prompt protection in relation to, and for restriction of, non-essential local produce, forest products (e.g. wild berries, wild mushrooms), milk from grazing animals, drinking water supplies, animal feed and commodities with contamination or possibly with contamination following a significant radioactive release, in accordance with the protection strategy (see Requirement 5). These arrangements shall include:

(a) Provision of instructions and advice:
    (i) To protect the food chain, water supply and commodities from contamination;
    (ii) To prevent ingestion of food, milk and drinking water with contamination or possibly with contamination;
    (iii) To prevent use of commodities with contamination or possibly with contamination;
(b) Prompt monitoring, sampling and analysis.

(c) Use of pre-established operational criteria in accordance with the protection strategy (see para. 4.28(4)).
(d) The means to enforce the restrictions.
(e) Provisions to expand monitoring and assessment and actions beyond this distance if necessary.

5.78. Within the emergency planning zones and the inner cordoned off area, arrangements shall be made for monitoring the levels of contamination of people, vehicles and goods moving out of areas with contamination, in order to control the spread of contamination and, as applicable, for the purposes of decontamination in accordance with the protection strategy (see Requirement 5). These arrangements shall include the use of pre-established operational criteria in accordance with the protection strategy (see para. 4.28(4)) and shall take into consideration that some vehicles and items potentially with contamination, as well as members of the public and emergency workers, might have left these areas before the establishment of contamination control points and boundaries.

5.79. Arrangements shall be made for access control and enforcing of restrictions for areas in which evacuations and relocations would be carried out within emergency planning zones, the extended planning distance and the inner cordoned off area, in accordance with the protection strategy (see Requirement 5). Returns to these areas for short periods of time shall be permitted if justified (e.g. to feed animals left behind) and provided that those individuals entering the area are:

(a) Subject to controls and to dose assessment while in the area;
(b) Instructed on how to protect themselves;
(c) Briefed on the associated health hazards.

5.80. Arrangements shall be made to test methods of decontamination before their general use and to assess their effectiveness in terms of dose reduction.

5.81. For a transnational emergency in category IV, arrangements shall be made for taking early protective actions and other response actions as appropriate for areas beyond category V, including promptly conducting monitoring and assessment of contamination (a) of food, milk and drinking water and, as appropriate, of commodities other than food, and (b) of vehicles and cargoes that are likely to have contamination, with the aim of mitigating the consequences of a nuclear or radiological emergency and reassurance of the public. These arrangements shall include the use of pre-established operational criteria in accordance with the protection strategy (see para. 4.28(4)).

5.82. Monitoring in response to a nuclear or radiological emergency shall be carried out on the basis of a strategy to be developed at the preparedness stage as part of the protection strategy. Arrangements shall be made to adjust the monitoring in the emergency response on the basis of prevailing conditions.

5.83. Arrangements shall be made to carry out retrospective assessment of exposure of members of the public in a nuclear or radiological emergency, and to make the results of these assessments publicly available. The assessments shall be based on the best available information, shall be put into perspective in terms of the associated health hazards (see para. 5.72) and shall be promptly updated in the light of information that would yield substantially more accurate results.

Requirement 15: Managing radioactive waste in an emergency

The government shall ensure that radioactive waste is managed safely and effectively in a nuclear or radiological emergency.

5.84. The national policy and strategy for radioactive waste management [19] shall apply for radioactive waste generated in a nuclear or radiological emergency, with account taken of paras 5.85 to 5.88.

5.85. The protection strategy (see Requirement 5) shall take into account radioactive waste that might arise from protective actions and other response actions that are to be taken.

5.86. Radioactive waste arising in a nuclear or radiological emergency, including radioactive waste arising from associated protective actions and other response actions taken, shall be identified, characterized and categorized in due time and shall be managed in a manner that does not compromise the protection strategy, with account taken of prevailing conditions as these evolve.

5.87. Arrangements shall be made for radioactive waste to be managed safely and effectively. These arrangements shall include:

(a) A plan to characterize waste, including in situ measurements and analysis of samples;
(b) Criteria for categorization of waste;
(c) Avoiding, to the extent possible, the mixing of waste of different categories;
(d) Minimizing the amount of material unduly declared as radioactive waste;

(e) A method for determining appropriate options for predisposal management of radioactive waste (including processing, storage and transport), with account taken of the interdependences between all steps as well as impacts on the anticipated end points (clearance, authorized discharge, reuse, recycling, disposal) [19, 20];
(f) A method of identifying appropriate storage options and sites;
(g) Consideration of non-radiological aspects of waste (e.g. chemical properties such as toxicity, and biological properties).

5.88. Consideration shall be given to the management of human remains and animal remains with contamination as a result of a nuclear or radiological emergency, with due account taken of religious practices and cultural practices.

Requirement 16: Mitigating non-radiological consequences of a nuclear or radiological emergency and of an emergency response

The government shall ensure that arrangements are in place for mitigation of non-radiological consequences of a nuclear or radiological emergency and of an emergency response.

5.89. Non-radiological consequences of a nuclear or radiological emergency and of an emergency response shall be taken into consideration in deciding on the protective actions and other response actions to be taken in the context of the protection strategy (see Requirement 5).

5.90. Arrangements shall be made for mitigating the non-radiological consequences of an emergency and those of an emergency response and for responding to public concern in a nuclear or radiological emergency. These arrangements shall include arrangements for providing the people affected with:

(a) Information on any associated health hazards and clear instructions on any actions to be taken (see Requirement 10 and Requirement 13);
(b) Medical and psychological counselling, as appropriate;
(c) Adequate social support, as appropriate.

5.91. Arrangements shall be made to mitigate the impacts on international trade of a nuclear or radiological emergency and associated protective actions and other response actions, with account taken of the generic criteria in Appendix II. These arrangements shall provide for issuing information to the public and interested parties (such as importing States) on controls put in place in relation to traded commodities, including food, and on vehicles and cargoes being shipped, and on any revisions of the relevant national criteria.

5.92. Arrangements shall be put in place for any actions taken, beyond those emergency response actions that are warranted, by members of the public and by commercial, industrial, infrastructural or other governmental or non-governmental bodies to be, to the extent practicable, promptly identified and appropriately addressed. This shall include the designation of organization(s) with the responsibility for monitoring for, identifying and addressing such actions.

Requirement 17: Requesting, providing and receiving international assistance for emergency preparedness and response

The government shall ensure that adequate arrangements are in place to benefit from, and to contribute to the provision of, international assistance for preparedness and response for a nuclear or radiological emergency.

5.93. Governments and international organizations shall put in place and shall maintain arrangements to respond in a timely manner to a request made by a State, in accordance with established mechanisms and respective mandates, for assistance in preparedness and response for a nuclear or radiological emergency.

5.94. Arrangements shall be put in place and maintained for requesting and obtaining international assistance from States or international organizations and for providing assistance to States (either directly or through the IAEA) in preparedness and response for a nuclear or radiological emergency, on the basis of international instruments (e.g. the Assistance Convention [13]), bilateral agreements or other mechanisms. These arrangements shall take due account of compatibility requirements for the capabilities to be obtained from and to be rendered to different States so as to ensure the usefulness of these capabilities.

Requirement 18: Terminating a nuclear or radiological emergency

The government shall ensure that arrangements are in place and are implemented for the termination of a nuclear or radiological emergency, with account taken of the need for the resumption of social and economic activity.

5.95. Adjustment of protective actions and other response actions and of other arrangements that are aimed at enabling the termination of an emergency shall be made by a formal process that includes consultation of interested parties.

5.96. Arrangements for communication with the public in a nuclear or radiological emergency (see Requirement 13) shall include arrangements for communication on the reasons for any adjustment of protective actions and other response actions and other arrangements aimed at enabling the termination of the emergency. This shall include providing the public with information on the need for any continuing protective actions following termination of the emergency and on any necessary modifications to their personal behaviour. Arrangements shall be made, during this period, to closely monitor public opinion and the reaction in the news media in order to ensure that any concerns can be promptly addressed. These arrangements shall ensure that any information provided to the public puts health hazards in perspective (see para. 5.72).

5.97. The termination of a nuclear or radiological emergency shall be based on a formal decision that is made public and shall include prior consultation with interested parties, as appropriate.

5.98. Both radiological consequences and non-radiological consequences shall be considered in deciding on the termination of an emergency as well as in the justification and optimization of further protection strategies as necessary.

5.99. The transition to an existing exposure situation or to a planned exposure situation shall be made in a coordinated and orderly manner, by making any necessary transfer of responsibilities and with the increased involvement of relevant authorities and interested parties.

5.100. The government shall ensure that, as part of its emergency preparedness, arrangements are in place for the termination of a nuclear or radiological emergency. The arrangements shall take into account that the termination of an emergency might be at different times in different geographical areas. The planning process shall include as appropriate:

(a) The roles and functions of organizations;
(b) Methods of transferring information;
(c) Means for assessing radiological consequences and non-radiological consequences;
(d) Conditions, criteria and objectives to be met for enabling the termination of a nuclear or radiological emergency (see Appendix II);
(e) A review of the hazard assessment and of the emergency arrangements;
(f) Establishment of national guidelines for the termination of an emergency;

(g) Arrangements for continued communication with the public, and for monitoring of public opinion and the reaction in the news media;
(h) Arrangements for consultation of interested parties.

5.101. Once the emergency is terminated, all workers undertaking relevant work shall be subject to the relevant requirements for occupational exposure in planned exposure situations [8], and individual monitoring, environmental monitoring and health surveillance shall be conducted subject to the requirements for planned exposure situations or existing exposure situations, as appropriate [8].

Requirement 19: Analysing the nuclear or radiological emergency and the emergency response

The government shall ensure that the nuclear or radiological emergency and the emergency response are analysed in order to identify actions to be taken to avoid other emergencies and to improve emergency arrangements.

5.102. Arrangements shall be made to document, protect and preserve, in an emergency response, to the extent practicable, data and information important for an analysis of the nuclear or radiological emergency and the emergency response. Arrangements shall be made to undertake a timely and comprehensive analysis of the nuclear or radiological emergency and the emergency response with the involvement of interested parties. These arrangements shall give due consideration to the need for making contributions to relevant internationally coordinated analyses and for sharing the findings of the analysis with relevant response organizations. The analysis shall give due consideration to:

(a) The reconstruction of the circumstances of the emergency;
(b) The root causes of the emergency;
(c) Regulatory controls including regulations and regulatory oversight;
(d) General implications for safety, including the possible involvement of other sources or devices (including those in other States);
(e) General implications for nuclear security, as appropriate;
(f) Necessary improvements to emergency arrangements;
(g) Necessary improvements to regulatory control.

5.103. Arrangements shall be made to enable comprehensive interviews on the circumstances of the nuclear or radiological emergency to be conducted with those involved.

5.104. Arrangements shall be made to acquire (e.g. from the IAEA, from another State or from the manufacturer of relevant equipment) the expertise necessary to conduct an analysis of the circumstances of the nuclear or radiological emergency.

5.105. Arrangements shall be made to take actions promptly on the basis of an analysis to avoid other emergencies, including provision of information to other operating organizations, as relevant, or to other States, directly or through the IAEA.

6. REQUIREMENTS FOR INFRASTRUCTURE

GENERAL

6.1. This section establishes the requirements for infrastructural elements that are essential to providing the capability for fulfilling the requirements established in Section 5 in accordance with the hazard assessment and the protection strategy.

Requirement 20: Authorities for emergency preparedness and response

The government shall ensure that authorities for preparedness and response for a nuclear or radiological emergency are clearly established.

6.2. The authorities for developing, maintaining and regulating arrangements, both on the site and off the site, for preparedness and response for a nuclear or radiological emergency shall be established by means of acts, legal codes or statutes.

6.3. All of the functions specified in Section 5 shall be assigned to the appropriate operating organizations and to local, regional and national response organizations. The involvement of all these organizations in the performance of these functions, or in support of their performance, shall be documented.[35] The documentation shall specify their roles, functions, authorities and responsibilities in emergency preparedness and response and shall

[35] Typically, the involvement of operating organizations and local, regional and national response organizations is documented as part of the appropriate facility, local, regional and national emergency plans.

assent to the authorities, roles and responsibilities of other response organizations. Conflicting or potentially conflicting and overlapping roles and responsibilities shall be identified and conflicts shall be resolved at the preparedness stage through the national coordinating mechanism (see para. 4.10).

6.4. The authority and responsibility for making decisions on response actions to be taken on the site and off the site (see para. 5.7) and the authority and responsibility for communication with the public shall be clearly assigned for each phase of the response.

6.5. The emergency arrangements shall include clear assignment of responsibilities and authorities, and shall provide for coordination and for communication in all phases of the response. These arrangements shall include:

— Ensuring that for each response organization a position in the response hierarchy has the authority and responsibility to direct and to coordinate its response actions;
— Clearly assigning the authority and responsibility for the direction and coordination of the entire response (see para. 5.7) and for the prevention and resolution of conflicts between response organizations;
— Assigning to an on-site position the authority and responsibility for notifying the appropriate response organization(s) of an emergency and for taking immediate on-site actions;
— Assigning to an on-site position the responsibility for directing the entire on-site emergency response (see paras 5.2 and 5.7).

These arrangements shall be such as to ensure that those personnel with authority and responsibility to perform critical response functions[36] in an emergency response are not assigned any other responsibilities in an emergency that would interfere with the prompt performance of the specified functions.

6.6. The arrangements for delegation and/or transfer of authority shall be specified in the relevant emergency plans, together with arrangements for notifying all appropriate parties of the transfer.

[36] Critical response functions are functions that must be performed promptly and correctly in order to classify, declare and notify an emergency, to activate an emergency response, to manage the response, to take mitigatory actions, to protect emergency workers and to take urgent protective actions on and off the site.

Requirement 21: Organization and staffing for emergency preparedness and response

The government shall ensure that overall organization for preparedness and response for a nuclear or radiological emergency is clearly specified and staffed with sufficient personnel who are qualified and are assessed for their fitness for their intended duties.

6.7. The organizational relationships for preparedness and response for a nuclear or radiological emergency and interfaces between all the response organizations shall be established.

6.8. The positions responsible within each operating organization and response organization for performance of the response functions specified in Section 5 shall be assigned in the emergency plans and procedures. The positions responsible in each operating organization, in each response organization and in the regulatory body for the performance of activities at the preparedness stage, in accordance with these requirements, shall be assigned as part of the routine organizational structures and shall be specified, as appropriate, in the emergency plans and procedures.

6.9. Personnel who are assigned to positions in all operating organizations and response organizations to perform the functions necessary to meet the requirements established in Section 5 shall be qualified and shall be assessed for their initial fitness and continuing fitness for their intended duties.

6.10. Appropriate numbers of suitably qualified personnel shall be available at all times (including during 24 hour a day operations) so that appropriate positions can be promptly staffed as necessary following the declaration and notification of a nuclear or radiological emergency. Appropriate numbers of suitably qualified personnel shall be available for the long term to staff the various positions necessary to take mitigatory actions, protective actions and other response actions.

6.11. For a site where multiple facilities in category I or II are collocated, an appropriate number of suitably qualified personnel shall be available to manage an emergency response at all facilities if each of the facilities is under emergency conditions simultaneously (see para. 5.4).

Requirement 22: Coordination of emergency preparedness and response

The government shall ensure that arrangements are in place for the coordination of preparedness and response for a nuclear or radiological emergency between the operating organization and authorities at the local, regional and national levels, and, where appropriate, at the international level.

6.12. Arrangements shall be developed, as appropriate, for the coordination of emergency preparedness and response and of protocols for operational interfaces between operating organizations and authorities at the local, regional and national levels, including those organizations and authorities responsible for the response to conventional emergencies and to nuclear security events (see paras 4.3, 4.10, 6.3 and Requirement 6). The arrangements shall be clearly documented and the documentation shall be made available to all relevant parties. Arrangements shall be put in place to ensure effective working relationships among these organizations, both at the preparedness stage and in an emergency.

6.13. When several different organizations of the State or of other States are expected to have or to develop tools, procedures or criteria for use in the response to an emergency, arrangements for coordination shall be put in place to improve the consistency of the assessments of the situation, including assessments of contamination, doses and radiation induced health effects and any other relevant assessments made in a nuclear or radiological emergency, so as not to give rise to confusion.

6.14. Arrangements shall be made to coordinate with other States in the event of a transnational emergency any protective actions and other response actions that are recommended to their citizens and to their embassies in order either to ensure that they are consistent with those recommended in other States, or to provide an opportunity for them to explain to the public the basis for any differences (see para. 5.73).

6.15. Arrangements shall be made to ensure that States with areas in category V are provided with appropriate information for developing their own preparedness to respond to a transboundary emergency and that appropriate coordination across national borders is in place. These arrangements shall include:

(a) Agreements and protocols to provide information necessary to develop a coordinated means for notification, classification schemes and criteria for taking and for adjusting protective actions and other response actions;

(b) Arrangements for communication with the public;
(c) Arrangements for the exchange of information between decision making authorities.

Requirement 23: Plans and procedures for emergency response

The government shall ensure that plans and procedures necessary for effective response to a nuclear or radiological emergency are established.

6.16. Plans, procedures and other arrangements for effective emergency response, including coordinating mechanisms, letters of agreement or legal instruments, shall be made for coordinating a national emergency response. The arrangements for a coordinated national emergency response:

— Shall specify the organization responsible for the development and maintenance of the arrangements;
— Shall describe the responsibilities of operating organizations and other response organizations;
— Shall describe the coordination effected between these arrangements and the arrangements for response to a conventional emergency and to a nuclear security event.

Consideration shall be given in these plans, procedures and other arrangements to the need to protect information that might be confidential.

6.17. Each response organization shall prepare an emergency plan or plans for coordinating and performing their assigned functions as specified in Section 5 and in accordance with the hazard assessment and the protection strategy. An emergency plan shall be developed at the national level that integrates all relevant plans for emergency response in a coordinated manner and consistently with an all-hazards approach. Emergency plans shall specify how responsibilities for managing operations in an emergency response are to be discharged on the site, off the site and across national borders, as appropriate. The emergency plans shall be coordinated with other plans and procedures that may be implemented in a nuclear or radiological emergency, to ensure that the simultaneous implementation of the plans would not reduce their effectiveness or cause conflicts. Such other plans and procedures include:

(a) Emergency plans for facilities in category I and for areas in category V;
(b) Security plans and contingency plans [9, 10];

(c) Procedures for the investigation of a nuclear security event, including identification, collection, packaging and transport of evidence contaminated with radionuclides, nuclear forensics and related activities [11];
(d) Evacuation plans;
(e) Plans for firefighting.

6.18. The appropriate responsible authorities shall ensure that:

(a) A 'concept of operations'[37] for emergency response is developed at the beginning of the preparedness stage.
(b) Emergency plans and procedures are prepared and, as appropriate, approved for any facility or activity, area or location that could give rise to an emergency warranting protective actions and other response actions.
(c) Response organizations and operating organizations, as appropriate, are involved in the preparation of emergency plans and procedures, as appropriate.
(d) Account is taken in the content, features and extent of emergency plans of the results of any hazard assessment and any lessons from operating experience and from past emergencies, including conventional emergencies (see paras 4.18–4.26).
(e) Emergency plans and procedures are periodically reviewed and updated (see paras 6.36 and 6.38).

6.19. The operating organization of a facility or for an activity in category I, II, III or IV shall prepare an emergency plan. This emergency plan shall be coordinated with those of all other bodies that have responsibilities in a nuclear or radiological emergency, including public authorities, and shall be submitted to the regulatory body for approval.

6.20. The operating organization and response organizations shall develop the necessary procedures and analytical tools to be able to perform the functions specified in Section 5 for the goals of emergency response to be achieved and for the emergency response to be effective.

[37] A concept of operations is a brief description of an ideal response to a postulated nuclear or radiological emergency, used to ensure that all those personnel and organizations involved in the development of a capability for emergency response share a common understanding.

6.21. Procedures and analytical tools shall be tested under simulated emergency conditions and shall be validated prior to initial use. Any arrangements for the use of analytical tools early in an emergency response for supporting decision making on protective actions and other response actions shall be made in due recognition of the limitations[38] of such analytical tools and in a way that would not reduce the effectiveness of response actions. These limitations shall be made clear to, and shall be recognized by, those responsible for decision making.

Requirement 24: Logistical support and facilities for emergency response

The government shall ensure that adequate logistical support and facilities are provided to enable emergency response functions to be performed effectively in a nuclear or radiological emergency.

6.22. Adequate tools, instruments, supplies, equipment, communication systems, facilities and documentation (such as documentation of procedures, checklists, manuals, telephone numbers and email addresses) shall be provided for performing the functions specified in Section 5. These items and facilities shall be selected or designed to be operational under the conditions (such as radiological conditions, working conditions and environmental conditions) that could be encountered in the emergency response, and to be compatible with other procedures and equipment for the response (e.g. compatible with the communication frequencies used by other response organizations), as appropriate. These support items shall be located or provided in a manner that allows their effective use under the emergency conditions postulated.

6.23. For facilities in categories I and II, as contingency measures, alternative supplies for taking on-site mitigatory actions, such as an alternative supply of water and an alternative electrical power supply, including any necessary equipment, shall be ensured. This equipment shall be located and maintained so that it can be functional and readily accessible when needed (see also Safety of Nuclear Power Plants: Design (SSR-2/1) [18]).

---

[38] An example of such limitations is that the timing and magnitude of radioactive releases in an emergency at a nuclear power plant that would warrant taking precautionary urgent protective actions and urgent protective actions off the site before, or shortly after, a radioactive release may not be predictable. In addition, the radioactive release could occur over several days, resulting in complex deposition patterns off the site.

6.24. Emergency response facilities or locations to support an emergency response under the full range of postulated hazardous conditions shall be designated and shall be assigned the following functions, as appropriate:

(a) Receiving notifications and initiating the response;
(b) Coordination and direction of on-site response actions;
(c) Providing technical and operational support to those personnel performing tasks at a facility and those personnel responding off the site;
(d) Direction of off-site response actions and coordination with on-site response actions;
(e) Coordination of national response actions;
(f) Coordination of communication with the public;
(g) Coordination of monitoring, sampling and analysis;
(h) Managing those people who have been evacuated (including reception, registration, monitoring and decontamination, as well as provision for meeting their personal needs, including for housing, food and sanitation);
(i) Managing the storage of necessary resources;
(j) Providing individuals who have undergone exposure or contamination with appropriate medical attention including medical treatment.

6.25. For facilities in category I, emergency response facilities[39] separate from the control room and supplementary control room shall be provided so that:

(a) Technical support can be provided to the operating personnel in the control room in an emergency (from a technical support centre).
(b) Operational control by personnel performing tasks at or near the facility can be maintained (from an operational support centre).
(c) The on-site emergency response is managed (from an emergency centre).

These emergency response facilities shall operate as an integrated system in support of the emergency response, without conflicting with one another's functions, and shall provide reasonable assurance of being operable and habitable under a range of postulated hazardous conditions, including conditions not considered in the design.

---

[39] Emergency response facilities may be collocated (i.e. these functions may be performed from a single emergency response facility or location) provided that it is ensured that they do not conflict with each other in performing their specified functions and provided that they are separated from the control rooms.

6.26. Arrangements shall be made for performing appropriate and reliable analyses of samples[40] and measurements of internal contamination for the purposes of emergency response and of health screening, as appropriate. Such arrangements shall include the designation of laboratories that would be operational under postulated emergency conditions.

6.27. Arrangements shall be made to obtain appropriate support from organizations responsible for providing support in conventional emergencies for logistics and communication, for social welfare and in other areas.

Requirement 25: Training, drills and exercises for emergency preparedness and response

The government shall ensure that personnel relevant for emergency response shall take part in regular training, drills and exercises to ensure that they are able to perform their assigned response functions effectively in a nuclear or radiological emergency.

6.28. The operating organization and response organizations shall identify the knowledge, skills and abilities necessary to perform the functions specified in Section 5. The operating organization and response organizations shall make arrangements for the selection of personnel and for training to ensure that the personnel selected have the requisite knowledge, skills and abilities to perform their assigned response functions. The arrangements shall include arrangements for continuing refresher training on an appropriate schedule and arrangements for ensuring that personnel assigned to positions with responsibilities in an emergency response undergo the specified training.

6.29. For facilities in category I, II or III, all personnel and all other persons on the site shall be instructed in the arrangements for them to be notified of an emergency and of their actions if notified of an emergency.

6.30. Exercise programmes shall be developed and implemented to ensure that all specified functions required to be performed for emergency response, all organizational interfaces for facilities in category I, II or III, and the national level programmes for category IV or V are tested at suitable intervals. These programmes shall include the participation in some exercises of, as appropriate

---

[40] Arrangements for analyses could include, for example, arrangements for performing analyses of environmental and biological samples as well as analyses of other samples taken from the facility for the purpose of assessing its operational status.

and feasible, all the organizations concerned, people who are potentially affected, and representatives of news media. The exercises shall be systematically evaluated (see para. 4.10(h)) and some exercises shall be evaluated by the regulatory body. Programmes shall be subject to review and revision in the light of experience gained (see paras 6.36 and 6.38).

6.31. The personnel responsible for critical response functions shall participate in drills and exercises on a regular basis so as to ensure their ability to take their actions effectively.

6.32. Officials off the site who are responsible for making decisions on protective actions and other response actions shall be trained and shall regularly participate in exercises. Officials off the site who are responsible for communication with the public in a nuclear or radiological emergency shall regularly participate in exercises.

6.33. The conduct of exercises shall be evaluated against pre-established objectives of emergency response to demonstrate that identification, notification, activation and response actions can be performed effectively to achieve the goals of emergency response (see para. 3.2).

Requirement 26: Quality management programme for emergency preparedness and response

The government shall ensure that a programme is established within an integrated management system to ensure the availability and reliability of all supplies, equipment, communication systems and facilities, plans, procedures and other arrangements necessary for effective response in a nuclear or radiological emergency.

6.34. The operating organization, as part of its management system (see Ref. [14]), and response organizations, as part of their emergency management system, shall establish a programme to ensure the availability and reliability of all supplies, equipment, communication systems and facilities, plans, procedures and other arrangements necessary to perform functions in a nuclear or radiological emergency as specified in Section 5 (see para. 6.22). The programme shall include arrangements for inventories, resupply, tests and calibrations, to ensure that these are continuously available and are functional for use in a nuclear or radiological emergency.

6.35. The programme shall also include periodic and independent appraisals against functions as specified in Section 5, including participation in international appraisals[41].

6.36. Arrangements shall be made to maintain, review and update emergency plans, procedures and other arrangements and to incorporate lessons from research, operating experience (such as in the response to emergencies) and emergency exercises.

6.37. The operating organization and response organizations shall establish and maintain adequate records in relation to both emergency arrangements and the response to a nuclear or radiological emergency, to include dose assessments, results of monitoring and inventory of radioactive waste managed, in order to allow for their review and evaluation. These records shall also provide for the identification of those persons requiring longer term medical actions, as necessary, and shall provide for the long term management of radioactive waste.

6.38. The operating organization and response organizations shall make arrangements to review and evaluate responses in actual events and in exercises, in order to record the areas in which improvements are necessary and to ensure that the necessary improvements are made (see Requirement 19).

6.39. Relevant international organizations shall review and update their applicable standards and guidelines and their relevant arrangements in emergency preparedness and response on the basis of research and lessons from the response to actual emergencies and in emergency exercises.

[41] Examples of international appraisals include those organized by the IAEA, such as Emergency Preparedness Review (EPREV) missions.

Safety through international standards

"Governments, regulatory bodies and operators everywhere must ensure that nuclear material and radiation sources are used beneficially, safely and ethically. The IAEA safety standards are designed to facilitate this, and I encourage all Member States to make use of them."

Yukiya Amano
Director General

INTERNATIONAL ATOMIC ENERGY AGENCY
VIENNA
ISBN 978–92–0–105715–0

APPENDIX

3

Human Reliability Assessment

DOI: https://doi.org/10.1016/B978-0-12-821319-3.00031-2

HRA generation type	Puplicly avilable	HRA method name	HRA acronym	Developed by	Originally developed for	Domain of application (aviation/ maritime/ indutry)	Main theoretical framework and area of focuse (design, operation, management, cognitive, behavioral, contextual)	Qualitative/ Quantitiave	Method procedures	Performance factors (PSFs), error producing conditions (EPCs), error forcing contexts (EFCs), common performance conditions (CPCs)	Validation	Method's strengths	Method's weakness	Acdemic references
(First Generation) The first generation HRA methods have been strongly influenced by the probabilistic safety assessment (PSA) and its aim is quantification the success/failure of the action, with less attention to the depth of the failure causes and reasons for human behavior, differentiation between omission and commission errors. The characteristics of a task represented by HEPs, are regarded as major factors. While, the context, which is represented by PSFs, is considered a minor factor in estimating the probability of human failure. PSFs in the first generation methods were mainly derived by focusing on the environmental impacts on operators.	Yes	Human Error Assessment and Reduction Technique	HEART	First developed by Williams (1985)	The Central Electricity Generating Board.	Nuclear sector, Chemical sector, Aviation sector, Transportation sector, Oil and gas sector, and Healthcare sector	Cognitive, (Human Task)	Quantitative	1) Classification of a task into one of the 9 generic categories (each with an associated nominal human error potential (HEP)); 2) assignment of nominal HEPs to the task; 3) determination of which error producing conditions (EPCS), which are about 38 EPCs; 4) calculating the task HEP based on the Assessed Proportion of Affect (APOA) for each EPC	38 EPCs	HEART has been empirically validated by many scholars such as Kirwan et al. (1997) Kennedy et al. (2000). Kirwan et al. (1997) HEART was validated with two other methods (THERP and JHEDI) with the support of 30 assessors and results found to achieve a reasonable level of accuracy and consistency with other methods	1- Simple to apply. 2- Empirically validated and showed a reasonable level of accuracy	1- HEART depends on the nature of the task and the "perfect condition," Thus, the human reliability will degrade as the task condition deviate from the perfect conditions of execution and will increase the probability of human errors. 2- highly subjective between generic tasks and their associated EPCs), error dependency is not considered. 3- double counting (some elements of EPCs	1- Williams, J.C. (1985). HEART – A Proposed Method for Achieving High Reliability in Process Operation by means of Human Factors Engineering Technology. In Proceedings of a Symposium on the Achievement of Reliability in Operating Plant, Safety and Reliability Society, 16 September 1985, Southport. 2- Kirwan, B. (1996). The validation of three human reliability quantification techniques, THERP, HEART and JHEDI: Part 1 technique descriptions and validation issues. Applied ergonomics, 27, (6), 359-373. 3- Kirwan, B., et al. (1997). The validation of three human reliability quantification techniques, THERP, HEART and JHEDI: Part II – results of validation exercise. Applied ergonomics, 28 (1), 17–25.

(continued on next page)

HRA generation type	Puplicly avilable	HRA method name	HRA acronym	Developed by	Originally developed for	Domain of application (aviation/ maritime/ indutry)	Main theoretical framework and area of focuse (design, operation, management, cognitive, behavioral, contextual)	Qualitative/ Quantitiave	Method procedures	Performance factors (PSFs), error producing conditions (EPCs), error forcing contexts (EFCs), common performance conditions (CPCs)	Validation	Method's strengths	Method's weakness	Acdemic references
	Yes	Technique for Human Error Rate Prediction	THERP	Developed by Swain and Guttmann (1983)	Sandia Laboratories for the US Nuclear Regulatory Commission	Nuclear industry, Aviation, Healthcare, Oil and Gas industry, Engineering Industry, Transportation industry	Behavioral, Factorial (consider that PSF have a direct impact on the task performance)	Quantitative	1- Decomposition of tasks into elements; 2- assignment of HEPs to each element; 3- determination of effects of PSF on each element; 4- calculation of effects of dependence between tasks; 5- modeling in HRA Event/Fault tree; 6- quantification of total task HEP	67 PSFs	Validated by Kirwan et al. (1997) with HEART and JHEDI and achieved a reasonable level of accuracy	Predictive and preventative tool that quantifies human error probabilities (HEPs) and system degradation using a fault-tree approach and human reliability analysis event tree (HRAET) taking into account the influence of shaping factors (PSFs) in the overall HEP calculation	Unclear guidance on modelling procedures and the impact of PSFs on performance.	Swain AD and Guttmann HE (1983). Handbook of human reliability analysis with emphasis on nuclear power plant applications. US Nuclear Regulatory Commission), Washington, DC. NUREG/CR-1278

(continued on next page)

HRA generation type	Puplicly avilable	HRA method name	HRA acronym	Developed by	Originally developed for	Domain of application (aviation/ maritime/ indutry)	Main theoretical framework and area of focuse (design, operation, management, cognitive, behavioral, contextual)	Qualitative/ Quantitiave	Method procedures	Performance factors (PSFs), error producing conditions (EPCs), error forcing contexts (EFCs), common performance conditions (CPCs)	Validation	Method's strengths	Method's weakness	Acdemic references
	Yes	Simplified Plant Analysis Risk Human Reliability Assessment	SPAR-H	USNRC in conjunction with the Idaho National Laboratory, INL in 1994	US Nuclear Research Commis-sion, Office	Mainly Nuclear Industry and Nuclear industry and few studies show some implementation of SPAR-H in industries including Gas Transmission Plant, Petroleum Industry , & Industrial maintenance of production process	Cognitive, (Task, System)	Quantitative	SPAR-H method comprises of five steps: 1-Categorizing the HFE either Diagnosis or Action based under 8 PSFs, which are (Available time, Stress and stressors, Experience and training, Complexity, Ergonomics, Procedures Fitness for duty,& Work processes). 2. Rating the PSFs. 3. Calculating PSF-Modified/ normalized HEP (HEP = [(NHEP Å PSFcompos-ite)/((NHEP) Å (PSFcomposite – 1) + 1)]. 4. Calculating the Dependence. 5. Calculating the Minimum Cutoff Value	8 PSFs, (Task, System)	NRC and other scholars such as German et al, 2004; and Forester et al, 2006 have validated the reliability of the SPAR-H and found that this method has incorporated old improvement in PSF elements that were adapted from HEART, CREAM, THERP, and ASEP) at that time as well as the decision made in relation to HEPs are not justified nor their bases are clear. Accordingly, the NRC group of reviewrs concluded that further improvement and clarification are required	Consideration of the Dependncy	SPAR H investigates the Dependency, which is the interaction between different components that contributed to human error. However, Dependency calculation is a critical issue in this method and may lead to incorrect quantification of human error probabilities if dependency is overused or applied frequently and if the link of dependency is not accurate or is latent, subjective, and dynamic. Taking into account that not all PSFs are directly measurable or observable. The basis for the selection of final values is not clear, PSFs may be inadequate for detailed analysis	1- Gertman, D., Blackman, H., Marble, J., Byers, and Smith, C. (2004a). The SPAR-H human reliability analysis method. NUREG/CR-6883 (https://www.nrc.gov/reading-rm/doc-collections/nuregs/contract/cr6883/cr6883.pdf). 2- Idaho National Laboratory, prepared for U. S. Nuclear Regulatory Commission Office of Nuclear Regulatory Research Washington, DC 205555-0001. 3- Byers, J. et al. (2000) Simplified Plant Analysis Risk (SPAR) Human Reliability Analysis (HRA) Methodology: Comparisons with Other HRA Methods.Human Factors and Ergonomics Society Annual Meeting Proceedings 44(18) DOI: 10.1177/154193120004401802

(continued on next page)

HRA generation type	Puplicly avilable	HRA method name	HRA acronym	Developed by	Originally developed for	Domain of application (aviation/ maritime/ indutry)	Main theoretical framework and area of focuse (design, operation, management, cognitive, behavioral, contextual)	Qualitative/ Quantitiave	Method procedures	Performance factors (PSFs), error producing conditions (EPCs), error forcing contexts (EFCs), common performance conditions (CPCs)	Validation	Method's strengths	Method's weakness	Acdemic references
	Yes	Success Likelihood Index Method	SLIM	Developed by Embrey in 1983	The US Nuclear Regulatory Commis-sion	Healthcare, engineering, nuclear, transportation, and manufacturing sector.	Behavioral	Quantitative	SLIM comprises mainly of 12 steps, which are 1. selection of the expert panel, 2. definition of situations and subsets, 3. elicitation of PSFs, 4. rating PSF, 5. scaling calculations, 6. independence checks, 7. weighting procedures, 8. calculation of the SLIs, 9. conversion of the SLIs into probabilities, 10. uncertainty-bound analysis, 11. error reduction through conducting sensitivity analysis, 12. documentation process.	Performance Shaping Factors (PSFs)Human, Task	Scholars such as Humphreys (1988) and Kirwan (1994) carried out a comparative validation between amongst other tools and found that its accuracy, validity, usefulness, effective is at a reasonably high level.	SLIM is a systematical analytical approach based on expert judgment under the MAUD (multi-attribute utility decom-position) model or SARAH (Systematic Approach to the Reliability Assessment of Humans)	Subjectivity that can be reduced if coupled with aggregating, fuzzification, defuzzification approaches to reduce uncertainty and quantify the degree of agreement.	1- Embrey, D.E., Humphreys, P., Rosa, E.A., Kirwan, B. and Rea, K. (1984) SLIM-MAUD: An approach to assessing human error probabilities using structured expert judgment. NUREG/CR-3518. 2- Embrey, D.E., Humphreys, P., Rosa, E.A., Kirwan, B. and Rea, K. (1984) SLIM-MAUD: An approach to assessing human error probabilities using structured expert judgment. Volume II: Detailed analysis of the technical issues. NUREG/CR-3518. Prepared for the US NRC

(*continued on next page*)

HRA generation type	Puplicly avilable	HRA method name	HRA acronym	Developed by	Originally developed for	Domain of application (aviation/ maritime/ indutry)	Main theoretical framework and area of focuse (design, operation, management, cognitive, behavioral, contextual)	Qualitative/ Quantitiave	Method procedures	Performance factors (PSFs), error producing conditions (EPCs), error forcing contexts (EFCs), common performance conditions (CPCs)	Validation	Method's strengths	Method's weakness	Acdemic references
	No	Human Reliability Management System	HRMS	Kirwan in the 1980s	Technical University of Delft, in the Netherlands, in the Safety Science Group	No much of references about its application in the industry	All aspects of the HRA process	Quantitative/ Qualitative	1- Task analysis module. 2- Error analysis module (e.g Skill, Rule and Knowledge (SRK) model of human error and performance, work on slips, lapses and mistakes, External Error Mode Performance Shaping Factor (PSF), Psychological Error Mechanism 3- Error analysis and quantification module, is based on actual data of the scenarios/ tasks, according to six major PSFs (the time - scale involved; the quality of the interface; training/experience/ familiarity;	HRMS is a fully-computerized HRA that uses Human Error Identification (HEI) and PSFs.	Accuracy of HRMS has not been properly validated	1- It is based on current theories and taxonomies of error identification 2- It incorporates factors known to influence performance 3- It has a taxonomic structure (for error identification and quantification) 4- It has error reduction functionality, based on PSF sensitivity analysis, and its error reduction approach is flexible and leads to tangible and practicable recommendations	1-Requires significant resources	1- Kirwan, B. and James, N.J. (1989). A Human Reliability Management System. In: Reliability Volume 89 (Brighton Metropole) 2- Kirwan, B. (1990). A resources flexible approach to human reliability assessment for PRA. In: Safety 3- KIRWAN, B., "A resources flexible approach to human reliability assessment for PRA", Safety and Reliability Symposium, Altrincham, Elsevier Applied Sciences, London (1990). 4- Kirwan B. (1997) The development of a nuclear chemical plant human reliability management approach: HRMS and JHEDI. Reliability Engineering & System Safety, 56, (2), 107–133.

(continued on next page)

HRA generation type	Puplicly avilable	HRA method name	HRA acronym	Developed by	Originally developed for	Domain of application (aviation/ maritime/ indutry)	Main theoretical framework and area of focuse (design, operation, management, cognitive, behavioral, contextual)	Qualitative/ Quantitiave	Method procedures	Performance factors (PSFs), error producing conditions (EPCs), error forcing contexts (EFCs), common performance conditions (CPCs)	Validation	Method's strengths	Method's weakness	Acdemic references
									the degree of adequacy of procedures; how the task is organized; and the degree of complexity of the task), like SLIM, the system can also carry out a PSF - based sensitivity analysis to determine how to reduce the likelihood of error and like HEART, the system provides error - reduction mechanisms. 4- Error reduction module, four ways to reduce error likelihood: Error Pathway Blocking, Consequence Reduction. Error Recovery Enhancement, and PSF-Based Reduction.			5. It is integrable within risk assessment (it has been used in a number of assessments and safety cases) 6. It is an integrated and computerized system, with face validity of those that have used the system 7. HRMS has a screening facility within the representation module, and feeds forward information to relevant operations departments 8. Its range is focused on nuclear reprocessing, though this would probably include chemical plant		

(continued on next page)

HRA generation type	Puplicly avilable	HRA method name	HRA acronym	Developed by	Originally developed for	Domain of application (aviation/ maritime/ indutry)	Main theoretical framework and area of focuse (design, operation, management, cognitive, behavioral, contextual)	Qualitative/ Quantitiave	Method procedures	Performance factors (PSFs), error producing conditions (EPCs), error forcing contexts (EFCs), common performance conditions (CPCs)	Validation	Method's strengths	Method's weakness	Acdemic references
									5- (Prediction of Human Operator Error using Numerical Index eXtrapolation–PHOENIX) quantification module, assign PSF, HEP gnertaed by PHOENIX extrapolation rules 6- Quality assurance module to ensure consistency, and to ensure assumptions are justified, usually done by the Human Reliability Manager 7- Representation module deals with aspects of recovery, screening, fault and event tree integration, and dependence.			operations, and it focuses on skill and rule-based tasks, with some limited treatment of violations, errors of commission, and knowledge-based failures 9. The system has been used primarily in the design stage assessments 10. The system is highly auditable, leaving an extensive audit trail.		

(continued on next page)

HRA generation type	Puplicly avilable	HRA method name	HRA acronym	Developed by	Originally developed for	Domain of application (aviation/ maritime/ indutry)	Main theoretical framework and area of focuse (design, operation, management, cognitive, behavioral, contextual)	Qualitative/ Quantitiave	Method procedures	Performance factors (PSFs), error producing conditions (EPCs), error forcing contexts (EFCs), common performance conditions (CPCs)	Validation	Method's strengths	Method's weakness	Acdemic references
	No, it is BNFL property	Justified Human Error Data Information (JHEDI)	JHEDI	Kirwan in the 1980s (Kirwan & James 1989, Kirwan 1990)	BNFL THORP (British Nuclear Fuels Ltd., Thermal Oxide Re-processing Plant)	Nuclear Industry	Design Process	Quantitative/ Qualitative	1- Identify a set of basic errors descriptors and empirically derived errors probabilities. 2- Assessor to answer a set of PSF questions. 3- Determine the HEP value using the set of PSF's answers by the assessor	PSFs	Kirwan (1997) and 30 HRA practitioners validated JHEDI with HEART and THERP, and found that JHEDI achieved higher accuracy and precision scores than HEART and THERP	1- faster screening technique based on the HRMS methodology 2- PSF rating process more straightfor-ward, as it asks factual questions rather than subjective judgement by the assessor 3- JHEDI is a auditable because it is a computerised tool	1- JHEDI involves a less detailed assessment than HRMS. 2- Nominal HEP which is increased according to levels of PSF. 3- The HEPs more conservative to allow for simplicity within JHEDI (i.e. account for a measure of dependence) 4- The multipliers used are fewer than those used in HRMS. it is not supporting complicated systems where no actual industry data are available	1- Kirwan, B. (1990). A resources flexible approach to human reliability assessment for PRA. In: Safety and Reliability Symposium, Elsevier Applied Sciences, pp. 114–135. 2- Kirwan, B. (1996). The validation of three human reliability quantification techniques, THERP, HEART and JHEDI: Part 1 technique descriptions and validation issues. Applied ergonomics, 27, (6), 359–373. 3- Kirwan, B. (1997) The development of a nuclear chemical plant human reliability management approach: HRMS and JHEDI. Reliability Engineering & System Safety, Volume 56, Issue 2, Pages 107–133. 4- Kirwan, B. The validation of three human reliability quantification techniques – THERP, HEART and JHEDI: Part III – Practical aspects of the usage of the techniques, Applied Ergonomics, Vol 28, No 1, pp. 27–39, 1997. 5- Kirwan, B., Kennedy, R., Taylor-Adams, S. and Lambert, B. (1997). The validation of three human reliability quantification techniques, THERP, HEART and JHEDI: Part II – results of validation exercise. Applied ergonomics, 28 (1), 17–25

(*continued on next page*)

HRA generation type	Puplicly avilable	HRA method name	HRA acronym	Developed by	Originally developed for	Domain of application (aviation/ maritime/ indutry)	Main theoretical framework and area of focuse (design, operation, management, cognitive, behavioral, contextual)	Qualitative/ Quantitiave	Method procedures	Performance factors (PSFs), error producing conditions (EPCs), error forcing contexts (EFCs), common performance conditions (CPCs)	Validation	Method's strengths	Method's weakness	Acdemic references
	No	Controller Action Reliability Assessment	CARA	Kirwan, B.	Air Traffic Manage-ment (ATM)	Air Traffic Management (ATM)	Human Error in ATM	Quantitative	1- Analyse task & context, Identify Error: Review the error to be quantified and understand the precursors and the controller activities that are undertaken (or fail) 2- Match to GTT: Select from CARA a Generic Task Type (GTT) that best reflects the activity 3- Baseline HEP: Select a set of contexts that provide Error Producing Conditions (EPC) 4-Calculate EPC Effects: Determine the extent to which an EPC will impact the GTT 5- Calculate HEP: Using a basic formula multiply the GTT error by the EPC(s) to provide the final HEP	EPCs	According to EUROPEAN ORGANISA-TION FOR THE SAFETY OF AIR NAVIGATION (2009). Controller Action Reliability Assessment (CARA). CARA User Manual that stated "CARA is based on HEART that used in a number of industries was selected for adaptation to ATM because it has been validated several times to show that it can be accurate in safety cases"	1-CARA is based on an existing HRA technique, HEART that has been tailored to the ATM and therefore the development of a set of generic ATM tasks, Review of available data and quantification of generic tasks where data are available, Development of error producing conditions, Review of Maximum Effect values and anchor points for error producing conditions. 2-CARA has same strengths of HEART	1-CARA has the same weakness of HEART. 2-CARA may not work well for comprehensively assessing 'emergent' contexts,human performance is likely to be impacted of complex ways by new systems. (situation awareness or complex decision making) 3- CARA is not a 'microscope' for providing a fully quantitative model of the "generic controller" it is an approach which puts generic controller activities into a quantitative context.	1- Gibson, W.H. and Kirwan, B. (2008) Application of the CARA HRA Tool to Air Traffic Management Safety Cases. http://www.eurocontrol.int/eec/public/standard_page/DOC_Conf_2008_002.html 2- Gibson, W.H. and Kirwan, B. (2007) CARA: A Human Reliability Assessment Technique for Air Traffic Management Safety Assurance. www.eurocontrol.int/eec/public/standard_page/DOC_Conf_2007_007.html 3- Kirwan, B. and Gibson, H. (2007) CARA: A Human Reliability Assessment Tool for Air Traffic Safety Management — Technical Basis and Preliminary Architecture. pp. 197 –214 in The Safety of Systems Proceedings of the Fifteenth Safety-critical Systems Symposium, Bristol, UK, 13–15 February 2007. 4- http://think.aero/HRA/Ref4.pdf

(continued on next page)

HRA generation type	Puplicly avilable	HRA method name	HRA acronym	Developed by	Originally developed for	Domain of application (aviation/ maritime/ indutry)	Main theoretical framework and area of focuse (design, operation, management, cognitive, behavioral, contextual)	Qualitative/ Quantitiave	Method procedures	Performance factors (PSFs), error producing conditions (EPCs), error forcing contexts (EFCs), common performance conditions (CPCs)	Validation	Method's strengths	Method's weakness	Acdemic references
(Second Generation) began in the 1990s and are on-going. They are cognitive and taxonomy-based model uses the contextual control model (COCOM) and aimed at the qualitative assessment of the operator's behavior and the search for models that describe the interaction with the production process. PSFs in the second generation were derived by focusing on the cognitive impacts on operators. They contain three main groups of HRA methods: Factorial, Contextual, and Based on expert judgment.	Yes	Cognitive Reliability and Error Analysis Method	CREAM	Erik Hollnagel (1993)		Nuclear Industry, Railway Industry	Cognitive, Contextual	Qualitative/ Quantitative	1- Identify the task and sub-activities. 2- Each of these tasks and sub-activities to be categorized under 9 Common Performance Conditions (CPCs), which are 1. Adequacy of organisation, 2. Working conditions, 3. Adequacy of the man-machine interface and operational support, 4. Availability of proce-dures/plans, 5. Number of simultaneous goals, 6. Available time, 7. Time of day, 8. Adequacy of training and experience, 9. Quality of crew collaboration. 3- These CPCS are given scores.	9 Common Performance Conditions (CPCs) (System)	Many scholars such as Marseguerra et al. (2007), Collier (2003) validated this method and they argue that CREAM Method is not a reliable tool for safety assessment in particular with HRA and even further development is limited due to several problems found with the technique and the data used for analysis.	Bidirectional (retrospective and Prediction)	Subjectivity associated with CPCS that are considered to be the key player in the occurrence of the likelihood of failure in these tasks. Based on the tasks nature and experts' decision, These CPCS are given scores and assessed for each activity based on how CPC will improved, or reduce or not significantly change the task failure probability	1- Hollnagel, E. (1998) Cognitive Reliability and Error Analysis Method. Elsevier. Hollnagel, E. (1993) Human reliability analysis: Context and control. Academic Press. 2- Marseguerra, M., Zio, E. and Librizzi, M. (2007) Human Reliability Analysis by Fuzzy "CREAM" Risk Analysis 27, (1), 137–154

(continued on next page)

HRA generation type	Puplicly avilable	HRA method name	HRA acronym	Developed by	Originally developed for	Domain of application (aviation/ maritime/ indutry)	Main theoretical framework and area of focuse (design, operation, management, cognitive, behavioral, contextual)	Qualitative/ Quantitiave	Method procedures	Performance factors (PSFs), error producing conditions (EPCs), error forcing contexts (EFCs), common performance conditions (CPCs)	Validation	Method's strengths	Method's weakness	Acdemic references
	Yes	A Technique for Human Event Analysis	ATHE-ANA	US Nuclear Regulatory Commis-sion in 2000	US Nuclear Regulatory Commis-sion	Nuclear industry	Behavioral, Cognitive, Contextual (error is mainly due to the context of the activity)	Qualitative/ Quantitative	ATHEANA process comprises of ten steps including 1) formulation of the task into the ATHEANA HRA/PRA framework; 2) determination of unsafe actions in the task and related human failure events; 3) correlating the causes that led for failure with events; 4) EFCS probability quantification; 5) calculation of the probability of each unsafe action and the probability of not recovering from the initial Unsafe Action and; 6) Results' evaluation and analysis based on expert elicitation and ATHEANA quantification approach.	Error Forcing Contexts (EFCs), combinations of PSFS (Contextual factors derived from: Human, Task, System, and Environment)	According to peer reviewers such as Hollnagel, et al. (1998) it has been found that ATHEANA is a useful tool that provides a significant improvement in HRA methodology.	1- Investigates psychology and human behaviors that are integrated with engineering disciplines. 2- it is capable of quantifying HRA for pre and post initiators using the quantification model for HFE probabilities based on the frequency estimation of Error Forcing Contexts (EFCs) 3- Provides a detailed procedure to find the reason why unsafe action happened. 4- has the ability to define human error type e.g error forcing context or error-prone situation.	1- ATHENA method is based on the HFEs from PSA model that the consequences of human error are limited to the preidentified PSA in the accident sequence. 2- ATHENA weak for predictive analysis and does not take advantage of cognitive identified errors.	1- US Nuclear Regulatory Commission (USNRC). A Technique for Human Event Analysis (ATHEANA)—Technical Basis and Methodological Description. NUREG/CR-6350 2- Brookhaven National Laboratory, Upton, NY, April. 1996. Prepared by Cooper, S.E., Ramey-Smith, A.M., Wreathall, J., Parry, G.W., Bley, D.C. Luckas, W.J., Taylor, J.H., Barriere, M.T. US 3- Nuclear Regulatory Commission (USNRC). Technical Basis and Implementation Guidelines for A Technique for Human Event Analysis (ATHEANA). NUREG- 1624. Division of Risk Analysis and Applications. Office of Nuclear Regulatory Research, Washington DC. May 2000.

(continued on next page)

HRA generation type	Puplicly avilable	HRA method name	HRA acronym	Developed by	Originally developed for	Domain of application (aviation/ maritime/ indutry)	Main theoretical framework and area of focuse (design, operation, management, cognitive, behavioral, contextual)	Qualitative/ Quantitiave	Method procedures	Performance factors (PSFs), error producing conditions (EPCs), error forcing contexts (EFCs), common performance conditions (CPCs)	Validation	Method's strengths	Method's weakness	Acdemic references
	NO. Electricité de France (EdF) property	Méthode d'Evaluation de la Réalisation des Missions Opérateur pour la Sûreté (MERMOS) (Name In English. EnglishAssessment method for the performance of safety operation)	MERMOS	Electricité de France (EdF).	Electricité de France (EdF). Nuclear Reactors	Nuclear Industry	(Contextual) Analysing emergency operation behaviour (Emergency operations system -EOS) once an incident has been triggered during the four hours after the incident initiator (post-incident behaviour) at the level of system operators, procedures and human-machine interface.	Qualitative/ Quantitative	1- A qualitative analysis to identify as many scenarios as possible leading to the HF mission failure in that acciden. The failure is a set of exclusive events (e.g., CICAs that refer to particular ways of operating the plant adopted by the EOS in the course of the emergency situation, action, diagnostic, strategy). 2-Then a quantitative analysis is performed to calculate the failure probability of the mission failure, which is the sum of the probabilities of failure each of the scenarios described for a mission.	Human Factors	Bieder, et al. (2000), reported that the probabilities calculated with the MERMOS process proved consistent with previous HRA results. Le Bot, P. (2004). Performed a retrospective analysis of the Three Mile Island (TMI) accident using MERMOS and concluded that EOS are logic upon which the design of the operating system was based and MERMOS is operating system focused method	MERMOS is not focusing on individual error, and instead considering the human erros in the operating system as a whole.	1- MERMOS is EdF proprietary and therefore not freely available for review and use 2- Subjective and high uncertainty because it depends on experts/simulation in the field of emergency operations. 3- It is a case by case method based on criteria. 4- only applicable for emergency operations	1- Le Bot, P., Desmares, E. Bieder, C. Cara, F., Bonnet, J-L. (1997). MERMOS: An EdF project to update the PHRA (Probabilistic Human Reliability Assessment) methodology, OECD Nuclear Energy Agency Specialists Meeting on Human Performance in Operational Events, Chattanooga, USA (October 13-17 1997). 2- Le Bot, P., Cara, F. and Bieder, C. (1999). A second generation HRA method: What it does and doesn't do. In: (Washington, DC: American Nuclear Society), Volume II, 852–860.

(*continued on next page*)

HRA generation type	Puplicly avilable	HRA method name	HRA acronym	Developed by	Originally developed for	Domain of application (aviation/ maritime/ indutry)	Main theoretical framework and area of focuse (design, operation, management, cognitive, behavioral, contextual)	Qualitative/ Quantitiave	Method procedures	Performance factors (PSFs), error producing conditions (EPCs), error forcing contexts (EFCs), common performance conditions (CPCs)	Validation	Method's strengths	Method's weakness	Acdemic references
Third Generation	NO. A nuclear specific version (A property of the nuclear power company, British Energy)	Nuclear Action Reliability Assessment	NARA	Kirwan, B. (Eurocon-trol) Gibson, H., & Kennedy, R. (University of Birming-ham) Edmunds, J., Cooksley, G. (Corporate Risk Associates) Umbers, I. (British Energy)	The nuclear power company, British Energy	Nuclear Industry	Cognitive	Quantitative	NARA process Like HEART because it was developed from HEART. The key elements of the approach are: 1- Classify the task for analysis into one of the Generic Task Types and assign the nominal Human Error Probability (HEP) to the task. 2- Decide which EPCs may affect task reliability. 3- Consider the assessed proportion of affect (APOA) for each EPC. 4- Calculate the task HEP.	14 Generic Task Types (GTTs) originally a subset of HEART and some are modified for nuclear operation, and 18 Error Producing Conditions (EPCs) (Task, System, Environment)	No scientific or empirical validation results have been found	1- Ongoing development to consider dependence. 2- Same strengths of HEART since it is developed from the HEART	1- Same weakness of HEART since it is developed from the HEART	1- Kirwan, B., Gibson, H., Kennedy, R., Edmunds, J., Cooksley, G. and Umbers, I. (2005), Nuclear action reliability assessment (NARA): a databased HRA tool. Safety & Reliability, vol. 25. no. 2 pp. 38–45 (2005) 2- Kirwan, B. and Gibson, H. (2007) CARA: A Human Reliability Assessment Tool for Air Traffic Safety Management — Technical Basis and Preliminary Architecture. pp. 197 –214 in The Safety of Systems Proceedings of the Fifteenth Safety-critical Systems Symposium, Bristol, UK, 13–15 February 2007.

(*continued on next page*)

HRA generation type	Puplicly avilable	HRA method name	HRA acronym	Developed by	Originally developed for	Domain of application (aviation/ maritime/ indutry)	Main theoretical framework and area of focuse (design, operation, management, cognitive, behavioral, contextual)	Qualitative/ Quantitiave	Method procedures	Performance factors (PSFs), error producing conditions (EPCs), error forcing contexts (EFCs), common performance conditions (CPCs)	Validation	Method's strengths	Method's weakness	Acdemic references
Third Generation	Yes	System-Theoretic Accident Model and Processes	STAMP	Nancy G. Leveson in 2011	MIT	Automotive industry, Commerical and Military Airline operations, Manufacturing industry, Mining, Oil & Gas industry, Chemical& petrochemical industry, Nuclear industry, Space systems, Military and Defense, Healthcare, Robotics industry, Cyber Security, Maritime industry, and transportation.	All	Qualitative	STAMP based on 2 main approaches, which are Causal Analysis Based on STAMP (CAST) System-Theoretic Process Analysis (STPA). Overall procedures are: 1- Establish the system engineering foundation for the analysis and for the system development. 2. Create a hierarchical control structure (HCS). 3. Define control actions. 4. Identify potentially unsafe control actions. 5. Use the identified unsafe control actions to create safety requirements and constraints.	Systemic factors (components and barri-ers)/Software errors/ Human behav-ior/System design errors	STAMP approach has been validated by many scholars and shows a high level of accuracy as a systemic, comprehensive, predictive, and preventative model compared with other methods.	1- STAMP is a hierarchical safety control structure. 2- Address component interaction (hardware, humans, software) between the system and its subsystem (direct or Indirect or non-linear or complex) 3- It provides structured guidance for hazard identification at the first stages of the de-sign/analysis (STPA method) 4- Advanced, comprehen-sive, and dynamic accident causality model based on systems theory.	1- STAMP considers system safety is mainly a control problem, not a reliability one 2- STAMP's error taxonom classifications need to be improved by using HFACS or Hierarchical Holographic Modeling or other useful tools. 3- STAMP depends on the experience and expertise of the analyst	1- Leveson, N. G. (2011). Engineering a Safer World. Engineering Systems. The MIT Press.

(*continued on next page*)

HRA generation type	Puplicly avilable	HRA method name	HRA acronym	Developed by	Originally developed for	Domain of application (aviation/ maritime/ indutry)	Main theoretical framework and area of focuse (design, operation, management, cognitive, behavioral, contextual)	Qualitative/ Quantitiave	Method procedures	Performance factors (PSFs), error producing conditions (EPCs), error forcing contexts (EFCs), common performance conditions (CPCs)	Validation	Method's strengths	Method's weakness	Acdemic references
									6. Determine how each potentially hazardous control action could occur to enable mitigation actions.			5-Identify Hazard of a complex e.g automated systems through: 1- Emergence and hierarchy, 2- Communi-cation and control, 3- Process models that helped designers and managment to improve the interactions among system and subsystems components that are not adequately handled by the control system, constraints, control loops and process models, and levels of control.		

(continued on next page)

HRA generation type	Puplicly avilable	HRA method name	HRA acronym	Developed by	Originally developed for	Domain of application (aviation/ maritime/ indutry)	Main theoretical framework and area of focuse (design, operation, management, cognitive, behavioral, contextual)	Qualitative/ Quantitiave	Method procedures	Performance factors (PSFs), error producing conditions (EPCs), error forcing contexts (EFCs), common performance conditions (CPCs)	Validation	Method's strengths	Method's weakness	Acdemic references
	Yes	Human Factors Analysis and Classi-fication System	HFACS	Developed by Dr Scott Shappell and Dr Doug Wiegmann	US Navy	Civil Aviation, Railway, Mining, Maritime, Shipping sector, healthcare, Construction	All Classification schemes are divided scheme into three main groups which describe: (i) the context of the incident, (ii) the production of the error (operator context), and (iii) the recovery of the incident	Qualitative	Classify human error into four levels of failure: 1.Unsafe acts of operators, 2.Preconditions for unsafe acts, 3. Unsafe supervision, and 4. Organizational influences	four levels of failure: 1. Unsafe acts of operators 2. Preconditions for unsafe acts, 3. Unsafe supervision, and 4. Organizational influences	HFACS has been empirically validated by many scholars such as Chen, S. et al., (2013); Reinach, S. et al. (2005); Soner, O. et al. (2015); Patterson, J.M. et al. (2010); Hale, A. et al. (2012) for different applications with some modification and results found to be reasonable.	1- Hierarchical Structure of human factors. 2- Provides useful Human error Taxonomy 3- identify active and latent failures. 4- Help to understand the underlying causal factors that lead to an accident 5- based on Reason's model	1- Industry restrictions because HFACS designed to analyze the causes of aviation accidents, some categories within this model are not applicable to other sectors. 2- HFACS presents a remarkable lack of versatility. 3- HFACS' psychological factors will be limited due to the subjectivity of interviews' results. 4-Scope of the investigation is limited to the predefined taxonomy and organization level; hence, it will be altered in each accident outcome or scenario being modelled.	Shappell, S. A., & Wiegmann, D. A. (2000). The Human Factors Analysis and Classification System–HFACS., (). Retrieved from https://commons.erau.edu/publication/7372- Shappell, S., Wiegmann, D., Fraser, J., Gregory, G., Kinsey, P., and Squier, H (1999b). Beyond mishap rates: A human factors analysis of U.S. Navy/ Marine Corps TACAIR and rotary wing mishaps using HFACS. Aviation, Space, and Environmental Medicine, 70, 416–17. 3- Shappell, S.A. and Wiegmann, D.A. (1997). A reliability analysis of the Taxonomy of Unsafe Operations. Aviation, Space, and Environmental Medicine, 68, 620. 4- Shappell, S., Wiegmann, D., (2000) The Human Factors Analysis and Classification System–HFACS. Final Report. U.S. Department of Transportation Federal Aviation Administration DOT/FAA/AM-00/7 Office of Aviation Medicine Washington, DC 20591

(*continued on next page*)

HRA generation type	Puplicly avilable	HRA method name	HRA acronym	Developed by	Originally developed for	Domain of application (aviation/ maritime/ indutry)	Main theoretical framework and area of focuse (design, operation, management, cognitive, behavioral, contextual)	Qualitative/ Quantitiave	Method procedures	Performance factors (PSFs), error producing conditions (EPCs), error forcing contexts (EFCs), common performance conditions (CPCs)	Validation	Method's strengths	Method's weakness	Acdemic references
	Yes	Software, Hardware, Environ-ment, Liveware Analysis	SHEL Analysis	Elwyn Edwards (1972)	Aviation	Health Care	ALL 1- Software (the rules, procedures, written documents operating procedures. etc. 2- Hardware (Instruments, devices. etc) 3- Environment (social and economic climate, natural environment) 4- Liveware (the human beings) According to the SHELL a mismatch between the Liveware and other 3 components contributes to human error	Qualitative	1- Identify all the components in each block of: Software (the rules, procedures, written documents operating procedures. etc.), Hardware (Instruments, devices. etc) Environment (social and economic climate, natural environment), Liveware (Involved people) 2- Investigate changes, interactions, and the causal	N.A	Limited validation studies found. For instance, Antunes, P. et al., (2011) conducted an empirical validation (a case study) of SHELL addressing risk assessment in a hospital unit and concluded SHELL had very positive results to address risk management with a focus on collaborative work settings and the SHELL model should be used more often.	1- SHELL model show how the four elements of these systems interact and affect one another and influence the decision-making process. 2- SHELL model considers both active and latent failures 3- Livware considers psychological and physiological factors 4-The SHELL model	1- This building block diagram does not cover the interfaces that are outside human factors. 2- It requires high-level experts in human factors and useful only if a reference scenario exists to compare it to the solution scenario.	1- Edwards, E., 1972. Man and machine: systems for safety. In: Proceedings of BritishAirline Pilots Associations Technical Symposium. British Airline Pilots Associations, London, pp. 21–36.

(continued on next page)

HRA generation type	Puplicly avilable	HRA method name	HRA acronym	Developed by	Originally developed for	Domain of application (aviation/ maritime/ indutry)	Main theoretical framework and area of focuse (design, operation, management, cognitive, behavioral, contextual)	Qualitative/ Quantitiave	Method procedures	Performance factors (PSFs), error producing conditions (EPCs), error forcing contexts (EFCs), common performance conditions (CPCs)	Validation	Method's strengths	Method's weakness	Acdemic references
									relationships between the changes and the components (L-E, L-H, L-L, and L-S model elements). 3- Draw causal relationships between the changes and the components (L-E, L-H, L-L, and L-S model elements). 4- Analysis of the relationship between causes and the changes that occurred in the components (L-E, L-H, L-L, and L-S model elements) and draw conclusions.			focuses on the fundamental drivers of change when inquiring about the changes, and produce streamlined explanations of the causal relationships between the changes and the L-E, L-H, L-L and L-S model elements		

APPENDIX

4

IAEA Nuclear Security Series No. 8-G (Rev. 1): Ppreventive and Protective Measures Against Insider Threats

Applications of Nuclear and Radioisotope Technology: The Atom for Peace and Sustainable Development
DOI: https://doi.org/10.1016/B978-0-12-821319-3.00067-1

IAEA Nuclear Security Series No. 8-G (Rev. 1)

Implementing Guide

Preventive and Protective Measures against Insider Threats

IAEA NUCLEAR SECURITY SERIES

Nuclear security issues relating to the prevention and detection of, and response to, criminal or intentional unauthorized acts involving, or directed at, nuclear material, other radioactive material, associated facilities or associated activities are addressed in the **IAEA Nuclear Security Series**. These publications are consistent with, and complement, international nuclear security instruments, such as the Convention on the Physical Protection of Nuclear Material and its Amendment, the International Convention for the Suppression of Acts of Nuclear Terrorism, United Nations Security Council resolutions 1373 and 1540, and the Code of Conduct on the Safety and Security of Radioactive Sources.

CATEGORIES IN THE IAEA NUCLEAR SECURITY SERIES

Publications in the IAEA Nuclear Security Series are issued in the following categories:

- **Nuclear Security Fundamentals** specify the objective of a State's nuclear security regime and the essential elements of such a regime. They provide the basis for the Nuclear Security Recommendations.
- **Nuclear Security Recommendations** set out measures that States should take to achieve and maintain an effective national nuclear security regime consistent with the Nuclear Security Fundamentals.
- **Implementing Guides** provide guidance on the means by which States could implement the measures set out in the Nuclear Security Recommendations. As such, they focus on how to meet the recommendations relating to broad areas of nuclear security.
- **Technical Guidance** provides guidance on specific technical subjects to supplement the guidance set out in the Implementing Guides. They focus on details of how to implement the necessary measures.

DRAFTING AND REVIEW

The preparation and review of Nuclear Security Series publications involves the IAEA Secretariat, experts from Member States (who assist the Secretariat in drafting the publications) and the Nuclear Security Guidance Committee (NSGC), which reviews and approves draft publications. Where appropriate, open-ended technical meetings are also held during drafting to provide an opportunity for specialists from Member States and relevant international organizations to review and discuss the draft text. In addition, to ensure a high level of international review and consensus, the Secretariat submits the draft texts to all Member States for a period of 120 days for formal review.

For each publication, the Secretariat prepares the following, which the NSGC approves at successive stages in the preparation and review process:

- An outline and work plan describing the intended new or revised publication, its intended purpose, scope and content;
- A draft publication for submission to Member States for comment during the 120 day consultation period;
- A final draft publication taking account of Member States' comments.

The process for drafting and reviewing publications in the IAEA Nuclear Security Series takes account of confidentiality considerations and recognizes that nuclear security is inseparably linked with general and specific national security concerns.

An underlying consideration is that related IAEA safety standards and safeguards activities should be taken into account in the technical content of the publications. In particular, Nuclear Security Series publications addressing areas in which there are interfaces with safety — known as interface documents — are reviewed at each of the stages set out above by relevant Safety Standards Committees as well as by the NSGC.

PREVENTIVE AND PROTECTIVE MEASURES AGAINST INSIDER THREATS

The following States are Members of the International Atomic Energy Agency:

AFGHANISTAN
ALBANIA
ALGERIA
ANGOLA
ANTIGUA AND BARBUDA
ARGENTINA
ARMENIA
AUSTRALIA
AUSTRIA
AZERBAIJAN
BAHAMAS
BAHRAIN
BANGLADESH
BARBADOS
BELARUS
BELGIUM
BELIZE
BENIN
BOLIVIA, PLURINATIONAL STATE OF
BOSNIA AND HERZEGOVINA
BOTSWANA
BRAZIL
BRUNEI DARUSSALAM
BULGARIA
BURKINA FASO
BURUNDI
CAMBODIA
CAMEROON
CANADA
CENTRAL AFRICAN REPUBLIC
CHAD
CHILE
CHINA
COLOMBIA
CONGO
COSTA RICA
CÔTE D'IVOIRE
CROATIA
CUBA
CYPRUS
CZECH REPUBLIC
DEMOCRATIC REPUBLIC OF THE CONGO
DENMARK
DJIBOUTI
DOMINICA
DOMINICAN REPUBLIC
ECUADOR
EGYPT
EL SALVADOR
ERITREA
ESTONIA
ESWATINI
ETHIOPIA
FIJI
FINLAND
FRANCE
GABON
GEORGIA
GERMANY
GHANA
GREECE
GRENADA
GUATEMALA
GUYANA
HAITI
HOLY SEE
HONDURAS
HUNGARY
ICELAND
INDIA
INDONESIA
IRAN, ISLAMIC REPUBLIC OF
IRAQ
IRELAND
ISRAEL
ITALY
JAMAICA
JAPAN
JORDAN
KAZAKHSTAN
KENYA
KOREA, REPUBLIC OF
KUWAIT
KYRGYZSTAN
LAO PEOPLE'S DEMOCRATIC REPUBLIC
LATVIA
LEBANON
LESOTHO
LIBERIA
LIBYA
LIECHTENSTEIN
LITHUANIA
LUXEMBOURG
MADAGASCAR
MALAWI
MALAYSIA
MALI
MALTA
MARSHALL ISLANDS
MAURITANIA
MAURITIUS
MEXICO
MONACO
MONGOLIA
MONTENEGRO
MOROCCO
MOZAMBIQUE
MYANMAR
NAMIBIA
NEPAL
NETHERLANDS
NEW ZEALAND
NICARAGUA
NIGER
NIGERIA
NORTH MACEDONIA
NORWAY
OMAN
PAKISTAN
PALAU
PANAMA
PAPUA NEW GUINEA
PARAGUAY
PERU
PHILIPPINES
POLAND
PORTUGAL
QATAR
REPUBLIC OF MOLDOVA
ROMANIA
RUSSIAN FEDERATION
RWANDA
SAINT LUCIA
SAINT VINCENT AND THE GRENADINES
SAN MARINO
SAUDI ARABIA
SENEGAL
SERBIA
SEYCHELLES
SIERRA LEONE
SINGAPORE
SLOVAKIA
SLOVENIA
SOUTH AFRICA
SPAIN
SRI LANKA
SUDAN
SWEDEN
SWITZERLAND
SYRIAN ARAB REPUBLIC
TAJIKISTAN
THAILAND
TOGO
TRINIDAD AND TOBAGO
TUNISIA
TURKEY
TURKMENISTAN
UGANDA
UKRAINE
UNITED ARAB EMIRATES
UNITED KINGDOM OF GREAT BRITAIN AND NORTHERN IRELAND
UNITED REPUBLIC OF TANZANIA
UNITED STATES OF AMERICA
URUGUAY
UZBEKISTAN
VANUATU
VENEZUELA, BOLIVARIAN REPUBLIC OF
VIET NAM
YEMEN
ZAMBIA
ZIMBABWE

The Agency's Statute was approved on 23 October 1956 by the Conference on the Statute of the IAEA held at United Nations Headquarters, New York; it entered into force on 29 July 1957. The Headquarters of the Agency are situated in Vienna. Its principal objective is "to accelerate and enlarge the contribution of atomic energy to peace, health and prosperity throughout the world".

IAEA NUCLEAR SECURITY SERIES No. 8-G (Rev. 1)

PREVENTIVE AND PROTECTIVE MEASURES AGAINST INSIDER THREATS

IMPLEMENTING GUIDE

COPYRIGHT NOTICE

Marketing and Sales Unit, Publishing Section
International Atomic Energy Agency
Vienna International Centre
PO Box 100
1400 Vienna, Austria
fax: +43 1 26007 22529
tel.: +43 1 2600 22417
email: sales.publications@iaea.org
www.iaea.org/publications

Printed by the IAEA in Austria
Janaury 2020
STI/PUB/1858

IAEA Library Cataloguing in Publication Data

Names: International Atomic Energy Agency.
Title: Preventive and protective measures against insider threats / International Atomic Energy Agency.
Description: Vienna : International Atomic Energy Agency, 2020. | Series: Preventive and protective measures against insider threats (IAEA Nuclear Security Series (Rev. 1)), ISSN 1816–9317 ; no. 8-G (rev. 1) | Includes bibliographical references.
Identifiers: IAEAL 19-01249 | ISBN 978–92–0–103419–9 (paperback : alk. paper)
Subjects: Nuclear facilities. | Radioactive substances. | Internal security.
Classification: UDC 341.67 | STI/PUB/1858

FOREWORD

The IAEA's principal objective under its Statute is "to accelerate and enlarge the contribution of atomic energy to peace, health and prosperity throughout the world." Our work involves both preventing the spread of nuclear weapons and ensuring that nuclear technology is made available for peaceful purposes in areas such as health and agriculture. It is essential that all nuclear and other radioactive materials, and the facilities at which they are held, are managed in a safe manner and properly protected against criminal or intentional unauthorized acts.

Nuclear security is the responsibility of each individual State, but international cooperation is vital to support States in establishing and maintaining effective nuclear security regimes. The central role of the IAEA in facilitating such cooperation and providing assistance to States is well recognized. The IAEA's role reflects its broad membership, its mandate, its unique expertise and its long experience of providing technical assistance and specialist, practical guidance to States.

Since 2006, the IAEA has issued Nuclear Security Series publications to help States to establish effective national nuclear security regimes. These publications complement international legal instruments on nuclear security, such as the Convention on the Physical Protection of Nuclear Material and its Amendment, the International Convention for the Suppression of Acts of Nuclear Terrorism, United Nations Security Council resolutions 1373 and 1540, and the Code of Conduct on the Safety and Security of Radioactive Sources.

Guidance is developed with the active involvement of experts from IAEA Member States, which ensures that it reflects a consensus on good practices in nuclear security. The IAEA Nuclear Security Guidance Committee, established in March 2012 and made up of Member States' representatives, reviews and approves draft publications in the Nuclear Security Series as they are developed.

The IAEA will continue to work with its Member States to ensure that the benefits of peaceful nuclear technology are made available to improve the health, well-being and prosperity of people worldwide.

CONTENTS

1. INTRODUCTION

BACKGROUND

1.1. The IAEA Nuclear Security Series provides guidance for States to assist them in implementing, reviewing and, when necessary, strengthening a national nuclear security regime. The series also provides guidance for States on fulfilling their obligations and commitments with respect to binding and non-binding international instruments. The Nuclear Security Fundamentals publication (IAEA Nuclear Security Series No. 20 [1]) provides the objective and essential elements for the entire nuclear security regime. Recommendations publications indicate what a nuclear security regime should address for the physical protection of nuclear material and nuclear facilities [2], radioactive material and associated facilities [3], and nuclear and other radioactive material out of regulatory control [4]. These publications, as well as many others in the IAEA Nuclear Security Series (Refs [5–12]), recognize the particular threats that could be posed by insiders, as well as the need to implement specific measures against insider threats and to evaluate those measures accordingly.

1.2. This publication is an update of IAEA Nuclear Security Series No. 8, Preventive and Protective Measures against Insider Threats, published by the IAEA in 2008[1]. This revision was undertaken to better align this Implementing Guide with the Nuclear Security Fundamentals and with the Recommendations that were published after 2008, to cross-reference other relevant Implementing Guides published since 2008, and to add further detail on certain topics based on the experience of the IAEA and Member States in using IAEA Nuclear Security Series No. 8.

OBJECTIVE

1.3. The objective of this Implementing Guide is to provide updated guidance to States, their competent authorities and operators[2], shippers, and carriers on

---

[1] INTERNATIONAL ATOMIC ENERGY AGENCY, Preventive and Protective Measures against Insider Threats, IAEA Nuclear Security Series No. 8, IAEA, Vienna (2008).

[2] The term 'operator' is used to describe an entity (person or organization) authorized to operate a nuclear or radiological facility or authorized to use, store or transport nuclear material and/or radioactive material. Such an entity would normally hold a licence or other document of authorization from a competent authority or be contractors of a holder of such an authorization.

selecting, implementing and evaluating measures for addressing insider threats. Threats to nuclear facilities can involve external or insider adversaries or both together in collusion (cooperation for an illegal or malicious purpose with another insider adversary or with an external adversary).

SCOPE

1.4. This publication applies to preventing and protecting against unauthorized removal of nuclear material and sabotage of nuclear material and facilities by insiders. This publication applies to any type of nuclear facility — notably nuclear power plants, research reactors and other nuclear fuel cycle facilities (e.g. enrichment plants, reprocessing plants, fuel fabrication plants, storage facilities) — whether in design, redesign, construction, commissioning, operation, shutdown or decommissioning.

1.5. The guidance in this publication on insider threats may also be applied to preventing and protecting against unauthorized removal and sabotage of radioactive material and associated facilities [3]; securing nuclear and radioactive materials undergoing transport [6, 13]; and the prevention and detection of, and response to, nuclear and other radioactive material out of regulatory control [4]. This guidance may also be applied to securing facility information held or obtained by other stakeholders, including the competent authority [8].

1.6. For the purposes of this publication, insider access to a facility includes physical access to locations and material; internal or authorized remote computer or network access; and access to sensitive information about the facility.

1.7. While safety considerations are not addressed in this publication, the preventive and protective measures described should be implemented in a balanced manner that is compatible with safety considerations and that considers worker radiation protection. Security measures and safety measures should be designed and implemented in an integrated manner to develop synergy between these two areas and in such a way that security measures do not compromise safety and safety measures do not compromise security [1].

STRUCTURE

1.8. After this introduction, this publication is separated into four sections. Section 2 introduces insider threats and ways to categorize insiders. Section 3 identifies the targets and facility systems to be protected against malicious acts by insiders. Section 4 discusses implementation at the facility level of preventive and protective measures to address insider threats. Section 5 discusses the evaluation of the measures discussed in Section 4.

2. IDENTIFICATION OF INSIDER THREATS

2.1. The term 'adversary' is used to describe any individual performing or attempting to perform a malicious act. An adversary could be an insider or could be external.

2.2. The term 'insider' is used to describe

> "an individual with authorized access to [nuclear material,] *associated facilities* or *associated activities* or to *sensitive information* or *sensitive information assets*, who could commit, or facilitate the commission of criminal or intentional unauthorized acts involving or directed at *nuclear material*, *other radioactive material*, *associated facilities* or *associated activities* or other acts determined by the State to have an adverse impact on nuclear security" [1].

The term 'external adversary' is used to describe an adversary other than an insider.

ATTRIBUTES OF INSIDERS

2.3. Insiders possess at least one of the following attributes that provide advantages over external adversaries when attempting malicious activities:

(a) Access: Insiders have authorized access to the areas, equipment and information needed to perform their work. Access includes physical access to nuclear facilities; nuclear materials and associated systems, components and equipment; and computer systems. Access also includes remote computer access to a facility, such as access to computer systems and networks that control processes, provide safety, contain sensitive information or otherwise contribute to nuclear security. The operator should not permit remote access to critical systems, such as systems relevant to safety.

(b) Authority: Insiders are authorized to conduct operations as part of their assigned duties and may also have the authority to direct other employees. This authority may be used to support malicious acts, including either physical or computer based acts such as digital file or process manipulation.
(c) Knowledge: Insiders may possess knowledge of the facility, associated activities or systems, ranging from limited to expert knowledge. This may include knowledge that could enable an insider to bypass or defeat dedicated physical protection systems and other facility systems that contribute to nuclear security, such as safety and nuclear material accounting and control (NMAC) systems, operating procedures and response capabilities.

These attributes may also include access to, or knowledge of, sensitive information or sensitive information assets, including information regarding the transport or movement of nuclear material [13].

2.4. An insider might not possess all three attributes but might still have sufficient capability to conduct a malicious act. For example, a headquarters manager may have limited physical access to a facility but could have the authority to issue a counterfeit delivery order to an outside location. Insider adversaries may use feigned authority or knowledge to facilitate or initiate a malicious act. An insider adversary may act independently or in collusion with another insider adversary or with an external adversary.

2.5. Owing to their access, authority and knowledge, insiders have the opportunity to select the most vulnerable target and the best time to attempt or perform a malicious act. To maximize the likelihood of success, an insider adversary might extend a malicious act over a long period of time. This tactic could consist of (a) tampering with physical protection equipment or safety equipment to prepare for an act of sabotage, (b) falsifying records so that the insider adversary is able to repeatedly remove without authorization small amounts of lower category nuclear material that has less robust protection than higher category nuclear material without being detected or (c) removing nuclear material without authorization in amounts below measurement system detection thresholds. Insider adversaries may have the opportunity to commit a malicious act during normal or abnormal conditions of a facility, including during maintenance, or during the movement of nuclear material, and may select the most favourable time to do so [14].

MOTIVATIONS OF INSIDERS

2.6. Insiders may have different motivations for initiating malicious acts, including money, ideology, revenge, ego, coercion or a combination of these motivations.

2.7. An insider may independently develop sufficient motivation to perform a malicious act, including as the result of a mental health issue. An insider may also be recruited by an external adversary seeking to exploit the insider's access, authority or knowledge. An insider could be forced to commit a malicious act through coercion (e.g. blackmail).

2.8. An insider could hold any position within an organization, from the highest level to the lowest. Insiders at all levels could have sufficient motivation to perform a malicious act. Other personnel not directly employed by the operator, shipper or carrier but who have authorized access on a periodic basis to the facility or its systems (e.g. vendors, first responders, contractors, inspectors from regulatory bodies or other competent authorities) should also be considered to be potential insider threats.

CATEGORIES OF INSIDERS

2.9. An unwitting insider is an insider without the intent and motivation to commit a malicious act who is exploited by an adversary without the unwitting insider's awareness. For example, in a computer based attack, an unwitting insider may not be aware that certain actions (e.g. clicking a malicious link in an email that is disguised as being from a trusted source) may provide information or authenticated access to an adversary.

2.10. An insider adversary is an insider that commits malicious activities with awareness, intent and motivation. An insider adversary may be passive or active, and an active insider adversary may be either violent or non-violent. This categorization is useful for assessment purposes, such as during the development of adversary profiles in the threat assessment or design basis threat (DBT), or when creating scenarios to be used to test nuclear security measures as part of an evaluation process for the nuclear security system.

2.11. A passive insider adversary assists another adversary by providing information to be used in performing a malicious act. A passive insider adversary would not participate in the malicious act in any other way and would likely cease involvement if there was a high probablility of being identified.

2.12. An active, non-violent insider adversary uses stealth or deceit to facilitate or conduct a malicious act and may provide information to another adversary. For example, an active, non-violent insider adversary may attempt an abrupt or protracted theft of nuclear material or may assist external adversaries in performing a malicious act by disabling or ignoring alarms or by opening doors. It is likely that an active, non-violent insider adversary would terminate the malicious act if there was a high probability of being identified (i.e. this type of insider adversary might risk being detected but would likely not risk being identified).

2.13. An active, violent insider adversary is similar to an active, non-violent insider adversary but is also willing to use physical force against personnel to facilitate or conduct a malicious act. Depending on the circumstances, an insider adversary may move from non-violent to violent.

IDENTIFICATION OF POTENTIAL INSIDER THREATS

2.14. The guidance contained in this section may be useful for the operator in identifying potential insider threats and should be used in conjunction with other insider threat identification processes, such as developing plausible scenarios as part of an evaluation of the nuclear security system.

2.15. Reference [2] recommends that "The appropriate State authorities, using various credible information sources, should define the *threat* and associated capabilities in the form of a *threat assessment* and, if appropriate, a *design basis threat*."[3] A State should consider the attributes, motivations and categories of insiders and describe any credible insider threats in the national threat assessment or DBT.

2.16. A threat and risk assessment may also help identify potential insider threats. In addition to the general information about insider threats contained in the national threat assessment or DBT, local threat information from the area around a particular facility should be considered in the facility specific assessment. This information may highlight relevant conditions (e.g. crime levels) or situations

[3] The DBT refers to the "attributes and characteristics of potential *insider* and/or external adversaries, who might attempt *unauthorized removal* or *sabotage*, against which a *physical protection system* is designed and evaluated" [2].

outside the facility (e.g. general attitude of the community, presence of organized hostile groups) that may be favourable to insider adversaries.

2.17. Potential insider threats may also be identified by determining which insiders have remote or on-site authorized access to facility systems through computer networks. Modern facility systems, including those that contribute to nuclear security, rely on computer based controls and networks. These systems should be protected against computer based attacks as described in Ref. [7]. Personnel with access to these systems should be considered when identifying insider threats.

3. TARGET IDENTIFICATION

3.1. Target identification, as described in Ref. [15], determines which material and equipment needs to be protected from an adversary. Targets may include nuclear material, associated areas, buildings, equipment, components, information, systems and functions. Guidance on target identification for facilities and for nuclear and radioactive material is provided in Refs [2–4, 8, 15, 16].

3.2. Assets (e.g. surveillance systems, portal monitors) that are not themselves identified as targets but are critical for the protection of identified targets may also require protection. An insider adversary could bypass or compromise these assets to conduct a malicious act.

TARGETS FOR UNAUTHORIZED REMOVAL

3.3. Nuclear material targets for unauthorized removal can be assigned to one of three categories (I–III) according to the relative attractiveness and characteristics of the nuclear material as well as the potential consequences if it were used in a nuclear explosive device. This categorization is defined in table 1 of Ref. [2]. The unauthorized removal of nuclear or radioactive material for the construction of a radiological dispersal device should also be considered [3]. In addition to nuclear and other radioactive material, theft targets may include sensitive information and sensitive information assets.

3.4. The identification of potential targets for unauthorized removal of nuclear material by an insider adversary should take into account the possibility of both abrupt and protracted theft. 'Abrupt theft' is the unauthorized removal of a target or a significant quantity of nuclear material during a single act. 'Protracted theft' is the repeated unauthorized removal of potentially small quantities of nuclear material from either a single location or multiple locations.

3.5. An insider adversary might use protracted theft of nuclear material to remain undetected by repeatedly removing small quantities of material that are within the detection limits of NMAC and physical protection systems. Protracted theft may be accomplished either by removing the nuclear material from the facility with each acquisition or by accumulating the nuclear material in a hidden location for later, possibly abrupt, removal from the facility. The possibility that an insider adversary could collect an amount of nuclear material equivalent to a higher category by collecting sufficient amounts of lower category nuclear material should be considered during target identification. Factors such as the element, the physical form of the material, how it is used, the quantity that is used during processing and the amount stored should also be considered during target identification when determining if protracted theft scenarios are possible and credible. Similar considerations should be made for abrupt theft scenarios as well.

SABOTAGE TARGETS

3.6. Sabotage targets in a facility are determined by analysing the potential for the facility's radioactive material inventory and waste, including nuclear material and radioactive sources [3], to result in unacceptable radiological consequences or high radiological consequences. Further details regarding nuclear security measures that should be taken to protect against sabotage as well as to perform an analysis of sabotage targets can be found in Refs [2, 15].

3.7. The identification of possible combinations of actions (scenarios) an insider adversary might take to degrade facility structures, systems and components that may result in unacceptable radiological consequences or high radiological consequences should be part of the target identification process.

IDENTIFICATION OF SYSTEMS THAT CONTRIBUTE TO NUCLEAR SECURITY

3.8. A target identification process should consider all systems that could require additional protection from insider threats. Physical protection systems, NMAC systems and safety and process control systems should be considered as potential targets for malicious acts, including those initiated by an insider adversary.

3.9. An insider adversary may have authorized access to the facility or to information about the facility and might attack other structures, systems or components to indirectly perpetrate an attack, mask malicious acts or aid an external adversary. Depending on the facility or operation, computer based systems may be exploited by the insider adversary (e.g. office networks or communication computers might be used to acquire sensitive information).

3.10. The compromise of computer based systems in a facility could adversely affect safety, the security of nuclear material or accident mitigation. The operator should evaluate and protect computer based systems that contain information related to safety or security in accordance with the risk and the potential consequences of the release of this information. This evaluation should aim to identify critical computer based systems that may be the most vulnerable to a malicious act and whose failure could result in a nuclear security event.

3.11. The operator should consider providing additional training to employees and contractors with access to sensitive systems to raise security awareness. External adversaries may target insiders with access to a facility, sensitive information, sensitive information assets or the facility's networks to gain assistance in facilitating or masking malicious activities.

4. MEASURES AGAINST POTENTIAL INSIDER THREATS

4.1. Nuclear security measures used to protect against insider threats should include both preventive and protective measures. The term 'preventive measures' refers to measures used to reduce the number of potential insiders before individuals are granted access, to minimize opportunities for an insider to undertake a malicious act if access is granted or to prevent a potential insider adversary from carrying out a malicious act. The term 'protective measures' refers to measures used to detect or delay malicious acts, respond to malicious acts or mitigate the consequences of a malicious act.

4.2. This guidance does not include all measures that could be used against an insider threat. However, the use of preventive and protective measures can help counter insider threats if the threat is properly defined, the target identification process is thorough and the measures are effectively implemented and evaluated.

4.3. Information regarding measures used against insider threats and incidents involving malicious acts by insider adversaries should be collected by competent authorities to analyse trends, weaknesses and good practices. If appropriate, the information should be shared with authorized international agencies to better understand the scope and nature of the security challenge posed by insider adversaries.

GENERAL APPROACH TO IMPLEMENTATION

4.4. As stated in Ref. [2], nuclear security requirements should be based on a graded approach, taking into account the current evaluation of the threat, the relative attractiveness and nature of the material, and the potential consequences associated with unauthorized removal of nuclear material or sabotage of nuclear material or nuclear facilities. General guidance on the implementation of a graded approach to protect nuclear materials and facilities against insider and external threats can be found in Ref. [15].

4.5. Implementing nuclear security measures to protect against insider threats involves selecting a combination of preventive and protective measures[4] and implementing them in accordance with a graded approach. It is important that the measures selected be implemented and evaluated effectively so that they perform as desired. Not all measures are appropriate for every facility or operation.

4.6. Layers of preventive and protective measures should be implemented in accordance with the concept of defence in depth, such that insider adversaries would need to overcome or circumvent multiple layers of measures or technologies to achieve their objectives. These layers may consist of administrative measures (e.g. procedures, instructions, access control rules, confidentiality rules), technical measures or a combination of both. Both types of measure should integrate people and equipment.

4.7. The operator should prepare a security plan as part of its application to obtain a licence, as described in Ref. [2], and ensure that it describes the measures needed to address insider threats, including measures that address insider threats to information and computer security (e.g. a cyber-attack conducted by an insider adversary [7, 8]). The operator should consider insider threats during the design, evaluation, implementation and maintenance of nuclear security systems at the facility level.

---

[4] Some measures may have both preventive and protective effects.

4.8. The security plan should define how nuclear security systems are implemented at the facility and identify the measures used to protect identified targets from insider threats. The plan should include information about these measures. For example, technical measures might include containment and surveillance measures intended to detect and delay an insider adversary, measures to monitor and harden networks and associated devices, and measures to enforce access control. Administrative measures might include procedures, instructions, administrative sanctions, the two person rule, confidentiality rules and administrative checks, as well as planned, unplanned or unannounced inspections of the implementation of preventive and protective measures. Inspections should be performed by the operator or by independent teams. The security plan should specify how the measures will be evaluated (see Section 5).

4.9. Security systems at existing operating facilities may need to be upgraded to respond to evolving insider threats.

IMPLEMENTING MEASURES AGAINST INSIDER THREATS

4.10. Preventive and protective measures should both be used to protect against potential insider threats. Preventive measures can be used as follows:

(a) To reduce potential insider threats before allowing individuals access by identifying undesirable behaviours or characteristics that may indicate motivation;
(b) To further reduce potential insider threats after insiders have gained access by identifying undesirable behaviours or characteristics that may indicate motivation;
(c) To minimize opportunities for malicious acts by limiting access, authority and knowledge of insiders.

Protective measures can be used as follows:

(1) To detect, delay and respond to malicious acts;
(2) To mitigate or minimize the consequences of a nuclear security event and, if necessary, locate or recover the material.

Figure 1 illustrates how these steps may be used to address insider threats.

4.11. Many of the measures listed in the two sections that follow can be considered as both preventive and protective. As part of the selection and evaluation process,

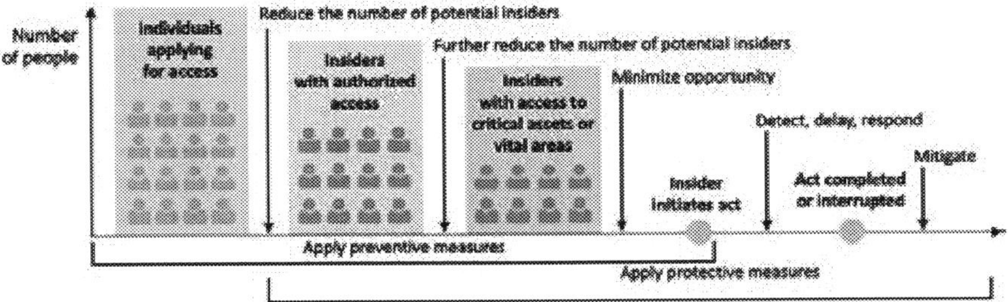

FIG. 1. Steps for using preventive and protective measures against potential insider threats.

the potential value of each proposed measure for both protection and prevention should be considered.

Implementing preventive measures

4.12. The goal of preventive measures is to reduce the number of potential insider threats and to minimize the opportunity for insiders to perform a malicious act. Preventive measures should be applied before employment, during employment and upon termination. In addition, preventive measures include quality assurance and specific computer security measures. Operators should apply the preventive measures described in this section.

Measures to be applied before employment

4.13. Individuals applying for work that requires access to a facility should be subject to identity verification, personal document verification and trustworthiness assessments.

4.14. Identity verification is used to confirm that the personal details of the individual in question are correct and genuine.

4.15. Personal document verification is used to authenticate the details of an applicant's work history, educational background and possession of the skill set required for the work to be performed. Verification and validation of documents and qualifications may be accomplished by contacting prior employers, educational institutions and references.

4.16. Trustworthiness assessments are used to provide an initial assessment (during the hiring process) and ongoing assessments (periodically throughout the employment period) of an individual's integrity, honesty and reliability. As recommended in Ref. [2]:

> "Taking into consideration State laws, regulations, or policies regarding personal privacy and job requirements, the State should determine the trustworthiness policy intended to identify the circumstances in which a trustworthiness determination is required and how it is made, using a *graded approach*."

4.17. The assessments should review the individual's observance of the law and adherence to facility rules, as well as any behaviours or motivational factors of concern. For example, the assessment should seek to identify motivational factors such as financial problems or pressures (e.g. debts, wage cuts), adherence to an ideology of concern, desire for revenge (e.g. a perceived injustice against the individual), physical dependency (e.g. drugs, alcohol, sex), psychological or psychiatric conditions, severe dissatisfaction with private or professional life, or other factors owing to which an individual could be coerced to commit a malicious act. These motivational factors may be identified by a review of information such as criminal records, personal and professional references, past work history, financial records, on-line and other social networks, medical records or job performance reports, as well as information from colleagues about observed behaviour.

4.18. National laws might restrict the scope or conduct of identity verification, personal document verification and trustworthiness assessments in a State.

Measures to be applied during employment

4.19. Insiders who have passed the pre-employment checks and have been granted authorized access, including access to critical assets, sensitive information and vital areas, should be subject to the measures described in the following paragraphs.

4.20. Escorting procedures should be developed and implemented. Persons whose trustworthiness has not been determined or whose duties do not require a trustworthiness assessment (e.g. temporary repair staff, administrative staff, maintenance staff, construction workers, visitors) should be escorted into vital areas or inner areas by persons who have authorized access and are not required to be themselves escorted. The escort should be knowledgeable about approved actions, including which areas and systems the escorted individual should be allowed access to and which activities he or she is authorized to perform.

4.21. Periodic reassessment of the trustworthiness of insiders should be conducted during employment. Certain behaviours and motivational factors of concern may not have previously been apparent or may develop over time. For example, random testing for drug or alcohol use during a work shift should be considered as a way to ensure a worker is reliable. The extent of the trustworthiness checks should be graded according to the level of access and authority the insider has to the facility and its assets. For example, insiders who perform network administration, who facilitate remote access to sensitive information assets and who work with nuclear material should be subject to more frequent and thorough trustworthiness checks than those who work in human resources.

4.22. Employees whose trustworthiness assessment has changed owing to personal circumstances might have their level of access temporarily demoted or they might be removed from management responsibilities until they are assessed again. Security awareness programmes and employee satisfaction and rewards, discussed below, may be used to maintain the trustworthiness of employees.

4.23. Sensitive information should be kept confidential so that only those who need to know the information are permitted to access it. Acquiring information on sensitive targets or regarding security procedures or measures (e.g. the location of the nuclear material inventory or transportation plans and schedules) might help insider adversaries successfully perform a malicious act. A record of persons accessing sensitive information, including the date and time at which the information was accessed, should be maintained and should also be protected against modification. Information addressing potential vulnerabilities in nuclear security systems should be highly protected and compartmentalized (as described in para. 4.30), since this information could facilitate a subsequent unauthorized removal or act of sabotage.

4.24. Access to nuclear facilities, nuclear material, nuclear facility systems and sensitive information should be controlled. A documented process for authorizing and revoking such access should be established and implemented. This process should apply to anyone who requires either remote or on-site access to a facility or its operations, including transportation. An individual's personal details could be verified through government issued identification documents and biometrics (e.g. retina, palm prints, finger prints, facial recognition). The process should apply strict need-to-know and need-to-access rules, as defined by the competent authority. Individuals should be permitted unescorted access only to the areas that they need to enter to complete assigned work. The number of persons with authorized access to designated areas should be kept to the minimum necessary.

4.25. The processing or movement of nuclear and other radioactive material should be authorized before processing or movement to minimize opportunities for the unauthorized removal of material and to detect unauthorized activities [6, 13]. For example, the facility operator should have a written procedure that specifies who can remove nuclear material from a storage vault for use in processing, when it can be removed and how the removal should be authorized and recorded. A daily or weekly schedule of activities coordinated and approved by operations staff may reduce opportunities for unauthorized activities by personnel who normally perform those activities.

4.26. Physical areas, duties, time and information should be compartmentalized so that one insider is unlikely to have sufficient access, authority or knowledge to complete a malicious act. Compartmentalization increases the effort that an insider would need to expend to complete a malicious act and increases the likelihood that an insider would need to exceed his or her normal authorized activities to complete a malicious act.

4.27. The facility operator should seek to ensure that physical areas are compartmentalized such that a single insider does not have access to all the systems, components and equipment that would enable him or her to complete a malicious act. The number of individuals with access to any area requiring protection should be limited. Rules should be defined to establish which personnel have a need to access compartmentalized areas; these rules should be applied to each compartmentalized area. These rules should be reviewed and changed when processes or configurations within the compartmentalized area are changed. Additionally, the number of persons permitted access to each of the compartmentalized areas should be strictly limited. Inspections and performance tests should be performed to ensure procedural adherence to the access rules.

4.28. Separation of duties compartmentalizes the work activities of insiders to limit an insider's ability to obtain sufficient authorized access, authority or knowledge to conduct a malicious act. Separation of duties includes applying the principle of least privilege to computer based systems, through which an insider is assigned only those privileges that are essential to his or her work.

4.29. Time should be compartmentalized by limiting authorized access during different periods of activity in a facility (e.g. working hours, maintenance, outages, non-routine conditions). For example, an insider's access to a critical area should be limited to his or her shifts.

4.30. Information should be compartmentalized by dividing information stored both in hard copy and electronically into separately controlled pieces and using administrative and technical measures to control access to the information. The purpose of compartmentalizing information is to prevent insiders from collecting all the information necessary to attempt a malicious act. Personnel need-to-know rules for sensitive information should be used when compartmentalizing information.

4.31. Standard operating procedures should be adhered to. Standard operating procedures are written instructions that govern recurring tasks according to approved specifications in order to produce a specified outcome. Standard operating procedures minimize variation and promote quality assurance through the consistent implementation of a process within an organization regardless of personnel changes. Standard operating procedures can assist in detecting, and thus preventing, an insider adversary's malicious act because they provide a baseline of predetermined activities from which deviations in procedure can be more readily detected and challenged by others.

4.32. A security awareness programme for staff and contractors should be developed and implemented. Such a programme contributes to the organization's nuclear security culture and can help prevent insider threats if security awareness of such threats is integrated into the facility's nuclear security culture. All personnel, regardless of job title or function, should be aware of the threats and potential consequences of malicious acts and their role in reducing the risk of a malicious act. Security awareness programmes may reduce the risk of blackmail, coercion, extortion or other threats to employees and their families, and should encourage the reporting of potential intimidation to the security management. Security awareness programmes should be developed in a coordinated manner with safety awareness programmes in order to establish effective and complementary safety and security cultures.

4.33. The security awareness programme should include clear security policies, the enforcement of security practices and continuous training. The purpose of training is to establish an environment in which all employees are aware of security policies and procedures so that they are able to aid in detecting and reporting suspicious or erroneous behaviour as well as unauthorized acts. Training should include methods to evaluate security awareness and training effectiveness as well as processes for continuous improvement or retraining. In addition to preparing personnel for the possibility of a physical incident at the facility or against its assets, the training should prepare personnel for the possibility of a cyber-attack.

4.34. A fitness for duty programme should be developed and implemented. Managers should be trained to identify concerns about an employee's behaviour and report them to the appropriate person. Fitness for duty programmes may be considered in order to monitor employees' health on a periodic basis. The facility operator may also consider offering assistance to employees who are in challenging situations (e.g. financial, medical, psychological).

4.35. Incidents of security concern (i.e. incidents at a facility that involve violations or irregularities associated with facility security policies, procedures or systems) should be reported and investigated. The reporting and investigation of incidents of security concern can help facilities develop corrective actions and prevent insider threats. An incident may be caused by an insider adversary as a precursor to a malicious act, either to prepare for the act or to test the response of a system. Thoroughly investigating these incidents might act as a deterrent to insiders and could identify personnel who might be insider adversaries.

4.36. Employees should be provided with good working conditions, rewards and recognition. Good working conditions, rewards and recognition are an important part of maintaining and increasing employee morale and loyalty, which contributes to an effective security culture.

4.37. Insiders should be made aware that deliberate violations of work instructions, regulations or laws will be sanctioned. The chance of disciplinary action or legal prosecution might deter insiders from committing malicious acts. In addition, requiring operators to inform the competent authority of attempted or completed malicious acts may provide, after proper evaluation, a basis for information sharing among operators as well as a source of needed modifications to regulatory requirements.

Measures to be applied upon termination

4.38. An individual's access and authority, including computer access, should be cancelled upon termination of the individual's position, employment or contract. Termination procedures should be established and should include revoking physical access to the facility; using a non-disclosure agreement to protect sensitive information; and changing encryption keys, passwords and access codes.

Quality assurance policy and programmes

4.39. The facility's quality assurance policy and programmes for nuclear security should address insider threats, as described in the threat assessment or DBT. As stated in para. 3.52 of Ref. [2]:

> "The quality assurance policy and programmes for physical protection should ensure that a *physical protection system* is designed, implemented, operated and maintained in a condition capable of effectively responding to the *threat assessment* or *design basis threat* and that it meets the State's regulations, including its prescriptive and/or performance based requirements."

4.40. The quality assurance programmes should include all facility systems that contribute to nuclear security to ensure adequate protection against insider threats. Quality assurance should require configuration management of the nuclear security systems to ensure that they continue to meet the desired performance criteria of these systems and to understand any potential consequences when changes are made to the systems, for example by an insider.

Measures for computer based systems

4.41. While certain measures, such as escorting, may be effective in limiting insider access to nuclear and radioactive material, they do not provide sufficient protection against potential insider threats to computer and network systems; such protection may be provided by information security measures [7, 8]. For example, third parties and vendors may have physical access on the site to sensitive information and assets during the development and maintenance of computer and network systems. While these third parties and vendors may wish to maintain remote access during all of the life cycle stages of the computer and network systems, such access should only be granted in accordance with the risk informed approach [1].

4.42. The facility operator should define and implement a policy addressing the acceptable use of computer based systems. This policy may define the approved use of computer based systems, outline employer expectations for monitoring approved use of these systems, provide for training and explicitly identify prohibited actions on computing systems. The facility operator should also consider the use of technical measures to enforce or enhance the systems policy. For example, the facility operator might define a social media policy and provide computer based training on the use of social media to reduce the likelihood of adversaries using employees as unwitting insiders.

Implementing protective measures

4.43. The purpose of protective measures against insider threats is to detect, delay and respond to a malicious act after it has been initiated and may include mitigation of consequences and recovery of nuclear or radioactive material. When designing and implementing protective measures, efforts should be made to ensure that these measures are supportive of and do not have an adverse effect on facility operations and safety. In case of conflict, particularly with safety, a solution should be reached in which the overall risk to the workers and the public is minimized and sufficient security is maintained.

4.44. Protective measures against insider threats should be applied using a graded approach for identified targets. In addition to protecting against unauthorized removal, as stated in para. 5.12 of Ref. [2], "The *operator* should design a *physical protection system* that is effective against the defined *sabotage* scenarios and complies with the required level of protection for the *nuclear facility* and *nuclear material*." Sabotage scenarios should include scenarios involving one or more insider adversaries. The following sections address protective measures against insider threats that should be considered during the design of a nuclear security system.

Detection measures

4.45. The detection of malicious acts attempted by external adversaries focuses on detecting the penetration of any one of a facility's protective measures. By contrast, insiders could bypass or defeat certain physical protection and NMAC measures owing to their authorized access, authority and knowledge. Operators should implement multiple and diverse protective measures for these systems to detect potential malicious acts performed by an insider and to provide the information needed for investigation and analysis. The facility operator should investigate all of the information provided by these detection measures in a comprehensive manner. Individual signals that seem insignificant might produce an indication of a malicious act when examined together.

4.46. An investigation might include reviewing recorded footage and network monitoring data, verifying tamper indicating devices or measurement data associated with nuclear materials, inspecting access logs or performing an emergency inventory. The personnel performing the investigation and analysis of the possible malicious act should be qualified. The time required to perform the investigation and analysis following detection directly affects the facility operator's ability to respond to a malicious act in a timely manner.

4.47. Suspicious or unauthorized activities should be detected and investigated because they might indicate that a malicious act is in the exploratory or preparatory phase. For example, an insider might attempt to bypass procedures (e.g. bringing prohibited items into an area), attempt to access an area that he or she is not authorized to access (e.g. entering through an emergency door), trigger an alarm to observe the timing and nature of the response, or attempt to obtain sensitive or otherwise need-to-know information to which the insider has not been granted access.

4.48. Protective measures to detect insider threats need to be designed to identify, correctly assess and report suspicious or malicious acts. Facility detection measures implemented against insider threats typically include measures related to access control, personnel tracking, detection of prohibited items, surveillance, NMAC systems and computer security. These types of measure are discussed in the following sections.

Access control

4.49. The operator should establish and document strict access control rules and procedures applicable to nuclear material, equipment used for processing or handling nuclear material, and data about nuclear material or systems relevant to safety or security. The robust implementation of access control rules and procedures minimizes insiders' access to material, systems and equipment. Access control rules and procedures may also act as a deterrent owing to the possibility of detection or identification if an insider attempts to access material, equipment or data for which he or she is not authorized.

4.50. Access control rules and procedures should be applicable to a variety of situations, including authorizing access to areas containing nuclear material and controlling nuclear material in routine and non-routine conditions, such as during actual or simulated emergency situations. For example, access control rules could apply to controlling and disseminating key and lock combinations in manual access control systems and to printing badges, enrolling personal identification numbers, gathering biometrics and controlling locks in electronic systems.

4.51. The operator should protect from unauthorized access (a) equipment that generates badges, (b) support equipment and associated spare parts and (c) systems used to grant access permissions. The facility operator should strictly control access to security equipment or equipment that contributes to security, calibration and maintenance. The operator should also establish procedures to ensure that this equipment remains intact. For example, to ensure that it has not been tampered with, equipment should be subject to testing by authorized personnel after maintenance has been performed and before the equipment is returned to service.

4.52. Access control rules should be defined for visitors and escorts and for abnormal conditions, such as response to emergencies and system outages.

4.53. Specific criteria, such as a personnel need-to-know and trustworthiness determination, should be verified before authorizing access to any area to which access is controlled. Establishment of rules for access control should be coordinated with NMAC, operations, safety and physical protection organizations.

4.54. Each access or attempted access to sensitive physical locations and computer systems should be recorded in access control records. Malicious acts committed by insider adversaries may be identified in the course of monitoring or inspecting these access control records. For example, inspections of access control records may identify events such as an unscheduled storage vault access, each failed personal identification number entry attempt, failed biometric authentification for an authorized badge or other indications of entry attempts by unauthorized individuals. Once identified, the irregularity or suspicious activity can be assessed as a potential malicious act. Detection measures and associated procedures used to monitor and inspect access control records should be considered as technical and administrative measures for access control during system design or upgrade.

4.55. Access control records should also be maintained of all persons who access vital areas or who have access to, or are in possession of, keys, key cards and other credentials relevant for accessing other systems, including computer systems that control access to inner areas, vital areas and other areas containing nuclear material [2].

4.56. If appropriately documented, access control records can be used during the investigation of a malicious act to determine a list of possible suspects. Requests for authorized access to security areas or systems relevant to safety or security, whether approved or denied, should also be reviewed and inspected to identify potential malicious activity.

Personnel tracking

4.57. Tracking the movement and location of personnel within a facility enables the operator to detect an attempted or actual violation of access control rules, such as multiple people exiting the facility using a single access control badge.

Existing technology makes it possible to track individuals either in real time or after the fact by recording the locations and areas they visit each day, along with the corresponding time and duration of each visit.

4.58. Awareness that a facility has a tracking system may deter insiders from carrying out unauthorized activities. Further, tracking records and access control records may be used during the investigation of a malicious act for assessment purposes or after an incident to generate an initial list of suspects.

Detection of prohibited items

4.59. As recommended in para. 4.43 of Ref. [2]:

> "Vehicles, persons and packages should be subject to search on entering both the *protected* and *inner areas* for *detection* and prevention of unauthorized access and of introduction of prohibited items. Vehicles, persons and packages leaving the *inner area* should be subject to search for *detection* and prevention of *unauthorized removal*."

4.60. The operator should identify and document prohibited items for limited areas, protected areas, inner areas and vital areas. Prohibited items may include unauthorized tools and material, such as computers, cell phones, tablets and other media or information technology devices with cameras; radiation shielding material; weapons; or explosives. These items could be used to gain access or cause damage to sensitive systems or equipment, or their components, or to enable the unauthorized removal or sabotage of nuclear material. Other prohibited items may be specifically identified by a facility to protect its physical protection, NMAC, safety and operational systems or to protect information against insider adversaries.

4.61. The operator should immediately investigate the detection of prohibited items entering or exiting an area as a potential malicious act performed by an insider. When preparing to perform a malicious act, an insider adversary might test the prohibited item detection system to ascertain the sensitivity of detectors or the strength of assessment procedures. Suspicious or repeated detections of prohibited items should be identified, assessed, reported and investigated.

4.62. Measures for the detection of prohibited items include manual searches of personnel, packages and vehicles (both periodic and random); use of metal detectors, X ray machines and radiation detectors; and use of dogs or other types of detector for chemicals and explosives. These measures should take into account the specifics of the facility and the threats against which protection is required according to the threat assessment or DBT, if applicable.

4.63. The operator should develop and implement policies identifying prohibited items and associated search and detection procedures. Personnel performing searches or using equipment to detect prohibited items should be trained to use the equipment and appropriately respond after identifying a prohibited item. Responses may include confirming an authorized exception, detaining the potential insider adversary or recording the event for the purpose of detecting potential malicious acts at a later date.

4.64. The stringency of searches and the determination of locations where they will be carried out should be commensurate with the sensitivity of the area where the search was triggered and the proximity of the area to the target. Searches should be carried out near the areas where the search was triggered. Periodic and random searches should be used to further deter the unauthorized removal or sabotage of nuclear and radioactive material. Searches should also be performed during emergency evacuation conditions, including exercises.

4.65. Monitoring procedures should be implemented during the detailed search of a transport vehicle before loading and shipment to ensure that those persons carrying out the search are not able to introduce prohibited items that would aid a malicious act.

4.66. Fixed or handheld radiation detectors should be used to detect the unauthorized removal of nuclear material on persons, in packages or in vehicles entering and leaving protected, inner and vital areas. Metal detectors should be placed in tandem with radiation detectors at pedestrian entrances and exits to enhance the effectiveness of the radiation detection, since shielding material might be used to block radioactive signatures from being detected if nuclear material is removed from the facility.

4.67. Procedures for approving exceptions to the introduction of prohibited or controlled items (e.g. radioactive calibration sources) into the facility should be specifically established [3].

Surveillance

4.68. Surveillance measures can be used to continuously monitor the activities of individuals inside the designated areas of the facility where a malicious act could occur so that unauthorized activities are identified, reported and assessed.

4.69. Surveillance includes visual observation, monitoring of live video footage or review of recorded footage gathered by automated surveillance systems. Surveillance can be useful not only as a detection measure but also for the deterrence and investigation of potential malicious acts performed by an insider.

4.70. Personnel performing surveillance activities should be capable of detecting authorized and unauthorized actions and should have the means to rapidly and safely report the observation of any unauthorized activity.

4.71. In the event of a reported unauthorized activity, recorded surveillance footage can be used to provide a correct assessment of a malicious act or identify possible suspects. Timely assessment of malicious acts may be difficult without surveillance information.

4.72. As recommended in para. 4.48 of Ref. [2], "whenever an *inner area* is occupied, *detection* of unauthorized action should be achieved by constant surveillance (e.g. the *two person rule*)." Surveillance measures should be considered for use during operations such as maintenance and particularly during packing, shipping and transfer operations [14]. Surveillance can be provided through co-workers, managers, automated surveillance systems or a combination thereof.

4.73. Periodic checks should be established and implemented by the operator to confirm that material control or other protective measures are applied according to the established procedures and that equipment is used correctly.

4.74. When the two person rule is the selected surveillance method in an area (e.g. in an area containing Category I material), the two authorized, knowledgeable persons should be physically located where they have an unobstructed view of each other and the nuclear material. Furthermore, each person should be trained and technically qualified to detect unauthorized activities or incorrect procedures. For visual surveillance to be effective, the persons observing need to be capable of recognizing unauthorized activities, correctly assessing the situation and reporting the activities to appropriate response personnel in time for them to prevent unauthorized removal. If the two person rule is applied in such surveillance, the two authorized individuals will both need to have appropriate training, have unobstructed views of the material and of each other, and be able to detect unauthorized or incorrect procedures [1].

4.75. In addition, the two person rule is only effective when the individuals do not become complacent, for example through long term friendship or association.

Whenever possible, managers should ensure that the members of each two person team are rotated. Enforcing the two person rule for access to designated areas may deter insider adversaries and assist with timely detection. In addition, the two person rule can help protect against insider adversaries tampering with physical protection systems. Attempts to defeat the two person rule should be reported and investigated.

Nuclear material accounting and control systems

4.76. The contribution of NMAC systems to nuclear security mainly derives from their ability to maintain precise knowledge of the types, quantities and locations of nuclear material at the facility; to conduct efficient physical inventory of the nuclear material; and, in some cases, to ensure that the activities performed in connection with the nuclear material have been properly authorized [9]. There are multiple measures through which an NMAC system can assist in detecting insider threats. These measures are described in more detail in Ref. [9].

4.77. NMAC and other detection measures should also be rigorously applied to prevent the unauthorized removal of additional nuclear material from a facility by, or with the assistance of, an insider adversary while an authorized shipment is in process. Other detection measures can include the use of (a) the two person rule during movement preparation, (b) material measurements, (c) tamper indicating devices, (d) document checks, (e) radiation monitors and (f) standard operating procedures.

Detection measures for computer based systems

4.78. Technical measures involving both hardware and software should be used to detect malicious acts. These measures may involve the following example activities:

(a) Establishing a baseline for and characterization of the network traffic of sensitive computer assets, and inspecting to the baseline.
(b) Implementing software intrusion detection tools to detect abnormal patterns of user behaviour.
(c) Monitoring, inspecting and assessing computer based systems to test for insider compliance with policies and procedures and to detect suspicious actions. For example, the operator might establish false targets and monitor them to detect attempts to gain unauthorized access to sensitive information, thus revealing a potential insider adversary while ensuring that no sensitive data are exposed.

(d) Restricting potential pathways that could be used to access data so that only authorized personnel are permitted to use those pathways, and ensuring that the pathways are controlled and monitored to protect against malicious use. This could include monitoring, physically blocking, prohibiting the use of removable media and mobile devices to limit access to sensitive systems by an insider, or using computer security zones to isolate nuclear security systems and their networks from other facility networks [7].

Delay measures

4.79. Multiple layers of different physical protection or procedural measures, including compartmentalization and separation of duties, can complicate the progress of an insider adversary by requiring a variety of tools and skills, thus providing additional time and opportunity for detection. By delaying the malicious act in this manner, an insider adversary could be detected and defeated. Delay may also deter insiders from attempting malicious acts.

4.80. Measures implemented close to equipment or nuclear material (e.g. tie-downs, restraints, locks) can be effective delay measures against insider adversaries when an area is under continuous surveillance or when other appropriate detection measures are in place. Such delay measures should be designed so that it is difficult for an insider adversary to use them to delay the response to a malicious act, particularly an act of sabotage.

4.81. Keeping nuclear material in a secure location can increase the delay for an insider adversary attempting to complete a malicious act. During production or usage, the minimum amount of nuclear material needed for production or usage should be removed from locked storage at one time, and measures should be taken to control the nuclear material between process steps. When material cannot be moved to a secure storage location during non-working hours, additional physical protection and surveillance measures should be implemented until the material is properly returned and stored in a normal secure location.

4.82. Certain types of delay measure may force insider adversaries to use more sophisticated tools, resources, logistics, training and skills to defeat the measure. Those sophisticated resources may not be available at the facility and may need to be introduced into the facility by the insider adversary or learned elsewhere.

4.83. System safety designs that provide for system self-protection (e.g. backup equipment, automatic equipment shutdown, automatic valve closure) may force the insider adversary to defeat multiple, redundant and dispersed equipment and systems. These features may delay a malicious act and prevent it from being successfully carried out. To the extent possible, access to information about the system safety designs should be restricted on a need-to-know basis to prevent it from being used to conduct a malicious act.

Delay measures in computer based systems

4.84. Physical security measures implemented to delay adversaries may not effectively protect computer based systems owing to the remote access to, and connectivity between, some computer based systems. For example, an insider with privileged access to sensitive computer based systems might be able to compromise physically separated assets remotely and simultaneously. Delays may also not be effective against an insider adversary who can use existing credentials to gain privileged access. Therefore, measures for computer based systems should emphasize prevention and, to a greater extent, detection and response.

4.85. The design and implementation of computer security zones and computer security levels at a facility can increase the complexity required to complete a malicious act using computer systems and provide security controls that may also increase the probability of detection [7].

Response measures

4.86. Both operations and security personnel may respond to an irregularity (e.g. an inventory difference, an opened door that should be locked). Typically, operations personnel respond to an irregularity to investigate its cause. If an irregularity is suspected to be due to a malicious act, security personnel should be notified and should respond if necessary. For example:

(a) Response to a passive insider adversary should depend on when detection occurs (when the information is obtained, when the information is passed on or when the investigation is completed).
(b) Response to an active, non-violent insider adversary should be by operations or security personnel depending on when detection occurs, since an active, non-violent insider adversary will stop a malicious act if confronted or challenged.
(c) Response to an active, violent insider adversary should be the same as for an external adversary.

4.87. Compared with an external adversary, an insider adversary is more difficult to identify and may not be easily identified as a threat anywhere within the facility. In addition, a malicious act committed by an insider adversary might consist of several acts separated by both time and space. Therefore, unless an insider adversary is identified when a suspicious or malicious act is detected, it may be difficult to identify him or her later among the other insiders.

4.88. To enable an effective response to be made, a protracted theft needs to be detected before the insider adversary accumulates a target quantity of material on or off the site. Scenarios should account for security systems and measures in place in the building and in any possible material balance areas, as well as for specific nuclear security procedures that could be used to detect unauthorized activities involving nuclear material early enough for an effective response to be made. For facilities where protracted theft might occur, scenarios should be analysed for the likelihood of detection if material were (a) taken off the site each time a quantity of the material was stolen or (b) accumulated in the facility or inside a process area to be taken off the site at one time in an abrupt theft.

4.89. An insider adversary might perform a set of acts ultimately intended to lead to unauthorized removal or sabotage in an unexpected order or with periods of inactivity between the individual acts. For example, an insider adversary might commit a single act and then wait to see if he or she is detected. This may complicate the security response necessary to identify and apprehend the insider adversary and increase the importance of investigation. Operations specialists may be needed to assist in the investigation by analysing the abnormal or irregular event to predict what further malicious acts might be attempted.

4.90. Insiders with access to a facility should be trained in detecting malicious acts and responding so that they protect themselves and transmit alarms according to a specified set of procedures. These procedures should be documented and used as part of the security awareness training provided to facility personnel by the operator. Response procedures should be based on the assumption that someone involved in response could be an adversary. For example, an insider adversary might report a fictitious emergency to distract others and prevent them from detecting a malicious act, or an insider adversary on the response team might use an emergency exercise or create an emergency to disguise a malicious act.

Response measures in computer based systems

4.91. For computer security incidents with the potential to adversely impact systems that contribute to nuclear security, response activities should be coordinated with nuclear security response personnel and documented. For example, the detection of unauthorized changes to access control by an insider should be responded to in a coordinated manner involving site security personnel and computer security personnel because such changes might facilitate unauthorized removal or sabotage. In the event of such a computer security incident, compensatory measures that involve site security and other appropriate facility organizations should also be considered.

COMPREHENSIVE ELEMENTS THAT REINFORCE PREVENTIVE AND PROTECTIVE MEASURES

Nuclear security culture

4.92. The foundation of nuclear security culture is the recognition that a credible threat exists and that nuclear security is important [11].

4.93. Nuclear security culture plays a key role in ensuring that individuals, organizations and institutions remain vigilant and that sustained measures are taken to counter insider threats. The effectiveness of preventive and protective measures against insider threats depends on the attitudes, behaviours and actions of individuals [17].

4.94. Management should promote a robust nuclear security culture to counter insider and external threats. The nuclear security culture creates the overall conditions for personnel to implement both preventive and protective measures. A facility's nuclear security culture should improve loyalty and adherence to security policies. For example, management should emphasize the employees' responsibility to report unusual activities or suspicious behaviour without fear of suffering disciplinary actions [11].

Contingency plans

4.95. As stated in para. 3.58 of Ref. [2]:

> "The State should establish a *contingency plan*. The State's *competent authority* should ensure that the *operator* prepares *contingency plans* to effectively counter the *threat assessment* or *design basis threat* taking actions of the *response forces* into consideration."

Paragraph 3.62 of Ref. [2] states that "The *operator* should initiate its *contingency plan* after *detection* and assessment of any *malicious act*." Paragraph 5.44 of Ref. [2] states that "The *contingency plan* should include measures which focus on preventing further damage, on securing the *nuclear facility* and on protecting emergency equipment and personnel."

4.96. The contingency plans developed by the State and the operator should address measures to respond to both insider and external threats. Protective measures against insider threats should be coordinated with contingency plans in accordance with agreed procedures. The contingency plan should require that personnel evacuating a building during a real or simulated emergency be controlled and examined for contamination and nuclear material to protect against insider threats.

4.97. Actions taken in response to suspected or confirmed malicious acts by an insider adversary may be different from the response to a malicious act by an external adversary.

System maintenance and recovery programme

4.98. A maintenance and recovery programme for all facility nuclear security systems that need to be protected may mitigate the consequences of a malicious act by an insider adversary. The maintenance programme should include the capability to rapidly repair operational and other vital systems, to rapidly replace parts that have been damaged and to implement compensatory measures as needed. Rapid repair and replacement limit the duration of the system outage and the time available for any subsequent malicious actions and may mitigate the consequences of the insider adversary's malicious act.

4.99. Operators should consider providing protection for spare parts (e.g. by installing barriers, storing the spare parts at a distance from the installed system and frequently monitoring the storage location) so that it would be difficult for an insider adversary to destroy or compromise both the installed parts and the spare parts for vital equipment.

4.100. Facility operating procedures and procedures for the recovery of security and operational systems should be validated and exercised to help ensure the rapid recovery of these systems, as well as to protect emergency equipment and personnel.

4.101. Procedures implemented for the protection of identified equipment should include the appropriate response to outages — such as implementing compensatory measures, investigating the cause of the outage and implementing a system for rapid repair (return to service) — to protect against the possibility of an unassessed and ongoing malicious act.

4.102. Secure backup and recovery processes should be implemented for sensitive computer based systems providing operation or security functions. System files used for recovery processes should be stored in a separate area with access control.

5. EVALUATION OF MEASURES

OBJECTIVES AND OVERVIEW OF THE EVALUATION PROCESS

5.1. Evaluating the effectiveness of preventive and protective measures against insider threats is a key component of a risk assessment that is intended to identify systems vulnerable to insider threats. The evaluation should use credible threat scenarios based on the threat assessment or DBT.

5.2. The results of the evaluation should be compared with previously established criteria for the effectiveness of preventive and protective measures. These criteria are usually established by the competent authority and are based on the potential consequences of a malicious act by an insider adversary and its likelihood of success. How the operator meets these criteria should be documented in the operator's comprehensive security plan, which includes plans for protecting both the NMAC and physical protection systems.

5.3. Evaluation of the effectiveness of the preventive and protective measures should be based on the operator's security plan. If the evaluation indicates that the preventive and protective measures defined in the security plan do not meet the criteria, upgrades should be implemented and the evaluation should be repeated until the criteria are met.

5.4. In the evaluation, consideration should be given by the operator to the relative ease of performing a malicious act and the level of risk associated with the potential malicious act. For example, a malicious act may have consequences that are deemed acceptable yet be relatively easy to perform (e.g. unauthorized alteration of the detection level of a radiation portal monitor); such an act may therefore be deemed unacceptable and require corrective action. Additionally, the risk may be deemed acceptable but may be close to the threshold beyond which the risk would no longer be acceptable. For example, an insider adversary might remove from a Category III process area small amounts of nuclear material that pose little risk, but if this unauthorized removal were repeated, the total quantity removed might reach a quantity that falls within a higher category. Such a case should not be disregarded, and prudent management practices would lead to additional protective measures.

5.5. The effectiveness of the preventive and protective measures should be re-evaluated periodically, particularly when there are changes in the threat assessment or DBT, in the preventive and protective measures or in the operating processes and conditions.

5.6. The criteria and performance requirements for an NMAC system are established in the overall context of nuclear security and can be useful in assessing the nuclear security system's effectiveness against insider threats. These criteria and performance requirements should address the different types of nuclear material and the time frames for the detection of unauthorized removal of nuclear material.

5.7. Different methods can be used to evaluate the effectiveness of the nuclear security system against insider threats (e.g. inspections and assessments, performance testing, measurement quality control, scenario analysis). Scenario analysis is an effective method of evaluation against insider threats. Performance testing supports the scenario analysis process by providing information such as the probability of detection and subsequent response. Plans for performance testing should be developed and implemented to test employee, facility and competent authority readiness for response to a potential malicious act by an insider adversary.

EVALUATION OF PREVENTIVE MEASURES

5.8. The implementation of preventive measures should be evaluated to ensure that they are implemented as designed. Although difficult to evaluate quantitatively, preventive measures can be effective in reducing the possibility of insider threats. Preventive measures should be evaluated by conducting performance testing on procedures to determine whether the procedures are adequate to address the threat and whether employees follow the procedures.

5.9. The opportunity for an insider adversary to perform a malicious act can be minimized by reducing the possibility for an insider to gain the access, authority or knowledge necessary to successfully carry out a malicious act. Credible scenarios for evaluation will incorporate the degree to which and the manner in which opportunity is minimized. A review should be performed to identify what preventive measures are in place and whether they are properly applied.

EVALUATION OF PROTECTIVE MEASURES

5.10. The effectiveness of the measures used to detect, delay and respond to malicious acts (protective measures) can be quantitatively or qualitatively analysed. The likelihood of detection and the timeliness of response are often quantifiable and can provide a basis for an evaluation of the effectiveness of the protective measures.

5.11. One way to evaluate the effectiveness of the protective measures against insider threats is to develop credible scenarios, including scenarios of collusion with other insider adversaries or with external adversaries, as appropriate. The effectiveness of the protective measures in countering these scenarios can then be evaluated.

5.12. The development of scenarios involves identifying the combination of actions necessary for an insider adversary to accomplish a malicious act. Operators should consider pairing identified targets (see Section 3) with a defined insider adversary (see Section 2) when developing scenarios. The set of actions that an insider adversary would need to take to achieve his or her goal should be defined, taking into account the threat assessment or DBT. These sets of actions should include the actions that would be performed and the locations where they would be performed, and all of the protective measures that could be encountered by insider adversaries while performing those actions should be identified. Because insider adversaries can perform the actions required for a malicious act over an extended period, and because the acts may not follow a predictable sequence, the concept of a path or timeline may or may not be relevant to the analysis.

5.13. For sabotage scenarios, the actions that need to be taken to initiate a sequence of events that would result in unacceptable radiological consequences should be identified. Sabotage scenarios should include attacks on both single and multiple targets.

5.14. For scenarios involving the unauthorized removal of nuclear material, the actions that need to be successfully taken to remove nuclear material from the facility should be identified. Scenarios involving unauthorized removal of nuclear material should consider both protracted and abrupt theft and should include situations in which the adversary leaves the facility directly with the nuclear material or hides material at the facility in order to remove it from the facility later under more favourable circumstances. Scenarios should consider attacks on, or the compromise of, computer based systems, combinations of physical attacks and cyber-attacks, and attacks by violent and non-violent insider adversaries.

5.15. Strategies that may be used by insider adversaries to defeat protective measures should also be considered as part of the scenario development process. The operator can develop such strategies by considering how access, authority and knowledge could enable an insider adversary to thwart the detection and delay measures. Possible efforts by insider adversaries to reduce the effectiveness of the response should also be considered. Emergency conditions that result in a facility evacuation may create opportunities for an insider adversary to complete a malicious act and should be considered during scenario development.

5.16. Once detailed scenarios involving insider threats have been developed, the effectiveness of the protective measures can be evaluated by considering the accumulated impact of detection and delay, as well as the response to and mitigation of the consequences of the scenario. For an active, non-violent insider adversary, the effectiveness of the response will depend on the probability of interrupting or neutralizing[5] a malicious act.

5.17. The evaluation process should be repeated for credible scenarios that require further analysis. Conclusions about the effectiveness of protective measures should be based on the results of all the evaluations conducted.

EVALUATION OF MEASURES AGAINST COLLUSION BETWEEN INSIDERS

5.18. The development of sufficient scenarios addressing collusion between two or more insider adversaries is challenging owing to the many combinations of insiders with different access, authority and knowledge that need to be considered. Evaluation of the effectiveness of the measures that help prevent collusion (e.g. compartmentalization, surveillance, preventive measures) may provide a good approach.

[5] 'Interruption' means the response occurs in time to prevent the completion of a malicious act. For an active, violent insider adversary, 'neutralization' means that the response force stops or prevents the attack permanently. For an active, non-violent insider adversary, neutralization occurs when the insider adversary is identified.

EVALUATION OF MEASURES AGAINST PROTRACTED THEFT

5.19. The evaluation of measures against protracted theft may be approached in the same manner as the evaluation of measures against abrupt theft. However, the evaluation of measures against protracted theft should also take into account additional challenges encountered by the insider adversary when attempting the unauthorized removal of small amounts of material over an extended period of time. These complexities include periodic inventory taking, the potential for inventory differences to be detected, record tracking, concealment of the amounts of material accumulated and defeat of radiation portal monitors. The evaluation method should also consider the increased probability of detection when the same action is repeated multiple times.

EVALUATION OF MEASURES AGAINST SABOTAGE

5.20. The evaluation of measures against sabotage by an insider adversary may use the same process as the evaluation of measures against abrupt and protracted theft and may reference the logic model approach (fault tree or event tree) provided in Ref. [16].

5.21. Sabotage scenarios to be evaluated should include scenarios for both direct sabotage of nuclear material and indirect sabotage (i.e. sabotage of facility systems) that could result in unacceptable radiological consequences. The evaluation of sabotage scenarios should consider scenarios by individuals who do not have direct access to material or equipment.

5.22. To perform an act of sabotage, the insider adversary would not necessarily need to leave the facility to complete the malicious act. Therefore, the evaluation of preventive and protective measures against any insider exiting the facility would be applicable.

EVALUATION OF A FACILITY FOR PROTECTION AGAINST INSIDER THREATS

5.23. The process of evaluating a facility for protection against insider threats begins with characterizing insiders according to attributes, motivations and categories to identify potential insider threats. The next step is target identification, which involves an evaluation of the assets that need to be protected from unauthorized removal or sabotage. The result of this evaluation is a prioritized list of targets.

5.24. Preventive measures should be implemented using the concept of defence in depth and a graded approach to minimize opportunities for the identified threats and targets to be subject to malicious acts.

5.25. Protective measures should be identified to protect targets in protected, inner or vital areas in a prioritized manner. The measures to detect, delay and respond to the insider threat should be increased in depth by using the results of the evaluation.

5.26. Preventive and protective measures against sabotage and unauthorized removal of nuclear material should be evaluated using a method such as the development of credible scenarios. Scenarios should be consistent with the threat assessment or DBT and may include physical attacks, cyber-attacks or a combination of both at the facility, along transport routes and within supply chains.

5.27. The system should be re-evaluated periodically to ensure that the measures are effectively implemented and sustained. The timing of the re-evaluation might be cyclic, or it might be triggered by changes to the threat or to the facility and its operation.

REFERENCES

[1] INTERNATIONAL ATOMIC ENERGY AGENCY, Objective and Essential Elements of a State's Nuclear Security Regime, IAEA Nuclear Security Series No. 20, IAEA, Vienna (2013).

[2] INTERNATIONAL ATOMIC ENERGY AGENCY, Nuclear Security Recommendations on Physical Protection of Nuclear Material and Nuclear Facilities (INFCIRC/225/Revision 5), IAEA Nuclear Security Series No. 13, IAEA, Vienna (2011).

[3] INTERNATIONAL ATOMIC ENERGY AGENCY, Nuclear Security Recommendations on Radioactive Material and Associated Facilities, IAEA Nuclear Security Series No. 14, IAEA, Vienna (2011).

[4] EUROPEAN POLICE OFFICE, INTERNATIONAL ATOMIC ENERGY AGENCY, INTERNATIONAL CIVIL AVIATION ORGANIZATION, INTERNATIONAL CRIMINAL POLICE ORGANIZATION–INTERPOL, UNITED NATIONS INTERREGIONAL CRIME AND JUSTICE RESEARCH INSTITUTE, UNITED NATIONS OFFICE ON DRUGS AND CRIME, WORLD CUSTOMS ORGANIZATION, Nuclear Security Recommendations on Nuclear and Other Radioactive Material out of Regulatory Control, IAEA Nuclear Security Series No. 15, IAEA, Vienna (2011).

[5] INTERNATIONAL ATOMIC ENERGY AGENCY, Security of Radioactive Sources, IAEA Nuclear Security Series No. 11, IAEA, Vienna (2009).

[6] INTERNATIONAL ATOMIC ENERGY AGENCY, Security in the Transport of Radioactive Material, IAEA Nuclear Security Series No. 9, IAEA, Vienna (2008).

[7] INTERNATIONAL ATOMIC ENERGY AGENCY, Computer Security at Nuclear Facilities, IAEA Nuclear Security Series No. 17, IAEA, Vienna (2011).

[8] INTERNATIONAL ATOMIC ENERGY AGENCY, Security of Nuclear Information, IAEA Nuclear Security Series No. 23-G, IAEA, Vienna (2015).

[9] INTERNATIONAL ATOMIC ENERGY AGENCY, Use of Nuclear Material Accounting and Control for Nuclear Security Purposes at Facilities, IAEA Nuclear Security Series No. 25-G, IAEA, Vienna (2015).

[10] INTERNATIONAL ATOMIC ENERGY AGENCY, Engineering Safety Aspects of the Protection of Nuclear Power Plants against Sabotage, IAEA Nuclear Security Series No. 4, IAEA, Vienna (2007).

[11] INTERNATIONAL ATOMIC ENERGY AGENCY, Nuclear Security Culture, IAEA Nuclear Security Series No. 7, IAEA, Vienna (2008).

[12] INTERNATIONAL ATOMIC ENERGY AGENCY, Development, Use and Maintenance of the Design Basis Threat, IAEA Nuclear Security Series No. 10, IAEA, Vienna (2009).

[13] INTERNATIONAL ATOMIC ENERGY AGENCY, Security of Nuclear Material in Transport, IAEA Nuclear Security Series No. 26-G, IAEA, Vienna (2015).

[14] INTERNATIONAL ATOMIC ENERGY AGENCY, Establishing a System for Control of Nuclear Material for Nuclear Security Purposes at a Facility during Use, Storage and Movement, IAEA Nuclear Security Series No. 32-T, IAEA, Vienna (2019).

[15] INTERNATIONAL ATOMIC ENERGY AGENCY, Physical Protection of Nuclear Material and Nuclear Facilities (Implementation of INFCIRC/225/Revision 5), IAEA Nuclear Security Series No. 27-G, IAEA, Vienna (2018).

[16] INTERNATIONAL ATOMIC ENERGY AGENCY, Identification of Vital Areas at Nuclear Facilities, IAEA Nuclear Security Series No. 16, IAEA, Vienna (2012).

[17] INTERNATIONAL ATOMIC ENERGY AGENCY, Self-assessment of Nuclear Security Culture in Facilities and Activities, IAEA Nuclear Security Series No. 28-T, IAEA, Vienna (2007).

No. 26

ORDERING LOCALLY

IAEA priced publications may be purchased from the sources listed below or from major local booksellers.

Orders for unpriced publications should be made directly to the IAEA. The contact details are given at the end of this list.

NORTH AMERICA

Bernan / Rowman & Littlefield
15250 NBN Way, Blue Ridge Summit, PA 17214, USA
Telephone: +1 800 462 6420 • Fax: +1 800 338 4550
Email: orders@rowman.com • Web site: www.rowman.com/bernan

Renouf Publishing Co. Ltd
22-1010 Polytek Street, Ottawa, ON K1J 9J1, CANADA
Telephone: +1 613 745 2665 • Fax: +1 613 745 7660
Email: orders@renoufbooks.com • Web site: www.renoufbooks.com

REST OF WORLD

Please contact your preferred local supplier, or our lead distributor:

Eurospan Group
Gray's Inn House
127 Clerkenwell Road
London EC1R 5DB
United Kingdom

Trade orders and enquiries:
Telephone: +44 (0)176 760 4972 • Fax: +44 (0)176 760 1640
Email: eurospan@turpin-distribution.com

Individual orders:
www.eurospanbookstore.com/iaea

For further information:
Telephone: +44 (0)207 240 0856 • Fax: +44 (0)207 379 0609
Email: info@eurospangroup.com • Web site: www.eurospangroup.com

Orders for both priced and unpriced publications may be addressed directly to:

Marketing and Sales Unit
International Atomic Energy Agency
Vienna International Centre, PO Box 100, 1400 Vienna, Austria
Telephone: +43 1 2600 22529 or 22530 • Fax: +43 1 26007 22529
Email: sales.publications@iaea.org • Web site: www.iaea.org/publications

19-00431-T

RELATED PUBLICATIONS

ENGINEERING SAFETY ASPECTS OF THE PROTECTION OF NUCLEAR POWER PLANTS AGAINST SABOTAGE
IAEA Nuclear Security Series No. 4
STI/PUB/1271 (58 pp.; 2007)
ISBN 92-0-109906-1 Price: €30.00

NUCLEAR SECURITY CULTURE
IAEA Nuclear Security Series No. 7
STI/PUB/1347 (37 pp.; 2008)
ISBN 978-92-0-107808-7 Price: €30.00

DEVELOPMENT, USE AND MAINTENANCE OF THE DESIGN BASIS THREAT
IAEA Nuclear Security Series No. 10
STI/PUB/1386 (30 pp.; 2009)
ISBN 978-92-0-102509-8 Price: €18.00

SECURITY OF RADIOACTIVE SOURCES
IAEA Nuclear Security Series No. 11
STI/PUB/1387 (66 pp.; 2009)
ISBN 978-92-0-102609-5 Price: €25.00

NUCLEAR SECURITY RECOMMENDATIONS ON PHYSICAL PROTECTION OF NUCLEAR MATERIAL AND NUCLEAR FACILITIES (INFCIRC/225/Revision 5)
IAEA Nuclear Security Series No. 13
STI/PUB/1481 (57 pp.; 2011)
ISBN 978-92-0-111110-4 Price: €28.00

USE OF NUCLEAR MATERIAL ACCOUNTING AND CONTROL FOR NUCLEAR SECURITY PURPOSES AT FACILITIES
IAEA Nuclear Security Series No. 25-G
STI/PUB/1685 (63 pp.; 2015)
ISBN 978-92-0-137810-1 Price: €20.00

PHYSICAL PROTECTION OF NUCLEAR MATERIAL AND NUCLEAR FACILITIES (IMPLEMENTATION OF INFCIRC/225/Revision 5)
IAEA Nuclear Security Series No. 27-G
STI/PUB/1760 (120 pp.; 2018)
ISBN 978-92-0-101915-8 Price: €30.00

ESTABLISHING A SYSTEM FOR CONTROL OF NUCLEAR MATERIAL FOR NUCLEAR SECURITY PURPOSES AT A FACILITY DURING USE, STORAGE AND MOVEMENT
IAEA Nuclear Security Series No. 32-T
STI/PUB/1786 (47 pp.; 2019)
ISBN 978-92-0-103017-7 Price: €38.00

COMPUTER SECURITY AT NUCLEAR FACILITIES
IAEA Nuclear Security Series No. 17
STI/PUB/1527 (69 pp.; 2011)
ISBN 978-92-0-120110-2 Price: €33.00

SECURITY OF NUCLEAR INFORMATION
IAEA Nuclear Security Series No. 23-G
STI/PUB/1677 (54 pp.; 2015)
ISBN 978-92-0-110614-8 Price: €30.00

COMPUTER SECURITY OF INSTRUMENTATION AND CONTROL SYSTEMS AT NUCLEAR FACILITIES
IAEA Nuclear Security Series No. 33-T
STI/PUB/1787 (58 pp.; 2018)
ISBN 978-92-0-103117-4 Price: €42.00

www.iaea.org/publications

This publication is an update of IAEA Nuclear Security Series No. 8, originally published in 2008. The revision was undertaken to better align this Implementing Guide with the Nuclear Security Fundamentals and with the Recommendations that were published after 2008, to cross-reference other relevant Implementing Guides published since 2008, and to add further detail on certain topics based on the experience of the IAEA and Member States in using IAEA Nuclear Security Series No. 8. This publication provides updated guidance to States, their competent authorities, and operators, shippers and carriers on selecting, implementing and evaluating measures for addressing insider threats. It applies to any type of nuclear facility, notably nuclear power plants, research reactors and other nuclear fuel cycle facilities (e.g. enrichment plants, reprocessing plants, fuel fabrication plants and storage facilities), whether in design, construction, commissioning, operation, shutdown or decommissioning.

INTERNATIONAL ATOMIC ENERGY AGENCY
VIENNA
ISBN 978–92–0–103419–9
ISSN 1816–9317

APPENDIX

5

Treaty on the Non-Proliferation of Nuclear Weapons-NPT

DOI: https://doi.org/10.1016/B978-0-12-821319-3.00082-8

INF

INFCIRC/140
22 April 1970

GENERAL Distr.
ENGLISH

International Atomic Energy Agency
INFORMATION CIRCULAR

TREATY ON THE NON-PROLIFERATION OF NUCLEAR WEAPONS

Notification of the entry into force

1. By letters addressed to the Director General on 5, 6 and 20 March 1970 respectively, the Governments of the United Kingdom of Great Britain and Northern Ireland, the United States of America and the Union of Soviet Socialist Republics, which are designated as the Depositary Governments in Article IX. 2 of the Treaty on the Non-Proliferation of Nuclear Weapons, informed the Agency that the Treaty had entered into force on 5 March 1970.

2. The text of the Treaty, taken from a certified true copy provided by one of the Depositary Governments, is reproduced below for the convenience of all Members.

TREATY
ON THE NON-PROLIFERATION OF NUCLEAR WEAPONS

The States concluding this Treaty, hereinafter referred to as the "Parties to the Treaty",

Considering the devastation that would be visited upon all mankind by a nuclear war and the consequent need to make every effort to avert the danger of such a war and to take measures to safeguard the security of peoples,

Believing that the proliferation of nuclear weapons would seriously enhance the danger of nuclear war,

In conformity with resolutions of the United Nations General Assembly calling for the conclusion of an agreement on the prevention of wider dis semination of nuclear weapons,

Undertaking to co-operate in facilitating the application of International Atomic Energy Agency safeguards on peaceful nuclear activities,

Expressing their support for research, development and other efforts to further the application, within the framework of the International Atomic Energy Agency safeguards system, of the principle of safeguarding effectively the flow of source and special fissionable materials by use of instruments and other techniques at certain strategic points,

Affirming the principle that the benefits of peaceful applications of nuclear technology, including any technological by-products which may be derived by nuclear-weapon States from the development of nuclear explosive devices, should be available for peaceful purposes to all Parties to the Treaty, whether nuclear-weapon or non-nuclear-weapon States,

Convinced that, in furtherance of this principle, all Parties to the Treaty are entitled to participate in the fullest possible exchange of scientific information for, and to contribute alone or in co-operation with other States to, the further development of the applications of atomic energy for peaceful purposes,

Declaring their intention to achieve at the earliest possible date the cessation of the nuclear arms race and to undertake effective measures in the direction of nuclear disarmament,

Urging the co-operation of all States in the attainment of this objective,

INFCIRC/140

Recalling the determination expressed by the Parties to the 1963 Treaty banning nuclear weapon tests in the atmosphere, in outer space and under water in its Preamble to seek to achieve the discontinuance of all test explosions of nuclear weapons for all time and to continue negotiations to this end,

Desiring to further the easing of international tension and the strengthening of trust between States in order to facilitate the cessation of the manufacture of nuclear weapons, the liquidation of all their existing stockpiles, and the elimination from national arsenals of nuclear weapons and the means of their delivery pursuant to a Treaty on general and complete disarmament under strict and effective international control,

Recalling that, in accordance with the Charter of the United Nations, States must refrain in their international relations from the threat or use of force against the territorial integrity or political independence of any State, or in any other manner inconsistent with the Purposes of the United Nations, and that the establishment and maintenance of international peace and security are to be promoted with the least diversion for armaments of the world's human and economic resources,

Have agreed as follows:

ARTICLE I

Each nuclear-weapon State Party to the Treaty undertakes not to transfer to any recipient whatsoever nuclear weapons or other nuclear explosive devices or control over such weapons or explosive devices directly, or indirectly; and not in any way to assist, encourage, or induce any non-nuclear-weapon State to manufacture or otherwise acquire nuclear weapons or other nuclear explosive devices, or control over such weapons or explosive devices.

ARTICLE II

Each non-nuclear-weapon State Party to the Treaty undertakes not to receive the transfer from any transferor whatsoever of nuclear weapons or other nuclear explosive devices or of control over such weapons or explosive devices directly, or indirectly; not to manufacture or otherwise acquire nuclear weapons or other nuclear explosive devices; and not to seek or receive any assistance in the manufacture of nuclear weapons or other nuclear explosive devices.

ARTICLE III

1. Each Non-nuclear-weapon State Party to the Treaty undertakes to accept safeguards, as set forth in an agreement to be negotiated and concluded with the International Atomic Energy Agency in accordance with the Statute of the International Atomic Energy Agency and the Agency's safeguards system, for the exclusive purpose of verification of the fulfilment of its obligations assumed under this Treaty with a view to preventing diversion of nuclear energy from peaceful uses to nuclear weapons or other nuclear explosive devices. Procedures for the safeguards required by this Article shall be followed with respect to source or special fissionable material whether it is being produced, processed or used in any principal nuclear facility or is outside any such facility. The safeguards required by this Article shall be applied on all source or special fissionable material in all peaceful nuclear activities within the territory of such State, under its jurisdiction, or carried out under its control anywhere.

INFCIRC/140

2. Each State Party to the Treaty undertakes not to provide: *(a)* source or special fissionable material, or *(b)* equipment or material especially designed or prepared for the processing, use or production of special fissionable material, to any non-nuclear-weapon State for peaceful purposes, unless the source or special fissionable material shall be subject to the safeguards required by this Article.

3. The safeguards required by this Article shall be implemented in a manner designed to comply with Article IV of this Treaty, and to avoid hampering the economic or technological development of the Parties or international co-operation in the field of peaceful nuclear activities, including the international exchange of nuclear material and equipment for the processing, use or production of nuclear material for peaceful purposes in accordance with the provisions of this Article and the principle of safeguarding set forth in the Preamble of the Treaty.

4. Non-nuclear-weapon States Party to the Treaty shall conclude agreements with the International Atomic Energy Agency to meet the requirements of this Article either individually or together with other States in accordance with the Statute of the International Atomic Energy Agency. Negotiation of such agreements shall commence within 180 days from the original entry into force of this Treaty. For States depositing their instruments of ratification or accession after the 180-day period, negotiation of such agreements shall commence not later than the date of such deposit. Such agreements shall enter into force not later than eighteen months after the date of initiation of negotiations.

ARTICLE IV

1. Nothing in this Treaty shall be interpreted as affecting the inalienable right of all the Parties to the Treaty to develop research, production and use of nuclear energy for peaceful purposes without discrimination and in conformity with Articles I and II of this Treaty.

2. All the Parties to the Treaty undertake to facilitate, and have the right to participate in. the fullest possible exchange of equipment, materials and scientific and technological information for the peaceful uses of nuclear energy. Parties to the Treaty in a position to do so shall also co-operate in contributing alone or together with other States or international organizations to the further development of the applications of nuclear energy for peaceful purposes, especially in the territories of non-nuclear-weapon States Party to the Treaty, with due consideration for the needs of the developing areas of the world.

ARTICLE V

Each Party to the Treaty undertakes to take appropriate measures to ensure that, in accordance with this Treaty, under appropriate international observation and through appropriate international procedures, potential benefits from any peaceful applications of nuclear explosions will be made available to non-nuclear-weapon States Party to the Treaty on a non-discriminatory basis and that the charge to such Parties for the explosive devices used will be as low as possible and exclude any charge for research and development. Non-nuclear-weapon States Party to the Treaty shall be able to obtain such benefits, pursuant to a special international agreement or agreements, through an appropriate international body with adequate representation of non-nuclear-weapon States. Negotiations on this subject shall commence as soon as possible after the Treaty enters into force. Non-nuclear-weapon States Party to the Treaty so desiring may also obtain such benefits pursuant to bilateral agreements.

INFCIRC/140

ARTICLE VI

Each of the Parties to the Treaty undertakes to pursue negotiations in good faith on effective measures relating to cessation of the nuclear arms race at an early date and to nuclear disarmament, and on a treaty on general and complete disarmament under strict and effective international control.

ARTICLE VII

Nothing in this Treaty affects the right of any group of States to conclude regional treaties in order to assure the total absence of nuclear weapons in their respective territories.

ARTICLE VIII

1. Any Party to the Treaty may propose amendments to this Treaty. The text of any proposed amendment shall be submitted to the Depositary Governments which shall circulate it to all Parties to the Treaty. Thereupon, if requested to do so by one-third or more of the Parties to the Treaty, the Depositary Governments shall convene a conference, to which they shall invite all the Parties to the Treaty, to consider such an amendment.

2. Any amendment to this Treaty must be approved by a majority of the votes of all the Parties to the Treaty, including the votes of all nuclear-weapon States Party to the Treaty and all other Parties which, on the date the amendment is circulated, are members of the Board of Governors of the International Atomic Energy Agency. The amendment shall enter into force for each Party that deposits its instrument of ratification of the amendment upon the deposit of such instruments of ratification by a majority of all the Parties, including the instruments of ratification of all nuclear-weapon States Party to the Treaty and all other Parties which, on the date the amendment is circulated, are members of the Board of Governors of the International Atomic Energy Agency. Thereafter, it shall enter into force for any other Party upon the deposit of its instrument of ratification of the amendment.

3. Five years after the entry into force of this Treaty, a conference of Parties to the Treaty shall be held in Geneva, Switzerland, in order to review the operation of this Treaty with a view to assuring that the purposes of the Preamble and the provisions of the Treaty are being realised. At intervals of five years thereafter. a majority of the Parties to the Treaty may obtain, by submitting a proposal to this effect to the Depositary Governments, the convening of further conferences with the same objective of reviewing the operation of the Treaty.

ARTICLE IX

1. This Treaty shall be open to all States for signature. Any State which does not sign the Treaty before its entry into force in accordance with paragraph 3 of this Article may accede to it at any time.

2. This Treaty shall be subject to ratification by signatory States. Instruments of ratification and instruments of accession shall be deposited with the Governments of the United Kingdom of Great Britain and Northern Ireland, the Union of Soviet Socialist Republics and the United States of America, which are hereby designated the Depositary Governments.

3. This Treaty shall enter into force after its ratification by the States, the Governments of which are designated Depositaries of the Treaty, and forty other States signatory to this Treaty and the deposit of their instruments of ratification. For the purposes of this Treaty, a nuclear-weapon State is one which has manufactured and exploded a nuclear weapon or other nuclear explosive device prior to 1 January, 1967.

INFCIRC/140

4. For States whose instruments of ratification or accession are deposited subsequent to the entry into force of this Treaty, it shall enter into force on the date of the deposit of their instruments of ratification or accession.

5. The Depositary Governments shall promptly inform all signatory and acceding States of the date of each signature, the date of deposit of each instrument of ratification or of accession, the date of the entry into force of this Treaty, and the date of receipt of any requests for convening a conference or other notices.

6. This Treaty shall be registered by the Depositary Governments pursuant to Article 102 of the Charter of the United Nations.

ARTICLE X

1. Each Party shall in exercising its national sovereignty have the right to withdraw from the Treaty if it decides that extraordinary events, related to the subject matter of this Treaty, have jeopardized the supreme interests of its country. It shall give notice of such withdrawal to all other Parties to the Treaty and to the United Nations Security Council three months in advance. Such notice shall include a statement of the extraordinary events it regards as having jeopardized its supreme interests.

2. Twenty-five years after the entry into force of the Treaty, a conference shall be convened to decide whether the Treaty shall continue in force indefinitely, or shall be extended for an additional fixed period or periods. This decision shall be taken by a majority of the Parties to the Treaty.

ARTICLE XI

This Treaty, the English, Russian, French, Spanish and Chinese texts of which are equally authentic, shall be deposited in the archives of the Depositary Governments. Duly certified copies of this Treaty shall be transmitted by the Depositary Governments to the Governments of the signatory and acceding States.

IN WITNESS WHEREOF the undersigned, duly authorised, have signed this Treaty.

DONE in triplicate, at the cities of London, Moscow and Washington, the first day of July, one thousand nine hundred and sixty-eight.

APPENDIX

6

Convention on the Physical Protection of Nuclear Material – CPPNM

Applications of Nuclear and Radioisotope Technology: The Atom for Peace and Sustainable Development
DOI: https://doi.org/10.1016/B978-0-12-821319-3.00092-0

International Atomic Energy Agency

INFORMATION CIRCULAR

IAEA - INFCIRC/274/Rev. 1
May 1980

GENERAL Distr.

Original: ENGLISH, FRENCH, RUSSIAN and SPANISH

THE CONVENTION ON THE PHYSICAL PROTECTION OF NUCLEAR MATERIAL

1. The Convention on the Physical Protection of Nuclear Material was opened for signature on 3 March 1980, pursuant to Article 18 1 thereof and following the conclusion of negotiations on 28 October 1979

2. The texts of the Convention[1] and of the Final Act of the Meeting of Governmental Representatives to Consider the Drafting of the Convention are reproduced in this document for the information of all Member States

3. Member States will be informed by an addendum to this document of the entry into force of the Convention pursuant to Article 19 1 thereof

---

[1] The text of the Convention was transmitted to the twenty-third (1979) regular session of the General Conference of the International Atomic Energy Agency, pursuant to paragraph 11 of the Final Act, as document INFCIRC/274

CONVENTION ON THE PHYSICAL PROTECTION OF NUCLEAR MATERIAL

THE STATES PARTIES TO THIS CONVENTION,

RECOGNIZING the right of all States to develop and apply nuclear energy for peaceful purposes and their legitimate interests in the potential benefits to be derived from the peaceful application of nuclear energy,

CONVINCED of the need for facilitating international co-operation in the peaceful application of nuclear energy,

DESIRING to avert the potential dangers posed by the unlawful taking and use of nuclear material,

CONVINCED that offences relating to nuclear material are a matter of grave concern and that there is an urgent need to adopt appropriate and effective measures to ensure the prevention, detection and punishment of such offences,

AWARE OF THE NEED FOR international co-operation to establish, in conformity with the national law of each State Party and with this Convention, effective measures for the physical protection of nuclear material,

CONVINCED that this Convention should facilitate the safe transfer of nuclear material,

STRESSING also the importance of the physical protection of nuclear material in domestic use, storage and transport,

RECOGNIZING the importance of effective physical protection of nuclear material used for military purposes, and understanding that such material is and will continue to be accorded stringent physical protection,

HAVE AGREED as follows:

Article 1

For the purposes of this Convention:

(a) "nuclear material" means plutonium except that with isotopic concentration exceeding 80% in plutonium-238; uranium-233; uranium enriched in the isotope 235 or 233; uranium containing the mixture of isotopes as occurring in nature other than in the form of ore or ore-residue; any material containing one or more of the foregoing;

(b) "uranium enriched in the isotope 235 or 233" means uranium containing the isotope 235 or 233 or both in an amount such that the abundance ratio of the sum of these isotopes to the isotope 238 is greater than the ratio of the isotope 235 to the isotope 238 occurring in nature;

(c) "international nuclear transport" means the carriage of a consignment of nuclear material by any means of transportation intended to go beyond the territory of the State where the shipment originates beginning with the departure from a facility of the shipper in that State and ending with the arrival at a facility of the receiver within the State of ultimate destination

Article 2

1. This Convention shall apply to nuclear material used for peaceful purposes while in international nuclear transport.

2. With the exception of articles 3 and 4 and paragraph 3 of article 5, this Convention shall also apply to nuclear material used for peaceful purposes while in domestic use, storage and transport.

3. Apart from the commitments expressly undertaken by States Parties in the articles covered by paragraph 2 with respect to nuclear material used for peaceful purposes while in domestic use, storage and transport, nothing in this Convention shall be interpreted as affecting the sovereign rights of a State regarding the domestic use, storage and transport of such nuclear material.

Article 3

Each State Party shall take appropriate steps within the framework of its national law and consistent with international law to ensure as far as practicable that, during international nuclear transport, nuclear material within its territory, or on board a ship or aircraft under its jurisdiction insofar as such ship or aircraft is engaged in the transport to or from that State, is protected at the levels described in Annex I.

Article 4

1. Each State Party shall not export or authorize the export of nuclear material unless the State Party has received assurances that such material will be protected during the international nuclear transport at the levels described in Annex I.

2. Each State Party shall not import or authorize the import of nuclear material from a State not party to this Convention unless the State Party has received assurances that such material will during the international nuclear transport be protected at the levels described in Annex I.

3. A State Party shall not allow the transit of its territory by land or internal waterways or through its airports or seaports of nuclear material between States that are not parties to this Convention unless the State Party has received assurances as far as practicable that this nuclear material will be protected during international nuclear transport at the levels described in Annex I.

4. Each State Party shall apply within the framework of its national law the levels of physical protection described in Annex I to nuclear material being transported from a part of that State to another part of the same State through international waters or airspace.

5. The State Party responsible for receiving assurances that the nuclear material will be protected at the levels described in Annex I according to paragraphs 1 to 3 shall identify and inform in advance States which the nuclear material is expected to transit by land or internal waterways, or whose airports or seaports it is expected to enter.

6. The responsibility for obtaining assurances referred to in paragraph 1 may be transferred, by mutual agreement, to the State Party involved in the transport as the importing State.

7. Nothing in this article shall be interpreted as in any way affecting the territorial sovereignty and jurisdiction of a State, including that over its airspace and territorial sea.

Article 5

1. States Parties shall identify and make known to each other directly or through the International Atomic Energy Agency their central authority and point of contact having responsibility for physical protection of nuclear material and for co-ordinating recovery and response operations in the event of any unauthorized removal, use or alteration of nuclear material or in the event of credible threat thereof.

2. In the case of theft, robbery or any other unlawful taking of nuclear material or of credible threat thereof, States Parties shall, in accordance with their national law, provide co-operation and assistance to the maximum feasible extent in the recovery and protection of such material to any State that so requests. In particular:

(a) a State Party shall take appropriate steps to inform as soon as possible other States, which appear to it to be concerned, of any theft, robbery or other unlawful taking of nuclear material or credible threat thereof and to inform, where appropriate, international organizations;

(b) as appropriate, the States Parties concerned shall exchange information with each other or international organizations with a view to protecting threatened nuclear material, verifying the integrity of the shipping container, or recovering unlawfully taken nuclear material and shall:

(i) co-ordinate their efforts through diplomatic and other agreed channels;

(ii) render assistance, if requested;

(iii) ensure the return of nuclear material stolen or missing as a consequence of the above-mentioned events.

The means of implementation of this co-operation shall be determined by the States Parties concerned.

3. States Parties shall co-operate and consult as appropriate, with each other directly or through international organizations, with a view to obtaining guidance on the design, maintenance and improvement of systems of physical protection of nuclear material in international transport.

Article 6

1. States Parties shall take appropriate measures consistent with their national law to protect the confidentiality of any information which they receive in confidence by virtue of the provisions of this Convention from another State Party or through participation in an activity carried out for the implementation of this Convention. If States Parties provide information to international organizations in confidence, steps shall be taken to ensure that the confidentiality of such information is protected.

2. States Parties shall not be required by this Convention to provide any information which they are not permitted to communicate pursuant to national law or which would jeopardize the security of the State concerned or the physical protection of nuclear material.

Article 7

1. The intentional commission of:

(a) an act without lawful authority which constitutes the receipt, possession, use, transfer, alteration, disposal or dispersal of nuclear material and which causes or is likely to cause death or serious injury to any person or substantial damage to property;

(b) a theft or robbery of nuclear material;

(c) an embezzlement or fraudulent obtaining of nuclear material;

(d) an act constituting a demand for nuclear material by threat or use of force or by any other form of intimidation;

(e) a threat:

(i) to use nuclear material to cause death or serious injury to any person or substantial property damage, or

(ii) to commit an offence described in sub-paragraph (b) in order to compel a natural or legal person, international organization or State to do or to refrain from doing any act;

(f) an attempt to commit any offence described in paragraphs (a), (b) or (c); and

(g) an act which constitutes participation in any offence described in paragraphs (a) to (f)

shall be made a punishable offence by each State Party under its national law.

2. Each State Party shall make the offences described in this article punishable by appropriate penalties which take into account their grave nature.

Article 8

1. Each State Party shall take such measures as may be necessary to establish its jurisdiction over the offences set forth in article 7 in the following cases:

(a) when the offence is committed in the territory of that State or on board a ship or aircraft registered in that State;

(b) when the alleged offender is a national of that State.

2. Each State Party shall likewise take such measures as may be necessary to establish its jurisdiction over these offences in cases where the alleged offender is present in its territory and it does not extradite him pursuant to article 11 to any of the States mentioned in paragraph 1.

3. This Convention does not exclude any criminal jurisdiction exercised in accordance with national law.

4. In addition to the States Parties mentioned in paragraphs 1 and 2, each State Party may, consistent with international law, establish its jurisdiction over the offences set forth in article 7 when it is involved in international nuclear transport as the exporting or importing State.

Article 9

Upon being satisfied that the circumstances so warrant, the State Party in whose territory the alleged offender is present shall take appropriate measures, including detention, under its national law to ensure his presence for the purpose of prosecution or extradition. Measures taken according to this article shall be notified without delay to the

States required to establish jurisdiction pursuant to article 8 and, where appropriate, all other States concerned.

Article 10

The State Party in whose territory the alleged offender is present shall, if it does not extradite him, submit, without exception whatsoever and without undue delay, the case to its competent authorities for the purpose of prosecution, through proceedings in accordance with the laws of that State.

Article 11

1. The offences in article 7 shall be deemed to be included as extraditable offences in any extradition treaty existing between States Parties. States Parties undertake to include those offences as extraditable offences in every future extradition treaty to be concluded between them.

2. If a State Party which makes extradition conditional on the existence of a treaty receives a request for extradition from another State Party with which it has no extradition treaty, it may at its option consider this Convention as the legal basis for extradition in respect of those offences. Extradition shall be subject to the other conditions provided by the law of the requested State.

3. States Parties which do not make extradition conditional on the existence of a treaty shall recognize those offences as extraditable offences between themselves subject to the conditions provided by the law of the requested State.

4. Each of the offences shall be treated, for the purpose of extradition between States Parties, as if it had been committed not only in the place in which it occurred but also in the territories of the States Parties required to establish their jurisdiction in accordance with paragraph 1 of article 8.

Article 12

Any person regarding whom proceedings are being carried out in connection with any of the offences set forth in article 7 shall be guaranteed fair treatment at all stages of the proceedings.

Article 13

1. States Parties shall afford one another the greatest measure of assistance in connection with criminal proceedings brought in respect of the offences set forth in article 7, including the supply of evidence at their disposal necessary for the proceedings. The law of the State requested shall apply in all cases.

2. The provisions of paragraph 1 shall not affect obligations under any other treaty, bilateral or multilateral, which governs or will govern, in whole or in part, mutual assistance in criminal matters.

Article 14

1. Each State Party shall inform the depositary of its laws and regulations which give effect to this Convention. The depositary shall communicate such information periodically to all States Parties.

2. The State Party where an alleged offender is prosecuted shall, wherever practicable, first communicate the final outcome of the proceedings to the States directly concerned. The State Party shall also communicate the final outcome to the depositary who shall inform all States.

3. Where an offence involves nuclear material used for peaceful purposes in domestic use, storage or transport, and both the alleged offender and the nuclear material remain in the territory of the State Party in which the offence was committed, nothing in this Convention shall be interpreted as requiring that State Party to provide information concerning criminal proceedings arising out of such an offence.

Article 15

The Annexes constitute an integral part of this Convention.

Article 16

1. A conference of States Parties shall be convened by the depositary five years after the entry into force of this Convention to review the implementation of the Convention and its adequacy as concerns the preamble, the whole of the operative part and the annexes in the light of the then prevailing situation.

2. At intervals of not less than five years thereafter, the majority of States Parties may obtain, by submitting a proposal to this effect to the depositary, the convening of further conferences with the same objective.

Article 17

1. In the event of a dispute between two or more States Parties concerning the interpretation or application of this Convention, such States Parties shall consult with a view to the settlement of the dispute by negotiation, or by any other peaceful means of settling disputes acceptable to all parties to the dispute.

2. Any dispute of this character which cannot be settled in the manner prescribed in paragraph 1 shall, at the request of any party to such dispute, be submitted to arbitration or referred to the International Court of Justice for decision. Where a dispute is submitted to arbitration, if, within six months from the date of the request, the parties to the dispute are unable to agree on the organization of the arbitration, a party may request the President of the International Court of Justice or the Secretary-General of the United Nations to appoint one or more arbitrators. In case of conflicting requests by the parties to the dispute, the request to the Secretary-General of the United Nations shall have priority.

3. Each State Party may at the time of signature, ratification, acceptance or approval of this Convention or accession thereto declare that it does not consider itself bound by either or both of the dispute settlement procedures provided for in paragraph 2. The other States Parties shall not be bound by a dispute settlement procedure provided for in paragraph 2, with respect to a State Party which has made a reservation to that procedure.

4. Any State Party which has made a reservation in accordance with paragraph 3 may at any time withdraw that reservation by notification to the depositary.

Article 18

1. This Convention shall be open for signature by all States at the Headquarters of the International Atomic Energy Agency in Vienna and at the Headquarters of the United Nations in New York from 3 March 1980 until its entry into force.

2. This Convention is subject to ratification, acceptance or approval by the signatory States.

3. After its entry into force, this Convention will be open for accession by all States.

4. (a) This Convention shall be open for signature or accession by international organizations and regional organizations of an integration or other nature, provided that any such organization is constituted by sovereign States and has competence in respect of the negotiation, conclusion and application of international agreements in matters covered by this Convention.

(b) In matters within their competence, such organizations shall, on their own behalf, exercise the rights and fulfil the responsibilities which this Convention attributes to States Parties.

(c) When becoming party to this Convention such an organization shall communicate to the depositary a declaration indicating which States are members thereof and which articles of this Convention do not apply to it

(d) Such an organization shall not hold any vote additional to those of its Member States.

5. Instruments of ratification, acceptance, approval or accession shall be deposited with the depositary.

Article 19

1. This Convention shall enter into force on the thirtieth day following the date of deposit of the twenty-first instrument of ratification, acceptance or approval with the depositary.

2. For each State ratifying, accepting, approving or acceding to the Convention after the date of deposit of the twenty-first instrument of ratification, acceptance or approval, the Convention shall enter into force on the thirtieth day after the deposit by such State of its instrument of ratification, acceptance, approval or accession.

Article 20

1. Without prejudice to article 16 a State Party may propose amendments to this Convention. The proposed amendment shall be submitted to the depositary who shall circulate it immediately to all States Parties. If a majority of States Parties request the depositary to convene a conference to consider the proposed amendments, the depositary shall invite all States Parties to attend such a conference to begin not sooner than thirty days after the invitations are issued. Any amendment adopted at the conference by a two-thirds majority of all States Parties shall be promptly circulated by the depositary to all States Parties.

2. The amendment shall enter into force for each State Party that deposits its instrument of ratification, acceptance or approval of the amendment on the thirtieth day after the date on which two thirds of the States Parties have deposited their instruments of ratification, acceptance or approval with the depositary. Thereafter, the amendment shall enter into force for any other State Party on the day on which that State Party deposits its instrument of ratification, acceptance or approval of the amendment.

Article 21

1. Any State Party may denounce this Convention by written notification to the depositary.

2 Denunciation shall take effect one hundred and eighty days following the date on which notification is received by the depositary.

Article 22

The depositary shall promptly notify all States of:

(a) each signature of this Convention;

(b) each deposit of an instrument of ratification, acceptance, approval or accession;

(c) any reservation or withdrawal in accordance with article 17;

(d) any communication made by an organization in accordance with paragraph 4(c) of article 18;

(e) the entry into force of this Convention;

(f) the entry into force of any amendment to this Convention; and

(g) any denunciation made under article 21.

Article 23

The original of this Convention, of which the Arabic, Chinese, English, French, Russian and Spanish texts are equally authentic, shall be deposited with the Director General of the International Atomic Energy Agency who shall send certified copies thereof to all States.

IN WITNESS WHEREOF, the undersigned, being duly authorized, have signed this Convention, opened for signature at Vienna and at New York on 3 March 1980.

ANNEX I

Levels of Physical Protection to be Applied in International Transport of Nuclear Material as Categorized in Annex II

1. Levels of physical protection for nuclear material during storage incidental to international nuclear transport include:

(a) For Category III materials, storage within an area to which access is controlled;

(b) For Category II materials, storage within an area under constant surveillance by guards or electronic devices, surrounded by a physical barrier with a limited number of points of entry under appropriate control or any area with an equivalent level of physical protection;

(c) For Category I material, storage within a protected area as defined for Category II above, to which, in addition, access is restricted to persons whose trustworthiness has been determined, and which is under surveillance by guards who are in close communication with appropriate response forces. Specific measures taken in this context should have as their object the detection and prevention of any assault, unauthorized access or unauthorized removal of material.

2. Levels of physical protection for nuclear material during international transport include:

(a) For Category II and III materials, transportation shall take place under special precautions including prior arrangements among sender, receiver, and carrier, and prior agreement between natural or legal persons subject to the jurisdiction and regulation of exporting and importing States, specifying time, place and procedures for transferring transport responsibility;

(b) For Category I materials, transportation shall take place under special precautions identified above for transportation of Category II and III materials, and in addition, under constant surveillance by escorts and under conditions which assure close communication with appropriate response forces;

(c) For natural uranium other than in the form of ore or ore-residue, transportation protection for quantities exceeding 500 kilograms uranium shall include advance notification of shipment specifying mode of transport, expected time of arrival and confirmation of receipt of shipment.

ANNEX II

TABLE: CATEGORIZATION OF NUCLEAR MATERIAL

Material	Form	Category		
		I	II	III[c]
1. Plutonium[a]	Unirradiated[b]	2 kg or more	Less than 2 kg but more than 500 g	500 g or less but more than 15 g
2. Uranium-235	Unirradiated[b]			
	- uranium enriched to 20% ^{235}U or more	5 kg or more	Less than 5 kg but more than 1 kg	1 kg or less but more than 15 g
	- uranium enriched to 10% ^{235}U but less than 20%		10 kg or more	Less than 10 kg but more than 1 kg
	- uranium enriched above natural, but less than 10% ^{235}U			10 kg or more
3. Uranium-233	Unirradiated[b]	2 kg or more	Less than 2 kg but more than 500 g	500 g or less but more than 15 g
4. Irradiated fuel			Depleted or natural uranium, thorium or low-enriched fuel (less than 10% fossile content)[d]/[e]	

a/ All plutonium except that with isotopic concentration exceeding 80% in plutonium-238

b/ Material not irradiated in a reactor or material irradiated in a reactor but with a radiation level equal to or less than 100 rads/hour at one metre unshielded.

c/ Quantities not falling in Category III and natural uranium should be protected in accordance with prudent management practice.

d/ Although this level of protection is recommended, it would be open to States, upon evaluation of the specific circumstances, to assign a different category of physical protection.

e/ Other fuel which by virtue of its original fissile material content is classified as Category I and II before irradiation may be reduced one category level while the radiation level from the fuel exceeds 100 rads/hour at one metre unshielded.

FINAL ACT

Meeting of Governmental Representatives to Consider the Drafting of a Convention on the Physical Protection of Nuclear Material

1. The Meeting of Governmental Representatives to Consider the Drafting of a Convention on the Physical Protection of Nuclear Material met in Vienna at the Headquarters of the International Atomic Energy Agency from 31 October to 10 November 1977, from 10 to 20 April 1978, from 5 to 16 February and from 15 to 26 October 1979. Informal consultations between Governmental Representatives took place from 4 to 7 September 1978 and from 24 to 25 September 1979.

2. Representatives of 58 States and one organization participated, namely, representatives of:

Algeria
Argentina
Australia
Austria
Belgium
Brazil
Bulgaria
Canada
Chile
Colombia
Costa Rica
Cuba
Czechoslovakia
Denmark
Ecuador
Egypt
Finland
France
German Democratic Republic
Germany, Federal Republic of
Greece
Guatemala
Holy See
Hungary
India
Indonesia
Ireland
Israel
Italy
Japan
Korea, Republic of
Libyan Arab Jamahiriya
Luxembourg
Mexico
Netherlands
Niger
Norway
Pakistan
Panama
Paraguay
Peru
Philippines
Poland
Qatar
Romania
South Africa
Spain
Sweden
Switzerland
Tunisia
Turkey
Union of Soviet Socialist Republics
United Arab Emirates
United Kingdom of Great Britain and Northern Ireland
United States of America
Venezuela
Yugoslavia
Zaire
European Atomic Energy Community

3. The following States and international organizations participated as observers:

Iran
Lebanon
Malaysia
Thailand
Nuclear Energy Agency of the Organisation for Economic Co-operation and Development

4. The Meeting elected Ambassador D. L. Siazon Jr. (Philippines) as Chairman. For the meetings in April 1978 and February 1979 Mr. R. A. Estrada-Oyuela (Argentina) was elected Chairman.

5 The Meeting elected as Vice-Chairmen:

Mr. K. Willuhn of the German Democratic Republic, who at the meeting in February 1979 was succeeded by Mr. H. Rabold of the German Democratic Republic,

Mr. R. J. S. Harry, Netherlands, who at the meeting of October 1979 was succeeded by Mr. G. Dahlhoff of the Federal Republic of Germany;

Mr. R. A. Estrada-Oyuela, Argentina, who at the meeting of October 1979 was succeeded by Mr. L. A. Olivieri of Argentina.

6. Mr. L. W. Herron (Australia) was elected Rapporteur. For the meeting in October 1979 Mr. N. R. Smith (Australia) was elected Rapporteur.

7. Secretariat services were provided by the International Atomic Energy Agency. The Director General of the Agency was represented by the Director of the Legal Division of the Agency, Mr. D. M. Edwards and, in succession to him, Mr. L. W. Herron.

8. The Meeting set up the following groups:

(a) Working Group on Technical Issues

Chairman: Mr. R. J. S. Harry, Netherlands

(b) Working Group on Legal Issues

Chairman: Mr. R. A. Estrada-Oyuela, Argentina

(c) Working Group on Scope of Convention

Chairman: Mr. K. Willuhn, German Democratic Republic

(d) Drafting Committee

Chairman: Mr. De Castro Neves, Brazil

Members: Representatives of Australia, Brazil, Canada, Chile, Czechoslovakia, Egypt, France, Federal Republic of Germany, Italy, Japan, Mexico, Qatar, Tunisia, Union of Soviet Socialist Republics, United States of America.

9. The Meeting had before it the following documents:

(a) Draft Convention on the Physical Protection of Nuclear Materials, Facilities and Transports, as contained in document CPNM/1;

(b) IAEA document INFCIRC/225/Rev. 1: The Physical Protection of Nuclear Material;

(c) IAEA document INFCIRC/254: Communications Received from Certain Member States regarding Guidelines for the Export of Nuclear Material, Equipment or Technology.

10. The Meeting completed consideration of a Convention, the text of which is attached as Annex I.[*] Certain delegations expressed reservations with regard to particular provisions in the text. These are recorded in the documents and in the Daily Reports of the Meeting. It was agreed that the text will be referred by delegations to their authorities for consideration.

11. The Meeting recommended that the text of the Convention be transmitted for information to the twenty-third General Conference of the International Atomic Energy Agency.

12. The Convention will, in accordance with its terms, be opened for signature from 3 March 1980 at the Headquarters of the International Atomic Energy Agency in Vienna and at the Headquarters of the United Nations in New York.

Vienna, 26 October 1979 (signed) D. L. Siazon Jr.

---

[*] Since the Convention has been opened for signature it is not attached here as Annex I; it is reproduced as the first part of this document.

APPENDIX

7

The Convention on Early Notification of a Nuclear Accident

Applications of Nuclear and Radioisotope Technology: The Atom for Peace and Sustainable Development
DOI: https://doi.org/10.1016/B978-0-12-821319-3.00028-2

International Atomic Energy Agency

INFORMATION CIRCULAR

INFCIRC/335
18 November 1986

GENERAL Distr.

Original: ARABIC, CHINESE, ENGLISH, FRENCH, RUSSIAN and SPANIS

CONVENTION ON EARLY NOTIFICATION OF A NUCLEAR ACCIDENT

1. The Convention on Early Notification of a Nuclear Accident was adopted by the General Conference at its special session, 24-26 September 1986, and was opened for signature at Vienna on 26 September 1986 and at New York on 6 October 1986. It entered into force on 27 October 1986, i.e. thirty days after the date (26 September 1986) on which three States expressed their consent to be bound by the Convention, as required under Article 12 thereof.

2. The text of the Convention, taken from a certified copy, is reproduced herein for the information of all Members.

CONVENTION ON EARLY NOTIFICATION OF A NUCLEAR ACCIDENT

THE STATES PARTIES TO THIS CONVENTION,

AWARE that nuclear activities are being carried out in a number of States,

NOTING that comprehensive measures have been and are being taken to ensure a high level o safety in nuclear activities, aimed at preventing nuclear accidents and minimizing the consequences of any such accident, should it occur,

DESIRING to strengthen further international co-operation in the safe development and use o nuclear energy,

CONVINCED of the need for States to provide relevant information about nuclear accidents as early as possible in order that transboundary radiological consequences can be minimized,

NOTING the usefulness of bilateral and multilateral arrangements on information exchange in this area,

HAVE AGREED as follows:

Article 1

Scope of application

1. This Convention shall apply in the event of any accident involving facilities or activities of a State Party or of persons or legal entities under its jurisdiction or control, referred to in paragraph 2 below, from which a release of radioactive material occurs or is likely to occur and which has resulted or may result in an international transboundary release that could be of radiological safety significance for another State.

2. The facilities and activities referred to in paragraph 1 are the following:

(a) any nuclear reactor wherever located;

(b) any nuclear fuel cycle facility;

(c) any radioactive waste management facility;

(d) the transport and storage of nuclear fuels or radioactive wastes;

(e) the manufacture, use, storage, disposal and transport of radioisotopes for agricultural, industrial, medical and related scientific and research purposes; and

(f) the use of radioisotopes for power generation in space objects.

Article 2

Notification and information

In the event of an accident specified in article 1 (hereinafter referred to as a "nuclear accident"), the State Party referred to in that article shall:

(a) forthwith notify, directly or through the International Atomic Energy Agency (hereinafter referred to as the "Agency"), those States which are or may be physically affected as specified in article 1 and the Agency of the nuclear accident, its nature, the time of its occurrence and its exact location where appropriate; and

(b) promptly provide the States referred to in sub-paragraph (a), directly or through the Agency, and the Agency with such available information relevant to minimizing the radiological consequences in those States, as specified in article 5.

Article 3

Other Nuclear Accidents

With a view to minimizing the radiological consequences, States Parties may notify in the event of nuclear accidents other than those specified in article 1.

Article 4

Functions of the Agency

The Agency shall:

(a) forthwith inform States Parties, Member States, other States which are or may be physically affected as specified in article 1 and relevant international intergovernmental organizations (hereinafter referred to as "international organizations") of a notification received pursuant to sub-paragraph (a) of article 2; and

(b) promptly provide any State Party, Member State or relevant international organization, upon request, with the information received pursuant to sub-paragraph (b) of article 2.

Article 5

Information to be provided

1. The information to be provided pursuant to sub-paragraph (b) of article 2 shall comprise the following data as then available to the notifying State Party:

(a) the time, exact location where appropriate, and the nature of the nuclear accident;

(b) the facility or activity involved;

(c) the assumed or established cause and the foreseeable development of the nuclear accident relevant to the transboundary release of the radioactive materials;

(d) the general characteristics of the radioactive release, including, as far as is practicable and appropriate, the nature, probable physical and chemical form and the quantity, composition and effective height of the radioactive release;

(e) information on current and forecast meteorological and hydrological conditions, necessary for forecasting the transboundary release of the radioactive materials;

(f) the results of environmental monitoring relevant to the transboundary release of the radioactive materials;

(g) the off-site protective measures taken or planned;

(h) the predicted behaviour over time of the radioactive release.

2. Such information shall be supplemented at appropriate intervals by further relevant information on the development of the emergency situation, including its foreseeable or actual termination.

3. Information received pursuant to sub-paragraph (b) of article 2 may be used without restriction, except when such information is provided in confidence by the notifying State Party.

Article 6

Consultations

A State Party providing information pursuant to sub-paragraph (b) of article 2 shall, as far as is reasonably practicable, respond promptly to a request for further information or consultations sought by an affected State Party with a view to minimizing the radiological consequences in that State.

Article 7

Competent authorities and points of contact

1. Each State Party shall make known to the Agency and to other States Parties, directly or through the Agency, its competent authorities and point of contact responsible for issuing and receiving the notification and information referred to in article 2. Such points of contact and a focal point within the Agency shall be available continuously.

2. Each State Party shall promptly inform the Agency of any changes that may occur in the information referred to in paragraph 1.

3. The Agency shall maintain an up-to-date list of such national authorities and points of contact as well as points of contact of relevant international organizations and shall provide it to States Parties and Member States and to relevant international organizations.

Article 8

Assistance to States Parties

The Agency shall, in accordance with its Statute and upon a request of a State Party which does not have nuclear activities itself and borders on a State having an active nuclear programme but not Party, conduct investigations into the feasibility and establishment of an appropriate radiation monitoring system in order to facilitate the achievement of the objectives of this Convention.

Article 9

Bilateral and multilateral arrangements

In furtherance of their mutual interests, States Parties may consider, where deemed appropriate, the conclusion of bilateral or multilateral arrangements relating to the subject matter of this Convention.

Article 10

Relationship to other international agreements

This Convention shall not affect the reciprocal rights and obligations of States Parties under existing international agreements which relate to the matters covered by this Convention, or under future international agreements concluded in accordance with the object and purpose of this Convention.

Article 11

Settlement of disputes

1. In the event of a dispute between States Parties, or between a State Party and the Agency, concerning the interpretation or application of this Convention, the parties to the dispute shall consult with a view to the settlement of the dispute by negotiation or by any other peaceful means of settling disputes acceptable to them.

2. If a dispute of this character between States Parties cannot be settled within one year from the request for consultation pursuant to paragraph 1, it shall, at the request of any party to such dispute, be submitted to arbitration or referred to the International Court of Justice for decision. Where a dispute is submitted to arbitration, if, within six months from the date of the request, the parties t the dispute are unable to agree on the organization of the arbitration, a party may request the President of the International Court of Justice or the Secretary-General of the United Nations to appoint one o more arbitrators. In cases of conflicting requests by the parties to the dispute, the request to th Secretary-General of the United Nations shall have priority.

3. When signing, ratifying, accepting, approving or acceding to this Convention, a State ma declare that it does not consider itself bound by either or both of the dispute settlement procedure provided for in paragraph 2. The other States Parties shall not be bound by a dispute settlement proce dure provided for in paragraph 2 with respect to a State Party for which such a declaration is in force.

4. A State Party which has made a declaration in accordance with paragraph 3 may at any time withdraw it by notification to the depositary.

Article 12

Entry into force

1. This Convention shall be open for signature by all States and Namibia, represented by the United Nations Council for Namibia, at the Headquarters of the International Atomic Energy Agency in Vienna and at the Headquarters of the United Nations in New York, from 26 September 1986 and 6 October 1986 respectively, until its entry into force or for twelve months, whichever period is longer.

2. A State and Namibia, represented by the United Nations Council for Namibia, may express its consent to be bound by this Convention either by signature, or by deposit of an instrument of ratification, acceptance or approval following signature made subject to ratification, acceptance or approval, or by deposit of an instrument of accession. The instruments of ratification, acceptance, approval or accession shall be deposited with the depositary.

3. This Convention shall enter into force thirty days after consent to be bound has been expressed by three States.

4. For each State expressing consent to be bound by this Convention after its entry into force, this Convention shall enter into force for that State thirty days after the date of expression of consent.

5. (a) This Convention shall be open for accession, as provided for in this article, by international organizations and regional integration organizations constituted by sovereign States, which have competence in respect of the negotiation, conclusion and application of international agreements in matters covered by this Convention.

(b) In matters within their competence such organizations shall, on their own behalf, exercise the rights and fulfil the obligations which this Convention attributes to States Parties.

(c) When depositing its instrument of accession, such an organization shall communicate to the depositary a declaration indicating the extent of its competence in respect of matters covered by this Convention.

(d) Such an organization shall not hold any vote additional to those of its Member States.

Article 13

Provisional application

A State may, upon signature or at any later date before this Convention enters into force for it, declare that it will apply this Convention provisionally.

Article 14

Amendments

1. A State Party may propose amendments to this Convention. The proposed amendment shall be submitted to the depositary who shall circulate it immediately to all other States Parties.

2. If a majority of the States Parties request the depositary to convene a conference to consider the proposed amendments, the depositary shall invite all States Parties to attend such a conference to begin not sooner than thirty days after the invitations are issued. Any amendment adopted at the conference by a two-thirds majority of all States Parties shall be laid down in a protocol which is open to signature in Vienna and New York by all States Parties.

3. The protocol shall enter into force thirty days after consent to be bound has been expressed by three States. For each State expressing consent to be bound by the protocol after its entry into force, the protocol shall enter into force for that State thirty days after the date of expression of consent.

Article 15

Denunciation

1. A State Party may denounce this Convention by written notification to the depositary.

2. Denunciation shall take effect one year following the date on which the notification is received by the depositary.

Article 16

Depositary

1. The Director General of the Agency shall be the depositary of this Convention.

2. The Director General of the Agency shall promptly notify States Parties and all other States of:

   (a) each signature of this Convention or any protocol of amendment;

   (b) each deposit of an instrument of ratification, acceptance, approval or accession concerning this Convention or any protocol of amendment;

   (c) any declaration or withdrawal thereof in accordance with article 11;

   (d) any declaration of provisional application of this Convention in accordance with article 13;

   (e) the entry into force of this Convention and of any amendment thereto; and

   (f) any denunciation made under article 15.

Article 17

Authentic texts and certified copies

The original of this Convention, of which the Arabic, Chinese, English, French, Russian and Spanish texts are equally authentic, shall be deposited with the Director General of the International Atomic Energy Agency who shall send certified copies to States Parties and all other States.

IN WITNESS WHEREOF the undersigned, being duly authorized, have signed this Convention, open for signature as provided for in paragraph 1 of article 12.

ADOPTED by the General Conference of the International Atomic Energy Agency meeting in special session at Vienna on the twenty-sixth day of September one thousand nine hundred and eighty-six.

APPENDIX

8

Convention on Assistance in the Case of a Nuclear Accident or Radiological Emergency

Applications of Nuclear and Radioisotope Technology: The Atom for Peace and Sustainable Development
DOI: https://doi.org/10.1016/B978-0-12-821319-3.00056-7

International Atomic Energy Agency

INFORMATION CIRCULAR

IAEA - INFCIRC/336
18 November 1986
GENERAL Distr.
Original: ARABIC, CHINESE, ENGLISH, FRENCH, RUSSIAN and SPANISH

CONVENTION ON ASSISTANCE IN THE CASE OF A NUCLEAR ACCIDENT OR RADIOLOGICAL EMERGENCY

1. The Convention on Assistance in the Case of a Nuclear Accident or Radiological Emergency was adopted by the General Conference at its special session, 24-26 September 1986, and was opened for signature at Vienna on 26 September 1986 and at New York on 6 October 1986.*/

2. The text of the Convention, taken from a certified copy, is reproduced herein for the information of all Members.

*/ The date of entry into force will be announced in an Addendum to this document.

CONVENTION ON ASSISTANCE IN THE CASE OF A NUCLEAR ACCIDENT OR RADIOLOGICAL EMERGENCY

THE STATES PARTIES TO THIS CONVENTION,

AWARE that nuclear activities are being carried out in a number of States,

NOTING that comprehensive measures have been and are being taken to ensure a high level o safety in nuclear activities, aimed at preventing nuclear accidents and minimizing the consequences of any such accident, should it occur,

DESIRING to strengthen further international co-operation in the safe development and use of nuclear energy,

CONVINCED of the need for an international framework which will facilitate the prompt provision of assistance in the event of a nuclear accident or radiological emergency to mitigate its consequences,

NOTING the usefulness of bilateral and multilateral arrangements on mutual assistance in this area,

NOTING the activities of the International Atomic Energy Agency in developing guidelines for mutual emergency assistance arrangements in connection with a nuclear accident or radiological emergency,

HAVE AGREED as follows:

Article 1

General provisions

1. The States Parties shall cooperate between themselves and with the International Atomic Energy Agency (hereinafter referred to as the "Agency") in accordance with the provisions of this Convention to facilitate prompt assistance in the event of a nuclear accident or radiological emergency to minimize its consequences and to protect life, property and the environment from the effects of radioactive releases.

2. To facilitate such cooperation States Parties may agree on bilateral or multilateral arrangements or, where appropriate, a combination of these, for preventing or minimizing injury and damage which may result in the event of a nuclear accident or radiological emergency.

3. The States Parties request the Agency, acting within the framework of its Statute, to use its best endeavours in accordance with the provisions of this Convention to promote, facilitate and support the cooperation between States Parties provided for in this Convention.

Article 2

Provision of assistance

1. If a State Party needs assistance in the event of a nuclear accident or radiological emergency, whether or not such accident or emergency originates within its territory, jurisdiction or control, it may call for such assistance from any other State Party, directly or through the Agency, and from the Agency, or, where appropriate, from other international intergovernmental organizations (hereinafter referred to as "international organizations").

2. A State Party requesting assistance shall specify the scope and type of assistance required and, where practicable, provide the assisting party with such information as may be necessary for that party to determine the extent to which it is able to meet the request. In the event that it is not practicable for the requesting State Party to specify the scope and type of assistance required, the requesting State Party and the assisting party shall, in consultation, decide upon the scope and type of assistance required.

3. Each State Party to which a request for such assistance is directed shall promptly decide and notify the requesting State Party, directly or through the Agency, whether it is in a position to render the assistance requested, and the scope and terms of the assistance that might be rendered.

4. States Parties shall, within the limits of their capabilities, identify and notify the Agency of experts, equipment and materials which could be made available for the provision of assistance to other States Parties in the event of a nuclear accident or radiological emergency as well as the terms, especially financial, under which such assistance could be provided.

5. Any State Party may request assistance relating to medical treatment or temporary relocation into the territory of another State Party of people involved in a nuclear accident or radiological emergency.

6. The Agency shall respond, in accordance with its Statute and as provided for in this Convention, to a requesting State Party's or a Member State's request for assistance in the event of a nuclear accident or radiological emergency by:

(a) making available appropriate resources allocated for this purpose;

(b) transmitting promptly the request to other States and international organizations which, according to the Agency's information, may possess the necessary resources; and

(c) if so requested by the requesting State, co-ordinating the assistance at the international level which may thus become available.

Article 3

Direction and control of assistance

Unless otherwise agreed:

(a) the overall direction, control, co-ordination and supervision of the assistance shall be the responsibility within its territory of the requesting State. The assisting party should, where the assistance involves personnel, designate in consultation with the requesting State, the person who should be in charge of and retain immediate operational supervision over the personnel and the equipment provided by it. The designated person should exercise such supervision in cooperation with the appropriate authorities of the requesting State;

(b) the requesting State shall provide, to the extent of its capabilities, local facilities and services for the proper and effective administration of the assistance. It shall also ensure the protection of personnel, equipment and materials brought into its territory by or on behalf of the assisting party for such purpose;

(c) ownership of equipment and materials provided by either party during the periods of assistance shall be unaffected, and their return shall be ensured;

(d) a State Party providing assistance in response to a request under paragraph 5 of article 2 shall co-ordinate that assistance within its territory.

Article 4

Competent authorities and points of contact

1. Each State Party shall make known to the Agency and to other States Parties, directly or through the Agency, its competent authorities and point of contact authorized to make and receive requests for and to accept offers of assistance. Such points of contact and a focal point within the Agency shall be available continuously.

2. Each State Party shall promptly inform the Agency of any changes that may occur in the information referred to in paragraph 1.

3. The Agency shall regularly and expeditiously provide to States Parties, Member States and relevant international organizations the information referred to in paragraphs 1 and 2.

Article 5

Functions of the Agency

The States Parties request the Agency, in accordance with paragraph 3 of article 1 and without prejudice to other provisions of this Convention, to:

(a) collect and disseminate to States Parties and Member States information concerning:

(i) experts, equipment and materials which could be made available in the event of nuclear accidents or radiological emergencies;

(ii) methodologies, techniques and available results of research relating to response to nuclear accidents or radiological emergencies;

(b) assist a State Party or a Member State when requested in any of the following or other appropriate matters:

(i) preparing both emergency plans in the case of nuclear accidents and radiological emergencies and the appropriate legislation;

(ii) developing appropriate training programmes for personnel to deal with nuclear accidents and radiological emergencies;

(iii) transmitting requests for assistance and relevant information in the event of a nuclear accident or radiological emergency;

(iv) developing appropriate radiation monitoring programmes, procedures and standards;

(v) conducting investigations into the feasibility of establishing appropriate radiation monitoring systems;

(c) make available to a State Party or a Member State requesting assistance in the event of a nuclear accident or radiological emergency appropriate resources allocated for the purpose of conducting an initial assessment of the accident or emergency;

(d) offer its good offices to the States Parties and Member States in the event of a nuclear accident or radiological emergency;

(e) establish and maintain liaison with relevant international organizations for the purposes of obtaining and exchanging relevant information and data, and make a list of such organizations available to States Parties, Member States and the aforementioned organizations.

Article 6

Confidentiality and public statements

1. The requesting State and the assisting party shall protect the confidentiality of any confidential information that becomes available to either of them in connection with the assistance in the event of a nuclear accident or radiological emergency. Such information shall be used exclusively for the purpose of the assistance agreed upon.

2. The assisting party shall make every effort to coordinate with the requesting State before releasing information to the public on the assistance provided in connection with a nuclear accident or radiological emergency.

Article 7

Reimbursement of costs

1. An assisting party may offer assistance without costs to the requesting State. When considering whether to offer assistance on such a basis, the assisting party shall take into account:

(a) the nature of the nuclear accident or radiological emergency;

(b) the place of origin of the nuclear accident or radiological emergency;

(c) the needs of developing countries;

(d) the particular needs of countries without nuclear facilities; and

(e) any other relevant factors.

2. When assistance is provided wholly or partly on a reimbursement basis, the requesting State shall reimburse the assisting party for the costs incurred for the services rendered by persons or organizations acting on its behalf, and for all expenses in connection with the assistance to the extent that such expenses are not directly defrayed by the requesting State. Unless otherwise agreed, reimbursement shall be provided promptly after the assisting party has presented its request for reimbursement to the requesting State, and in respect of costs other than local costs, shall be freely transferrable.

3. Notwithstanding paragraph 2, the assisting party may at any time waive, or agree to the postponement of, the reimbursement in whole or in part. In considering such waiver or postponement, assisting parties shall give due consideration to the needs of developing countries.

Article 8

Privileges, immunities and facilities

1. The requesting State shall afford to personnel of the assisting party and personnel acting on its behalf the necessary privileges, immunities and facilities for the performance of their assistance functions.

2. The requesting State shall afford the following privileges and immunities to personnel of the assisting party or personnel acting on its behalf who have been duly notified to and accepted by the requesting State:

(a) immunity from arrest, detention and legal process, including criminal, civil and administrative jurisdiction, of the requesting State, in respect of acts or omissions in the performance of their duties; and

(b) exemption from taxation, duties or other charges, except those which are normally incorporated in the price of goods or paid for services rendered, in respect of the performance of their assistance functions.

3. The requesting State shall:

(a) afford the assisting party exemption from taxation, duties or other charges on the equipment and property brought into the territory of the requesting State by the assisting party for the purpose of the assistance; and

(b) provide immunity from seizure, attachment or requisition of such equipment and property.

4. The requesting State shall ensure the return of such equipment and property. If requested by the assisting party, the requesting State shall arrange, to the extent it is able to do so. for the necessary decontamination of recoverable equipment involved in the assistance before its return.

5. The requesting State shall facilitate the entry into, stay in and departure from its national territory of personnel notified pursuant to paragraph 2 and of equipment and property involved in the assistance.

6. Nothing in this article shall require the requesting State to provide its nationals or permanent residents with the privileges and immunities provided for in the foregoing paragraphs.

7. Without prejudice to the privileges and immunities, all beneficiaries enjoying such privileges and immunities under this article have a duty to respect the laws and regulations of the requesting State. They shall also have the duty not to interfere in the domestic affairs of the requesting State.

8. Nothing in this article shall prejudice rights and obligations with respect to privileges and immunities afforded pursuant to other international agreements or the rules of customary international law.

9. When signing, ratifying, accepting, approving or acceding to this Convention, a State may declare that it does not consider itself bound in whole or in part by paragraphs 2 and 3.

10. A State Party which has made a declaration in accordance with paragraph 9 may at any time withdraw it by notification to the depositary.

Article 9

Transit of personnel, equipment and property

Each State Party shall, at the request of the requesting State or the assisting party, seek to facilitate the transit through its territory of duly notified personnel, equipment and property involved in the assistance to and from the requesting State.

Article 10

Claims and compensation

1. The States Parties shall closely cooperate in order to facilitate the settlement of legal proceedings and claims under this article.

2. Unless otherwise agreed, a requesting State shall in respect of death or of injury to persons, damage to or loss of property, or damage to the environment caused within its territory or other area under its jurisdiction or control in the course of providing the assistance requested:

   (a) not bring any legal proceedings against the assisting party or persons or other legal entities acting on its behalf;

   (b) assume responsibility for dealing with legal proceedings and claims brought by third parties against the assisting party or against persons or other legal entities acting on its behalf;

   (c) hold the assisting party or persons or other legal entities acting on its behalf harmless in respect of legal proceedings and claims referred to in sub-paragraph (b), and

   (d) compensate the assisting party or persons or other legal entities acting on its behalf for:

      (i) death of or injury to personnel of the assisting party or persons acting on its behalf;

      (ii) loss of or damage to non-consumable equipment or materials related to the assistance;

except in cases of wilful misconduct by the individuals who caused the death, injury, loss or damage.

3. This article shall not prevent compensation or indemnity available under any applicable international agreement or national law of any State.

4. Nothing in this article shall require the requesting State to apply paragraph 2 in whole or in part to its nationals or permanent residents.

5. When signing, ratifying, accepting, approving or acceding to this Convention, a State may declare:

(a) that it does not consider itself bound in whole or in part by paragraph 2;

(b) that it will not apply paragraph 2 in whole or in part in cases of gross negligence by the individuals who caused the death, injury, loss or damage.

6. A State Party which has made a declaration in accordance with paragraph 5 may at any time withdraw it by notification to the depositary.

Article 11

Termination of assistance

The requesting State or the assisting party may at any time, after appropriate consultations and by notification in writing, request the termination of assistance received or provided under this Convention. Once such a request has been made, the parties involved shall consult with each other to make arrangements for the proper conclusion of the assistance.

Article 12

Relationship to other international agreements

This Convention shall not affect the reciprocal rights and obligations of States Parties under existing international agreements which relate to the matters covered by this Convention, or under future international agreements concluded in accordance with the object and purpose of this Convention.

Article 13

Settlement of disputes

1. In the event of a dispute between States Parties, or between a State Party and the Agency, concerning the interpretation or application of this Convention, the parties to the dispute shall consult with a view to the settlement of the dispute by negotiation or by any other peaceful means of settling disputes acceptable to them.

2. If a dispute of this character between States Parties cannot be settled within one year from the request for consultation pursuant to paragraph 1, it shall, at the request of any party to such dispute, be submitted to arbitration or referred to the International Court of Justice for decision. Where a dispute is submitted to arbitration, if, within six months from the date of the request, the parties to the dispute are unable to agree on the organization of the arbitration, a party may request the President of the International Court of Justice or the Secretary-General of the United Nations to appoint one or more arbitrators. In cases of conflicting requests by the parties to the dispute, the request to the Secretary-General of the United Nations shall have priority.

3. When signing, ratifying, accepting, approving or acceding to this Convention, a State may declare that it does not consider itself bound by either or both of the dispute settlement procedures

provided for in paragraph 2. The other States Parties shall not be bound by a dispute settlement procedure provided for in paragraph 2 with respect to a State Party for which such a declaration is in force.

4. A State Party which has made a declaration in accordance with paragraph 3 may at any time withdraw it by notification to the depositary.

Article 14

Entry into force

1. This Convention shall be open for signature by all States and Namibia, represented by the United Nations Council for Namibia, at the Headquarters of the International Atomic Energy Agency in Vienna and at the Headquarters of the United Nations in New York, from 26 September 1986 and 6 October 1986 respectively, until its entry into force or for twelve months, whichever period is longer.

2. A State and Namibia, represented by the United Nations Council for Namibia, may express its consent to be bound by this Convention either by signature, or by deposit of an instrument of ratification, acceptance or approval following signature made subject to ratification, acceptance or approval, or by deposit of an instrument of accession. The instruments of ratification, acceptance, approval or accession shall be deposited with the depositary.

3. This Convention shall enter into force thirty days after consent to be bound has been expressed by three States.

4. For each State expressing consent to be bound by this Convention after its entry into force, this Convention shall enter into force for that State thirty days after the date of expression of consent.

5. (a) This Convention shall be open for accession, as provided for in this article, by international organizations and regional integration organizations constituted by sovereign States, which have competence in respect of the negotiation, conclusion and application of international agreements in matters covered by this Convention.

   (b) In matters within their competence such organizations shall, on their own behalf, exercise the rights and fulfil the obligations which this Convention attributes to States Parties.

   (c) When depositing its instrument of accession, such an organization shall communicate to the depositary a declaration indicating the extent of its competence in respect of matters covered by this Convention.

   (d) Such an organization shall not hold any vote additional to those of its Member States.

Article 15

Provisional application

A State may, upon signature or at any later date before this Convention enters into force for it, declare that it will apply this Convention provisionally.

Article 16

Amendments

1. A State Party may propose amendments to this Convention. The proposed amendment shall be submitted to the depositary who shall circulate it immediately to all other States Parties.

2. If a majority of the States Parties request the depositary to convene a conference to consider the proposed amendments, the depositary shall invite all States Parties to attend such a conference to begin not sooner than thirty days after the invitations are issued. Any amendment adopted at the conference by a two-thirds majority of all States Parties shall be laid down in a protocol which is open to signature in Vienna and New York by all States Parties.

3. The protocol shall enter into force thirty days after consent to be bound has been expressed by three States. For each State expressing consent to be bound by the protocol after its entry into force, the protocol shall enter into force for that State thirty days after the date of expression of consent.

Article 17

Denunciation

1. A State Party may denounce this Convention by written notification to the depositary.

2. Denunciation shall take effect one year following the date on which the notification is received by the depositary.

Article 18

Depositary

1. The Director General of the Agency shall be the depositary of this Convention.

2. The Director General of the Agency shall promptly notify States Parties and all other States of:

   (a) each signature of this Convention or any protocol of amendment;

   (b) each deposit of an instrument of ratification, acceptance, approval or accession concerning this Convention or any protocol of amendment;

   (c) any declaration or withdrawal thereof in accordance with articles 8, 10 and 13;

   (d) any declaration of provisional application of this Convention in accordance with article 15;

   (e) the entry into force of this Convention and of any amendment thereto; and

   (f) any denunciation made under article 17.

Article 19

Authentic texts and certified copies

The original of this Convention, of which the Arabic, Chinese, English, French, Russian and Spanish texts are equally authentic, shall be deposited with the Director General of the International Atomic Energy Agency who shall send certified copies to States Parties and all other States.

IN WITNESS WHEREOF the undersigned, being duly authorized, have signed this Convention, open for signature as provided for in paragraph 1 of article 14.

ADOPTED by the General Conference of the International Atomic Energy Agency meeting in special session at Vienna on the twenty-sixth day of September one thousand nine hundred and eighty-six.

APPENDIX

9

Convention on Nuclear Safety

Applications of Nuclear and Radioisotope Technology: The Atom for Peace and Sustainable Development
DOI: https://doi.org/10.1016/B978-0-12-821319-3.00061-0

IAEA - INFCIRC/449
5 July 1994

International Atomic Energy Agency
INFORMATION CIRCULAR

GENERAL Distr.
Original: ARABIC, CHINESE, ENGLISH, FRENCH, RUSSIAN, SPANISH

CONVENTION ON NUCLEAR SAFETY

1. The Convention on Nuclear Safety was adopted on 17 June 1994 by a Diplomatic Conference convened by the International Atomic Energy Agency at its Headquarters from 14 to 17 June 1994. The Convention will be opened for signature on 20 September 1994 during the thirty-eighth regular session of the Agency's General Conference and will enter into force on the ninetieth day after the date of deposit with the Depositary (the Agency's Director General) of the twenty-second instrument of ratification, acceptance or approval, including the instruments of seventeen States, having each at least one nuclear installation which has achieved criticality in a reactor core.

2. The text of the Convention as adopted is reproduced in the Annex hereto for the information of all Member States.

CONVENTION ON NUCLEAR SAFETY

PREAMBLE

THE CONTRACTING PARTIES

(i) Aware of the importance to the international community of ensuring that the use of nuclear energy is safe, well regulated and environmentally sound;

(ii) Reaffirming the necessity of continuing to promote a high level of nuclear safety worldwide;

(iii) Reaffirming that responsibility for nuclear safety rests with the State having jurisdiction over a nuclear installation;

(iv) Desiring to promote an effective nuclear safety culture;

(v) Aware that accidents at nuclear installations have the potential for transboundary impacts;

(vi) Keeping in mind the Convention on the Physical Protection of Nuclear Material (1979), the Convention on Early Notification of a Nuclear Accident (1986), and the Convention on Assistance in the Case of a Nuclear Accident or Radiological Emergency (1986);

(vii) Affirming the importance of international co-operation for the enhancement of nuclear safety through existing bilateral and multilateral mechanisms and the establishment of this incentive Convention;

(viii) Recognizing that this Convention entails a commitment to the application of fundamental safety principles for nuclear installations rather than of detailed safety standards and that there are internationally formulated safety guidelines which are updated from time to time and so can provide guidance on contemporary means of achieving a high level of safety;

(ix) Affirming the need to begin promptly the development of an international convention on the safety of radioactive waste management as soon as the ongoing process to develop waste management safety fundamentals has resulted in broad international agreement;

(x) Recognizing the usefulness of further technical work in connection with the safety of other parts of the nuclear fuel cycle, and that this work may, in time, facilitate the development of current or future international instruments;

HAVE AGREED as follows:

CHAPTER 1. OBJECTIVES, DEFINITIONS AND SCOPE OF APPLICATION

ARTICLE 1. OBJECTIVES

The objectives of this Convention are:

(i) to achieve and maintain a high level of nuclear safety worldwide through the enhancement of national measures and international co-operation including, where appropriate, safety-related technical co-operation;

(ii) to establish and maintain effective defences in nuclear installations against potential radiological hazards in order to protect individuals, society and the environment from harmful effects of ionizing radiation from such installations;

(iii) to prevent accidents with radiological consequences and to mitigate such consequences should they occur.

ARTICLE 2. DEFINITIONS

For the purpose of this Convention:

(i) "nuclear installation" means for each Contracting Party any land-based civil nuclear power plant under its jurisdiction including such storage, handling and treatment facilities for radioactive materials as are on the same site and are directly related to the operation of the nuclear power plant. Such a plant ceases to be a nuclear installation when all nuclear fuel elements have been removed permanently from the reactor core and have been stored safely in accordance with approved procedures, and a decommissioning programme has been agreed to by the regulatory body.

(ii) "regulatory body" means for each Contracting Party any body or bodies given the legal authority by that Contracting Party to grant licences and to regulate the siting, design, construction, commissioning, operation or decommissioning of nuclear installations.

(iii) "licence" means any authorization granted by the regulatory body to the applicant to have the responsibility for the siting, design, construction, commissioning, operation or decommissioning of a nuclear installation.

ARTICLE 3. SCOPE OF APPLICATION

This Convention shall apply to the safety of nuclear installations.

CHAPTER 2. OBLIGATIONS

(a) General Provisions

ARTICLE 4. IMPLEMENTING MEASURES

Each Contracting Party shall take, within the framework of its national law, the legislative, regulatory and administrative measures and other steps necessary for implementing its obligations under this Convention.

ARTICLE 5. REPORTING

Each Contracting Party shall submit for review, prior to each meeting referred to in Article 20, a report on the measures it has taken to implement each of the obligations of this Convention.

ARTICLE 6. EXISTING NUCLEAR INSTALLATIONS

Each Contracting Party shall take the appropriate steps to ensure that the safety of nuclear installations existing at the time the Convention enters into force for that Contracting Party is reviewed as soon as possible. When necessary in the context of this Convention, the Contracting Party shall ensure that all reasonably practicable improvements are made as a matter of urgency to upgrade the safety of the nuclear installation. If such upgrading cannot be achieved, plans should be implemented to shut down the nuclear installation as soon as practically possible. The timing of the shut-down may take into account the whole energy context and possible alternatives as well as the social, environmental and economic impact.

(b) Legislation and regulation

ARTICLE 7. LEGISLATIVE AND REGULATORY FRAMEWORK

1. Each Contracting Party shall establish and maintain a legislative and regulatory framework to govern the safety of nuclear installations.

2. The legislative and regulatory framework shall provide for:

(i) the establishment of applicable national safety requirements and regulations;

(ii) a system of licensing with regard to nuclear installations and the prohibition of the operation of a nuclear installation without a licence;

(iii) a system of regulatory inspection and assessment of nuclear installations to ascertain compliance with applicable regulations and the terms of licences;

(iv) the enforcement of applicable regulations and of the terms of licences, including suspension, modification or revocation.

ARTICLE 8. REGULATORY BODY

1. Each Contracting Party shall establish or designate a regulatory body entrusted with the implementation of the legislative and regulatory framework referred to in Article 7, and provided with adequate authority, competence and financial and human resources to fulfil its assigned responsibilities.

2. Each Contracting Party shall take the appropriate steps to ensure an effective separation between the functions of the regulatory body and those of any other body or organization concerned with the promotion or utilization of nuclear energy.

ARTICLE 9. RESPONSIBILITY OF THE LICENCE HOLDER

Each Contracting Party shall ensure that prime responsibility for the safety of a nuclear installation rests with the holder of the relevant licence and shall take the appropriate steps to ensure that each such licence holder meets its responsibility.

(c) General Safety Considerations

ARTICLE 10. PRIORITY TO SAFETY

Each Contracting Party shall take the appropriate steps to ensure that all organizations engaged in activities directly related to nuclear installations shall establish policies that give due priority to nuclear safety.

ARTICLE 11. FINANCIAL AND HUMAN RESOURCES

1. Each Contracting Party shall take the appropriate steps to ensure that adequate financial resources are available to support the safety of each nuclear installation throughout its life.

2. Each Contracting Party shall take the appropriate steps to ensure that sufficient numbers of qualified staff with appropriate education, training and retraining are available for all safety-related activities in or for each nuclear installation, throughout its life.

ARTICLE 12. HUMAN FACTORS

Each Contracting Party shall take the appropriate steps to ensure that the capabilities and limitations of human performance are taken into account throughout the life of a nuclear installation.

ARTICLE 13. QUALITY ASSURANCE

Each Contracting Party shall take the appropriate steps to ensure that quality assurance programmes are established and implemented with a view to providing confidence that specified requirements for all activities important to nuclear safety are satisfied throughout the life of a nuclear installation.

ARTICLE 14. ASSESSMENT AND VERIFICATION OF SAFETY

Each Contracting Party shall take the appropriate steps to ensure that:

(i) comprehensive and systematic safety assessments are carried out before the construction and commissioning of a nuclear installation and throughout its life. Such assessments shall be well documented, subsequently updated in the light of operating experience and significant new safety information, and reviewed under the authority of the regulatory body;

(ii) verification by analysis, surveillance, testing and inspection is carried out to ensure that the physical state and the operation of a nuclear installation continue to be in accordance with its design, applicable national safety requirements, and operational limits and conditions.

ARTICLE 15. RADIATION PROTECTION

Each Contracting Party shall take the appropriate steps to ensure that in all operational states the radiation exposure to the workers and the public caused by a nuclear installation shall be kept as low as reasonably achievable and that no individual shall be exposed to radiation doses which exceed prescribed national dose limits.

ARTICLE 16. EMERGENCY PREPAREDNESS

1. Each Contracting Party shall take the appropriate steps to ensure that there are on-site and off-site emergency plans that are routinely tested for nuclear installations and cover the activities to be carried out in the event of an emergency.

For any new nuclear installation, such plans shall be prepared and tested before it commences operation above a low power level agreed by the regulatory body.

2. Each Contracting Party shall take the appropriate steps to ensure that, insofar as they are likely to be affected by a radiological emergency, its own population and the competent authorities of the States in the vicinity of the nuclear installation are provided with appropriate information for emergency planning and response.

3. Contracting Parties which do not have a nuclear installation on their territory, insofar as they are likely to be affected in the event of a radiological emergency at a nuclear installation in the vicinity, shall take the appropriate steps for the preparation and testing of emergency plans for their territory that cover the activities to be carried out in the event of such an emergency.

(d) Safety of Installations

ARTICLE 17. SITING

Each Contracting Party shall take the appropriate steps to ensure that appropriate procedures are established and implemented:

(i) for evaluating all relevant site-related factors likely to affect the safety of a nuclear installation for its projected lifetime;

(ii) for evaluating the likely safety impact of a proposed nuclear installation on individuals, society and the environment;

(iii) for re-evaluating as necessary all relevant factors referred to in sub-paragraphs (i) and (ii) so as to ensure the continued safety acceptability of the nuclear installation;

(iv) for consulting Contracting Parties in the vicinity of a proposed nuclear installation, insofar as they are likely to be affected by that installation and, upon request providing the necessary information to such Contracting Parties, in order to enable them to evaluate and make their own assessment of the likely safety impact on their own territory of the nuclear installation.

ARTICLE 18. DESIGN AND CONSTRUCTION

Each Contracting Party shall take the appropriate steps to ensure that:

(i) the design and construction of a nuclear installation provides for several reliable levels and methods of protection (defense in depth) against the release of radioactive materials, with a view to preventing the occurrence of accidents and to mitigating their radiological consequences should they occur;

(ii) the technologies incorporated in the design and construction of a nuclear installation are proven by experience or qualified by testing or analysis;

(iii) the design of a nuclear installation allows for reliable, stable and easily manageable operation, with specific consideration of human factors and the man-machine interface.

ARTICLE 19. OPERATION

Each Contracting Party shall take the appropriate steps to ensure that:

(i) the initial authorization to operate a nuclear installation is based upon an appropriate safety analysis and a commissioning programme demonstrating that the installation, as constructed, is consistent with design and safety requirements;

(ii) operational limits and conditions derived from the safety analysis, tests and operational experience are defined and revised as necessary for identifying safe boundaries for operation;

(iii) operation, maintenance, inspection and testing of a nuclear installation are conducted in accordance with approved procedures;

(iv) procedures are established for responding to anticipated operational occurrences and to accidents;

(v) necessary engineering and technical support in all safety-related fields is available throughout the lifetime of a nuclear installation;

(vi) incidents significant to safety are reported in a timely manner by the holder of the relevant licence to the regulatory body;

(vii) programmes to collect and analyse operating experience are established, the results obtained and the conclusions drawn are acted upon and that existing mechanisms are used to share important experience with international bodies and with other operating organizations and regulatory bodies;

(viii) the generation of radioactive waste resulting from the operation of a nuclear installation is kept to the minimum practicable for the process concerned, both in activity and in volume, and any necessary treatment and storage of spent fuel and waste directly related to the operation and on the same site as that of the nuclear installation take into consideration conditioning and disposal.

CHAPTER 3. MEETINGS OF THE CONTRACTING PARTIES

ARTICLE 20. REVIEW MEETINGS

1. The Contracting Parties shall hold meetings (hereinafter referred to as "review meetings") for the purpose of reviewing the reports submitted pursuant to Article 5 in accordance with the procedures adopted under Article 22.

2. Subject to the provisions of Article 24 sub-groups comprised of representatives of Contracting Parties may be established and may function during the review meetings as deemed necessary for the purpose of reviewing specific subjects contained in the reports.

3. Each Contracting Party shall have a reasonable opportunity to discuss the reports submitted by other Contracting Parties and to seek clarification of such reports.

ARTICLE 21. TIMETABLE

1. A preparatory meeting of the Contracting Parties shall be held not later than six months after the date of entry into force of this Convention.

2. At this preparatory meeting, the Contracting Parties shall determine the date for the first review meeting. This review meeting shall be held as soon as possible, but not later than thirty months after the date of entry into force of this Convention.

3. At each review meeting, the Contracting Parties shall determine the date for the next such meeting. The interval between review meetings shall not exceed three years.

ARTICLE 22. PROCEDURAL ARRANGEMENTS

1. At the preparatory meeting held pursuant to Article 21 the Contracting Parties shall prepare and adopt by consensus Rules of Procedure and Financial Rules. The Contracting Parties shall establish in particular and in accordance with the Rules of Procedure:

(i) guidelines regarding the form and structure of the reports to be submitted pursuant to Article 5;

(ii) a date for the submission of such reports;

(iii) the process for reviewing such reports.

2. At review meetings the Contracting Parties may, if necessary, review the arrangements established pursuant to sub-paragraphs (i)-(iii) above, and adopt revisions by consensus unless otherwise provided for in the Rules of Procedure. They may also amend the Rules of Procedure and the Financial Rules, by consensus.

ARTICLE 23. EXTRAORDINARY MEETINGS

An extraordinary meeting of the Contracting Parties shall be held:

(i) if so agreed by a majority of the Contracting Parties present and voting at a meeting, abstentions being considered as voting; or

(ii) at the written request of a Contracting Party, within six months of this request having been communicated to the Contracting Parties and notification having been received by the secretariat referred to in Article 28, that the request has been supported by a majority of the Contracting Parties.

ARTICLE 24. ATTENDANCE

1. Each Contracting Party shall attend meetings of the Contracting Parties and be represented at such meetings by one delegate, and by such alternates, experts and advisers as it deems necessary.

2. The Contracting Parties may invite, by consensus, any intergovernmental organization which is competent in respect of matters governed by this Convention to attend, as an observer, any meeting, or specific sessions thereof. Observers shall be required to accept in writing, and in advance, the provisions of Article 27.

ARTICLE 25. SUMMARY REPORTS

The Contracting Parties shall adopt, by consensus, and make available to the public a document addressing issues discussed and conclusions reached during a meeting.

ARTICLE 26. LANGUAGES

1. The languages of meetings of the Contracting Parties shall be Arabic, Chinese, English, French, Russian and Spanish unless otherwise provided in the Rules of Procedure.

2. Reports submitted pursuant to Article 5 shall be prepared in the national language of the submitting Contracting Party or in a single designated language to be agreed in the Rules of Procedure. Should the report be submitted in a national language other than the designated

language, a translation of the report into the designated language shall be provided by the Contracting Party.

3. Notwithstanding the provisions of paragraph 2, if compensated, the secretariat will assume the translation into the designated language of reports submitted in any other language of the meeting.

ARTICLE 27. CONFIDENTIALITY

1. The provisions of this Convention shall not affect the rights and obligations of the Contracting Parties under their law to protect information from disclosure. For the purposes of this Article, "information" includes, inter alia, (i) personal data; (ii) information protected by intellectual property rights or by industrial or commercial confidentiality; and (iii) information relating to national security or to the physical protection of nuclear materials or nuclear installations.

2. When, in the context of this Convention, a Contracting Party provides information identified by it as protected as described in paragraph 1, such information shall be used only for the purposes for which it has been provided and its confidentiality shall be respected.

3. The content of the debates during the reviewing of the reports by the Co..tracting Parties at each meeting shall be confidential.

ARTICLE 28. SECRETARIAT

1. The International Atomic Energy Agency, (hereinafter referred to as the "Agency") shall provide the secretariat for the meetings of the Contracting Parties.

2. The secretariat shall:

(i) convene, prepare and service the meetings of the Contracting Parties;

(ii) transmit to the Contracting Parties information received or prepared in accordance with the provisions of this Convention.

The costs incurred by the Agency in carrying out the functions referred to in sub-paragraphs i) and (ii) above shall be borne by the Agency as part of its regular budget.

3. The Contracting Parties may, by consensus, request the Agency to provide other services in support of meetings of the Contracting Parties. The Agency may provide such services if they can be undertaken within its programme and regular budget. Should this not be possible, the Agency may provide such services if voluntary funding is provided from another source.

CHAPTER 4. FINAL CLAUSES AND OTHER PROVISIONS

ARTICLE 29. RESOLUTION OF DISAGREEMENTS

In the event of a disagreement between two or more Contracting Parties concerning the interpretation or application of this Convention, the Contracting Parties shall consult within the framework of a meeting of the Contracting Parties with a view to resolving the disagreement.

ARTICLE 30. SIGNATURE, RATIFICATION, ACCEPTANCE, APPROVAL, ACCESSION

1. This Convention shall be open for signature by all States at the Headquarters of the Agency in Vienna from 20 September 1994 until its entry into force.

2. This Convention is subject to ratification, acceptance or approval by the signatory States.

3. After its entry into force, this Convention shall be open for accession by all States.

4. (i) This Convention shall be open for signature or accession by regional organizations of an integration or other nature, provided that any such organization is constituted by sovereign States and has competence in respect of the negotiation, conclusion and application of international agreements in matters covered by this Convention.

   (ii) In matters within their competence, such organizations shall, on their own behalf, exercise the rights and fulfil the responsibilities which this Convention attributes to States Parties.

   (iii) When becoming party to this Convention, such an organization shall communicate to the Depositary referred to in Article 34, a declaration indicating which States are members thereof, which articles of this Convention apply to it, and the extent of its competence in the field covered by those articles.

   (iv) Such an organization shall not hold any vote additional to those of its Member States.

5. Instruments of ratification, acceptance, approval or accession shall be deposited with the Depositary.

ARTICLE 31. ENTRY INTO FORCE

1. This Convention shall enter into force on the ninetieth day after the date of deposit with the Depositary of the twenty-second instrument of ratification, acceptance or approval, including the instruments of seventeen States, each having at least one nuclear installation which has achieved criticality in a reactor core.

2. For each State or regional organization of an integration or other nature which ratifies, accepts, approves or accedes to this Convention after the date of deposit of the last instrument required to satisfy the conditions set forth in paragraph 1, this Convention shall enter into force on the ninetieth day after the date of deposit with the Depositary of the appropriate instrument by such a State or organization.

ARTICLE 32. AMENDMENTS TO THE CONVENTION

1. Any Contracting Party may propose an amendment to this Convention. Proposed amendments shall be considered at a review meeting or an extraordinary meeting.

2. The text of any proposed amendment and the reasons for it shall be provided to the Depositary who shall communicate the proposal to the Contracting Parties promptly and at least ninety days before the meeting for which it is submitted for consideration. Any comments received on such a proposal shall be circulated by the Depositary to the Contracting Parties.

3. The Contracting Parties shall decide after consideration of the proposed amendment whether to adopt it by consensus, or, in the absence of consensus, to submit it to a Diplomatic Conference. A decision to submit a proposed amendment to a Diplomatic Conference shall require a two-thirds majority vote of the Contracting Parties present and voting at the meeting, provided that at least one half of the Contracting Parties are present at the time of voting. Abstentions shall be considered as voting.

4. The Diplomatic Conference to consider and adopt amendments to this Convention shall be convened by the Depositary and held no later than one year after the appropriate decision taken in accordance with paragraph 3 of this Article. The Diplomatic Conference shall make every effort to ensure amendments are adopted by consensus. Should this not be possible, amendments shall be adopted with a two-thirds majority of all Contracting Parties.

5. Amendments to this Convention adopted pursuant to paragraphs 3 and 4 above shall be subject to ratification, acceptance, approval, or confirmation by the Contracting Parties and shall enter into force for those Contracting Parties which have ratified, accepted, approved or confirmed them on the ninetieth day after the receipt by the Depositary of the relevant instruments by at least three fourths of the Contracting Parties. For a Contracting Party which subsequently ratifies, accepts, approves or confirms the said amendments, the amendments will enter into force on the ninetieth day after that Contracting Party has deposited its relevant instrument.

ARTICLE 33. DENUNCIATION

1. Any Contracting Party may denounce this Convention by written notification to the Depositary.

2. Denunciation shall take effect one year following the date of the receipt of the notification by the Depositary, or on such later date as may be specified in the notification.

ARTICLE 34. DEPOSITARY

1. The Director General of the Agency shall be the Depositary of this Convention.

2. The Depositary shall inform the Contracting Parties of:

(i) the signature of this Convention and of the deposit of instruments of ratification, acceptance, approval or accession, in accordance with Article 30;

(ii) the date on which the Convention enters into force, in accordance with Article 31;

(iii) the notifications of denunciation of the Convention and the date thereof, made in accordance with Article 33;

(iv) the proposed amendments to this Convention submitted by Contracting Parties, the amendments adopted by the relevant Diplomatic Conference or by the meeting of the Contracting Parties, and the date of entry into force of the said amendments, in accordance with Article 32.

ARTICLE 35. AUTHENTIC TEXTS

The original of this Convention of which the Arabic, Chinese, English, French, Russian and Spanish texts are equally authentic, shall be deposited with the Depositary, who shall send certified copies thereof to the Contracting Parties.

Index

Page numbers followed by "*f*" and "*t*" indicate, figures and tables respectively.

I

J

K

L

T

U

W

X

Printed in the United States
by Baker & Taylor Publisher Services